T0134508

METHODS IN MOLECULAR BIOLOGY

Series Editor
John M. Walker
School of Life and Medical Sciences
University of Hertfordshire
Hatfield, Hertfordshire, AL10 9AB, UK

For further volumes:
http://www.springer.com/series/7651

Campylobacter jejuni

Methods and Protocols

Edited by

James Butcher and Alain Stintzi

University of Ottawa, Ottawa, Canada

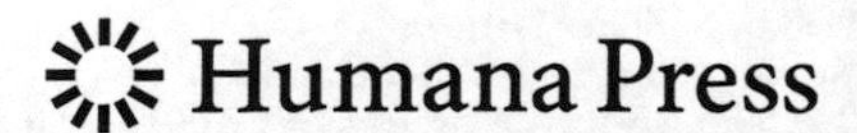

Editors
James Butcher
University of Ottawa
Ottawa, Canada

Alain Stintzi
University of Ottawa
Ottawa, Canada

Videos to this book can be accessed http://www. link.springer.com/videos/[978-1-4939-6536-6]

ISSN 1064-3745 ISSN 1940-6029 (electronic)
Methods in Molecular Biology
ISBN 978-1-4939-8229-5 ISBN 978-1-4939-6536-6 (eBook)
DOI 10.1007/978-1-4939-6536-6

Printed on acid-free paper

This Humana Press imprint is published by Springer Nature
The registered company is Springer Science+Business Media LLC
The registered company address is: 233 Spring Street, New York, NY 10013, U.S.A.

Preface

Campylobacter jejuni is one of the most common causes of bacterial gastroenteritis, threatening human health worldwide. *C. jejuni* infection results in Campylobacteriosis: an acute diarrhoeal disease which ranges from a day of mild watery or bloody diarrhea to severe abdominal pain that can linger for several weeks. *C. jejuni* is also involved in the development of Guillain–Barré syndrome, a debilitating autoimmune disease that causes muscle weakness/paralysis. *C. jejuni* colonizes poultry as a commensal organism, and poultry farms must implement extremely stringent biosecurity protocols to prevent flocks from *Campylobacter* infection. *C. jejuni* can also colonize bovine and porcine hosts who then shed *C. jejuni* into waterways and farm fields. *C. jejuni* has been identified in river streams contaminated with livestock run-off, on houseflies, in raw milk, and may even be hidden within protozoa hosts such as Acanthamoeba.

Paradoxically, despite these various infection vectors and *C. jejuni*'s presence in a variety of different hosts, *C. jejuni* is a fastidious organism that is difficult to grow and study in vitro. Thus, the mechanisms underlying how *C. jejuni* causes disease in humans are still poorly understood as compared to what is known for other common bacteria pathogens such as *Helicobacter*, *Salmonella*, and *Listeria*. To accelerate the pace of unraveling *C. jejuni*'s pathogenic mechanisms, we have assembled a series of protocols detailing key techniques used by leading experts in the field to study this pathogen. These techniques span from methods to isolate *Campylobacter* from environmental samples, molecular biology approaches to study *Campylobacter*, methods to study *Campylobacter*–host interactions, and systems biology approaches to study *Campylobacter* on a genome-wide scale.

We begin with different approaches to isolate *Campylobacter* from various types of environmental sources (Chapters 1 and 2) and also detail how to isolate *Campylobacter*-specific bacteriophages as well (Chapter 3). We subsequently present methods to study *Campylobacter* using molecular biology approaches. These span from detailing *Campylobacter* interactions with various ligands (Chapters 6 and 10) to classical approaches to delete/knock-down genes and assess their impact using in vitro assays (Chapters 4, 5 and 7). We also include approaches to study *Campylobacter*–bacteriophage interactions (Chapter 11) and to crystallize *Campylobacter* proteins for detailed structural biology studies (Chapter 8). Next, we catalogue methods to study *Campylobacter*–host interactions. These methods include in vitro studies using cell culture (Chapter 11–13), ex vivo studies using polarized in vitro organ culture from intestinal biopsies (Chapter 12), and in vivo studies using chick (Chapters 13 and 21), larvae (Chapter 14), and mouse (Chapter 15) models. We end with systems biology approaches to study *Campylobacter*. These include methods for characterizing the *Campylobacter* metabolome (Chapter 16), methylome (Chapter 17), proteome (Chapter 19), glycoproteome (Chapter 18), and transcriptome (Chapter 20). We also describe a systems biology approach for studying the effect of all potential Campylobacter gene mutants at the same time in vivo by generating insertion sequencing compatible mutant libraries (Chapter 21).

We aspire that the availability of these demonstrated techniques in Campylobacter will encourage existing *Campylobacter* researchers to employ novel methods to further their own research and also encourage new researchers to include *Campylobacter* in their future research initiatives. We are grateful for the various laboratories across the world that contributed to this series in Methods in Molecular biology and hope that this compendium of *Campylobacter* techniques will be well received in the *Campylobacter* field and beyond.

Ottawa, ON, Canada

James Butcher
Alain Stintzi

Contents

Contributors

LUIS ALVAREZ • *Conway Institute, School of Medicine and Medical Science, University College Dublin, Dublin, Ireland; National Children's Research Center, Our Lady's Children's Hospital Crumlin, Dublin, Ireland*

MOMEN ASKOURA • *Department of Biochemistry, Microbiology and Immunology, Ottawa Institute of Systems Biology, University of Ottawa, Ottawa, ON, Canada*

SOPHIE BÉRUBÉ • *Ottawa Institute of Systems Biology, Department of Biochemistry, Microbiology and Immunology, University of Ottawa, Ottawa, ON, Canada*

TINA BIRK • *National Food Institute, Technical University of Denmark, Søborg, Denmark*

BILLY BOURKE • *Conway Institute, School of Medicine and Medical Science, University College Dublin, Dublin, Ireland*

LONE BRØNDSTED • *Department of Veterinary Disease Biology, Faculty of Health and Medical Sciences, University of Copenhagen, Frederiksberg C, Denmark*

JAMES BUTCHER • *Ottawa Institute of Systems Biology, Department of Biochemistry, Microbiology and Immunology, University of Ottawa, Ottawa, ON, Canada*

CATHERINE D. CARRILLO • *Canadian Food Inspection Agency, Ottawa, ON, Canada*

TYSON A. CLARK • *Pacific Biosciences, Menlo Park, CA, USA*

MARGUERITE CLYNE • *Conway Institute, School of Medicine and Medical Science, University College Dublin, Dublin, Ireland*

JEAN-FRANÇOIS COUTURE • *Ottawa Institute of Systems Biology, University of Ottawa, Ottawa, ON, Canada; Department of Biochemistry, Microbiology and Immunology, University of Ottawa, Ottawa, ON, Canada*

MATTHEW P. DAVEY • *Department of Molecular Biology and Biotechnology, The University of Sheffield, Sheffield, UK*

CHRISTOPHER J. DAY • *Institute for Glycomics, Griffith University, Gold Coast Campus, Gold Coast, Australia*

VICTOR J. DIRITA • *Department of Microbiology and Molecular Genetics, Michigan State University, East Lansing, MI, USA*

BRENDAN DOLAN • *Conway Institute, School of Medicine and Medical Science, University College Dublin, Dublin, Ireland*

GINA DUGGAN • *Conway Institute, School of Medicine and Medical Science, University College Dublin, Dublin, Ireland*

CIARA DUNNE • *Conway Institute, School of Medicine and Medical Science, University College Dublin, Dublin, Ireland*

YILMAZ EMRE GENCAY • *Department of Veterinary Disease Biology, Faculty of Health and Medical Sciences, University of Copenhagen, Frederiksberg C, Denmark*

FRANZISKA A. GRAEF • *Division of Gastroenterology, Department of Pediatrics, Child and Family Research Institute, University of British Columbia, Vancouver, BC, Canada*

NABILA HADDAD • *INRA, UMR 1014 Secalim, Nantes, France; LUNAM Université, Nantes, France*

LAUREN E. HARTLEY-TASSELL • *Institute for Glycomics, Griffith University, Gold Coast Campus, Gold Coast, Australia*

KELLI L. HIETT • *Poultry Microbiological Safety and Processing Research Unit, United States National Poultry Research Center, Agricultural Research Service, U.S. Department of Agriculture, Athens, GA, USA*
ROBERT M. HOWLETT • *Department of Molecular Biology and Biotechnology, The University of Sheffield, Sheffield, UK*
IRENE IUGOVAZ • *Health Canada, Québec Region, Longueuil, QC, Canada*
BYEONGHWA JEON • *School of Public Health, University of Alberta, Edmonton, AB, Canada*
JEREMIAH G. JOHNSON • *Department of Microbiology and Molecular Genetics, Michigan State University, East Lansing, MI, USA; Department of Microbiology, University of Tennessee, Knoxville, TN, USA*
DAVID J. KELLY • *Department of Molecular Biology and Biotechnology, The University of Sheffield, Sheffield, UK*
ROBYN KENWELL • *Bureau of Microbial Hazards, Health Canada, Ottawa, ON, Canada*
REBECCA M. KING • *Institute for Glycomics, Griffith University, Gold Coast Campus, Gold Coast, Australia*
MICHAEL E. KONKEL • *School of Molecular Biosciences, College of Veterinary Medicine, Washington State University, Pullman, WA, USA*
VICTORIA KOROLIK • *Institute for Glycomics, Griffith University, Gold Coast Campus, Gold Coast, Australia*
JUN LIN • *Department of Animal Science, University of Tennessee, Knoxville, TN, USA*
KATHY T. MOU • *Department of Veterinary Microbiology and Preventive Medicine, College of Veterinary Medicine, Iowa State University, Ames, IA, USA*
USHA K. MUPPIRALA • *Genome Informatics Facility, Office of Biotechnology, Iowa State University, Ames, IA, USA*
JULIE NAUGHTON • *Conway Institute, School of Medicine, University College Dublin, Dublin, Ireland*
NICHOLAS M. NEGRETTI • *School of Molecular Biosciences, College of Veterinary Medicine, Washington State University, Pullman, WA, USA*
EUNA OH • *School of Public Health, University of Alberta, Edmonton, AB, Canada*
OMAR A. OYARZABAL • *University of Vermont, St. Albans, VT, UK*
PAUL J. PLUMMER • *Department of Veterinary Microbiology and Preventive Medicine, College of Veterinary Medicine, Iowa State University, Ames, IA, USA; Department of Veterinary Diagnostic and Production Animal Medicine, College of Veterinary Medicine, Iowa State University, Ames, IA, USA*
RAMILA C. RODRIGUES • *INRA, UMR 1014 Secalim, Nantes, France; LUNAM Université, Nantes, France*
ORHAN SAHIN • *Department of Veterinary Microbiology and Preventive Medicine, Iowa State University, Ames, IA, USA; Department of Veterinary Diagnostic and Production Animal Medicine, Iowa State University, Ames, IA, USA*
SABINA SARVAN • *Ottawa Institute of Systems Biology, University of Ottawa, Ottawa, ON, Canada; Department of Biochemistry, Microbiology and Immunology, University of Ottawa, Ottawa, ON, Canada*
NICHOLLAS E. SCOTT • *Department of Microbiology and Immunology, Doherty Institute, The University of Melbourne, Melbourne, Victoria, Australia*
ANDREW J. SEVERIN • *Genome Informatics Facility, Office of Biotechnology, Iowa State University, Ames, IA, USA*
ZHANGQI SHEN • *Department of Veterinary Microbiology and Preventive Medicine, Iowa State University, Ames, IA, USA*

MARTINE CAMILLA HOLST SØRENSEN • *Department of Veterinary Disease Biology, Faculty of Health and Medical Sciences, University of Copenhagen, Frederiksberg C, Denmark*
MARTIN STAHL • *Division of Gastroenterology, Department of Pediatrics, Child and Family Research Institute, University of British Columbia, Vancouver, BC, Canada*
ALAIN STINTZI • *Department of Biochemistry, Microbiology and Immunology, Ottawa Institute of Systems Biology, University of Ottawa, Ottawa, ON, Canada*
ODILE TRESSE • *INRA, UMR 1014 Secalim, Nantes, France; LUNAM Université, Nantes, France*
BRUCE A. VALLANCE • *Division of Gastroenterology, Department of Pediatrics, Child and Family Research Institute, University of British Columbia, Vancouver, BC, Canada*
XIMIN ZENG • *Department of Animal Science, University of Tennessee, Knoxville, TN, USA*
QIJING ZHANG • *Department of Veterinary Microbiology and Preventive Medicine, Iowa State University, Ames, IA, USA*

Chapter 1

Campylobacter jejuni Isolation/Enumeration from Environmental Samples

Kelli L. Hiett

Abstract

Currently, there is no universally accepted standard media or method for the recovery of *Campylobacter* species. This is likely due to the ubiquity of the organism in nature, the complex sample matrices from which the organism is often recovered, as well as the fragile/viable-but nonculturable state the organism assumes in response to stress. The use of a sterile filter placed upon a nonselective Brucella Agar Blood Plate (BAB), followed by incubation at 37 °C in a hydrogen-containing atmosphere (Campycheck), is one method to recover stressed and emerging *Campylobacter* spp. from complex environmental matrices; however, this technique does not currently allow for the enumeration of the recovered organisms. Enumeration is performed using serial dilutions of sample homogenate plated onto modified Campy-Cefex media followed by incubation at either 37 °C or 42 °C in a microaerobic atmosphere.

Key words *Campylobacter*, Cape Town, Campycheck, VBNC, Hydrogen, Environmental, Zoonotic pathogen

1 Introduction

A multitude of investigations focused on delineating *Campylobacter* spp. epidemiology have been conducted; however, a complete understanding of the critical sources for *Campylobacter* spp. transmission remains elusive [1]. A significant contributor to the current knowledge gap regarding *Campylobacter* spp. transmission is the use of several highly selective, cultivation media that demonstrate suboptimal recovery [2–4]. The majority of culture media widely in use rely upon the incorporation of antimicrobials (greater than 17 used either singularly or in some combination) for growth inhibition of other bacteria and fungi [5, 6]. The use of varied antimicrobials likely introduces bias in recovery, as isolates that are resistant to the antimicrobial used will preferentially survive;

Electronic supplementary material: The online version of this chapter (doi:10.1007/978-1-4939-6536-6_1) contains supplementary material, which is available to authorized users. Videos can also be accessed at http://www.springerimages.com/videos/[doi:10.1007/978-1-4939-6536-6_1]

James Butcher and Alain Stintzi (eds.), *Campylobacter jejuni: Methods and Protocols*, Methods in Molecular Biology, vol. 1512, DOI 10.1007/978-1-4939-6536-6_1, © Springer Science+Business Media New York 2017

additionally, these antimicrobials often have deleterious effects on the recovery of isolates, especially from environments where sublethally injured cells (viable-but nonculturable [VBNC]) are of concern [2, 7]. The use of an enrichment step is often considered a solution for recovery of these injured cells [4, 8, 9], but there remains a concern that an enrichment (using both selective and nonselective broths) step will only allow for faster growing, healthier, or adaptive strains to dominate [10–12]. Other variables often encountered in methodology include the use of various temperatures, 37 °C or 42 °C or a combination thereof, for incubation of both enrichment cultures as well as plates.

Moreover, during a majority of prevalence and epidemiologic studies, only one colony, exhibiting "typical" *C. jejuni* morphology, is often chosen for further analyses. Published investigations analyzing subtype variability of multiple isolates recovered relative to different media formulations and incubation conditions revealed that different *C. jejuni* subtypes are recovered with different media [10, 13, 14]. A recent concern is the demonstration that the use of the aforementioned selective media has an inherent bias for the recovery of *C. jejuni* and *C. coli* over other species, while newly emerging or novel species remain undetected [15–17]. When molecular methods (such as PCR, quantitative PCR, 16S rDNA sequence analyses, or high-throughput sequencing) are used for direct (noncultural) detection, several additional *Campylobacter* species are often identified [18, 19]. It is evident that large knowledge gaps exist related to the cultural recovery of *Campylobacter* spp. The cultural recovery of pathogenic organisms is not only required for regulatory agencies and public health agencies, but recovery is critical for future investigations including phenotypic expression studies, subtype analyses, and antimicrobial resistance testing.

Novel cultural methods that take advantage of both the small size (~0.2–0.8 × 0.5–5 μm) and motile nature of *Campylobacter* spp., relative to other organisms, are currently being evaluated and refined to facilitate the isolation and tracking of novel *Campylobacter jejuni* subtypes and emerging *Campylobacter* spp. [13, 20]. One such method, the Campycheck procedure (EC FP5-CAMPYCHECK project QLK CT 2002 02201; also referred to as the modified Cape Town procedure [21]), employs size filtration onto nonselective plates followed by incubation in a hydrogen containing atmosphere (7.5% H_2, 2.5% O_2, 10% CO_2, and 80% N_2). While this method is currently tedious to perform in large-scale settings, it is of great benefit for the recovery of novel/emerging *Campylobacter* spp. for subsequent characterization and investigations.

2 Materials

2.1 Campycheck

1. Campycheck Transport Media (CCTM; 1 l): Brucella Broth—28 g (dissolve into 950 mL purified water; mix thoroughly, autoclave 20 min at 121 °C, cool to 50 °C in water bath). Supplement with either *Campylobacter* Growth Supplement liquid (Oxoid—#SR0232E—two vials) **OR** 0.25 g sodium pyruvate, 0.25 g sodium bisulfite, 0.25 g ferrous sulfate. Add 50 mL lysed horse blood.
2. Brucella Agar with Blood Plates (BAB; 1 l): Brucella Agar—43 g (dissolve into 950 mL purified water; autoclave 20 min at 121 °C, cool to 50 °C in water bath). Supplement with either *Campylobacter* Growth Supplement liquid (Oxoid—#SR0232E—2 vials) **OR** 0.25 g sodium pyruvate, 0.25 g sodium bisulfite, 0.25 g ferrous sulfate. Add 50 mL lysed horse blood.
3. Petri Plates—sterile 100 mm × 15 mm polystyrene.
4. Hot plate.
5. Sterile 50 mm, 0.6 μm mixed cellulose ester filter.
6. Pipettors.
7. Sterile pipette tips.
8. ZipTop bag (*see* **Note 1**) or microaerobic gas generation system (e.g., Anoxomat™ System).
9. Pulsifier or Stomacher.
10. Filtered stomacher bags.
11. Sterile tweezers.
12. Gas Tank—7.5% H_2, 2.5% O_2, 10% CO_2, and 80% N_2.
13. 37 °C incubator.

2.2 Modified Campy-Cefex

1. Sterile 1× PBS—8 g NaCl, 0.2 g KCl, 1.44 g Na_2HPO_4, 0.24 g KH_2PO_4 dissolved into 800 mL distilled water. Adjust pH to 7.4 with HCl. Adjust final volume, with distilled water, to 1 L. Autoclave for 20 min at 121 °C. May be stored at room temperature or at 4 °C.
2. Cefoperazone stock—1 g added to 10 mL sterile water. Shake vigorously to dissolve. Dispense 1 mL aliquots into sterile cryovial tubes and store at −20 °C or −80 °C. Add 333 μL for 1 L of Campy-Cefex Agar.
3. Cycloheximide stock—2 g dissolved into 10 mL of 1:1 dH_2O/methanol (*see* **Note 2**). Filter sterilize. Add 1 mL for 1 L of Campy-Cefex Agar. Make fresh as needed.

4. Modified Campy-Cefex Agar Plates (1 L): Brucella Agar—43 g, 0.5 g ferrous sulfate, 0.2 g sodium bisulfite, 0.5 g pyruvic acid (dissolve into 950 mL purified water and bring to a boil, autoclave 20 min at 121 °C, cool to 50 °C in water bath). Add 50 mL lysed horse blood, 333 μL cefoperazone stock, 1 mL cycloheximide stock.
5. Petri Plates—sterile 100 mm × 15 mm polystyrene.
6. Hot Plate.
7. Sterile loops.
8. Pipettors.
9. Sterile pipette tips.
10. Turntable.
11. ZipTop bag, anaerobic chamber, or microaerobic gas generation system.
12. Pulsifier or Stomacher.
13. Filtered stomacher bags.
14. Blood-Gas Gas Tank—5.0% O_2, 10% CO_2, and 85% N_2.
15. 37 °C or 42 °C incubator.

2.3 Serial Dilutions for Enumeration

1. 1× PBS.
2. Sterile, 24-well, flat bottom, polystyrene, tissue culture plates.
3. Single channel pipettors or 8-channel pipettor, 15–1250 μL volumes.
4. Sterile pipette tips.

2.4 Long-Term Storage

1. Sterile cryovials.
2. Sterile 80% Glycerol (*see* **Note 3**)—80 mL of glycerol plus 20 mL sterile water. Autoclave 15 min at 121 °C.
3. Brucella Broth (1 L)—Dissolve 28 g of Brucella Broth into 1 L of purified water; mix thoroughly, autoclave 15 min at 121 °C.
4. 16% glycerol stocks—Add 800 μL of sterile Brucella Broth and 200 μL of sterile 80% glycerol into a sterile cryovial.

3 Methods

3.1 Sampling Methodology

3.1.1 Soil Sample Resuspension (1:3 w:v Soil:Sterile Campy Check Transport Media [CCTM])

1. Using gloves, weigh out 1 g of the collected soil sample.
2. Place the sample into a filtered stomacher bag.
3. Add 3 mL of sterile CCTM to the sample.
4. Place into a stomacher and homogenize for ~1 min.

3.1.2 Fecal Sample Resuspension (1:3 w:v Feces:Sterile CCTM)

1. Using gloves, weigh out 1 g of the collected fecal sample.
2. Place the sample into a filtered stomacher bag.
3. Add 3 mL of sterile CCTM to the sample.
4. Place into a stomacher and homogenize for ~1 min.

3.1.3 Water Sample

1. Place 500 mL of the water sample into a sterile 1 L centrifuge bottle and centrifuge at 4 °C for 20 min, at 8500 rpm.
2. Carefully decant the supernatant and resuspend the pellet in 5 mL of sterile CCTM.

3.1.4 Sediment Sample (1:3 w:v Sediment:Sterile CCTM)

1. Using gloves, weigh out 1 g of the collected sediment sample.
2. Place the sample into a filtered stomacher bag.
3. Add 3 mL of sterile CCTM to the sample.
4. Place into a stomacher and homogenize for ~1 min.

3.2 Campycheck Cultural Method

1. Using sterile tweezers, aseptically place a 50 mm, 0.6 μm mixed cellulose ester filter (*see* **Note 4**) onto the middle surface of a Brucella Agar with Blood (BAB) Plate (Video S1).
2. Cover the plate and allow the filter to set for 5 min at room temperature. The filter will uniformly absorb moisture from the plate so that no bubbles are present.
3. Pipette 4 × 50 μL aliquots of homogenized sample, or resuspended water sample, onto the filter surface at distinct locations. Avoid seepage/spreading of the homogenate over the filter edge (Video S2).
4. Cover the plate and allow it to incubate at room temperature for 15 min.
5. Using sterile tweezers, aseptically remove the inoculated filter from the plate. Use caution and avoid excess homogenate from dripping onto the plate. Place inoculated filter in biohazard disposal container (Video S3).
6. Invert plates and place into a ZipTop Bag. Partially seal the bag, push out the ambient atmosphere, and flush twice with hydrogen-enriched gas (7.5% H_2, 2.5% O_2, 10% CO_2, and 80% N_2) prior to the final fill (Video S4). Alternatively, a gas dispensing system may be used to introduce the proper atmosphere (Video S5).
7. Incubate at 37 °C for 24–72 h (*see* **Note 5**).
8. Colonies will primarily grow within the boundaries of where each aliquot was placed (Fig. 1).

Fig. 1 Campycheck colony recovery: Colony growth occurs within the boundaries aliquot placement

9. Sub-culture suspect *Campylobacter* spp. colonies on BAB plates, and incubate at 37 °C in a hydrogen-enriched atmosphere (7.5 % H_2, 2.5 % O_2, 10 % CO_2, and 80 % N_2) as described in **steps 6** and 7.
10. Remove a loopful of culture, resuspend in 16 % glycerol stock vial and store at −80 °C for further identification and analyses.

3.3 Modified Campy-Cefex, Serial Dilutions, and Enumeration

1. Remove 100 μL of sample homogenate and directly plate onto a modified Campy-Cefex Agar plate.
2. Place inoculated plate onto a turntable, and use a sterile loop or spreader to spread the inoculum across the plate while spinning (*see* **Note 6**)
   (a) For samples diluted 1:3 w:v (soil, feces, sediment), this will be a 2.5×10^{-2} dilution.
   (b) For water samples, pelleted and resuspended in 5 mL, this will be a direct plating.
3. Place 900 μL of sterile 1X PBS into five individual, consecutive wells of a sterile 24-well tissue culture plate.
   (a) When processing soil, fecal, or sediment samples, label the first well 10^{-3}, the second well 10^{-4}, the third well 10^{-5}, etc.
   (b) When processing water samples, label the first well 10^{-1}, the second well 10^{-2}, the third well 10^{-3}, etc.
4. Place 100 μL of the original sample homogenate (stomacher bag) into the first well (10^{-3} for soil/feces/sediment; 10^{-1} water). Mix thoroughly by gently pipetting the mixture up and down (*see* **Note 7**).
5. Dispose of the old tip.

6. Remove 100 μL of the 10^{-3} soil, fecal, sediment mixture, add to the second well (10^{-4}), and mix thoroughly as previously described.
7. Dispose of the old tip.
8. Continue dilution series.
9. Remove 100 μL from each serial dilution and plate onto a modified Campy-Cefex Agar plate.
10. Place inoculated plate onto a turntable and use a sterile loop to spread the inoculum across the plate while spinning (*see* **Note 6**).
11. Continue for each dilution series.
12. Invert plates and place into a ZipTop Bag. Partially seal the bag, push out the ambient atmosphere, and flush twice with Blood-Gas Mixture—5.0% O_2, 10% CO_2, and 85% N_2 prior to the final fill (Video S4). Alternatively, a gas-dispensing system may be used to introduce the proper atmosphere (Video S5).
13. Incubate at 42 °C for 36–48 h in a microaerobic atmosphere (5% O_2, 10% CO_2, 85% N_2). subsequently stored at −80 °C in 16% glycerol stocks for further identification and analyses.
14. Select the plate with 50–200 colonies to count. As you count, use a permanent pen to mark the colonies. Multiply the number of colonies by your dilution factor for enumeration.
15. Sub-culture suspect *Campylobacter* spp. colonies on BAB plates, and incubate at 42 °C in a microaerobic atmosphere (5.0% O_2, 10% CO_2, and 85% N_2) as described in **steps 12** and **13**.
16. Remove a loopful of culture, resuspend in 16% glycerol stock vial and store at −80 °C for further identification and analyses.

4 Notes

1. ZipTop bags are an inexpensive technique for maintaining defined atmospheres during incubation.
2. The cyclohexamide stock solution is best dissolved first in methanol followed by the addition of water.
3. The preparation of an 80% stock glycerol solution facilitates aseptic pipetting of small amounts of viscous liquid.
4. Filters are individually separated by blue paper. Be sure that the filter is placed on the BAB plate. Discard the blue papers.
5. During extended incubation periods, the ZipTop bags may deflate. It is best to check bags every 24 h and refill with fresh gas mixture.
6. Be careful to not gouge the agar with the loop while spreading homogenates.
7. Use filter tips to avoid contamination of the pipettor.

References

1. Agunos A, Waddell L, Leger D et al (2014) A systematic review characterizing on-farm sources of *Campylobacter* spp. for broiler chickens. PLoS One 9(8):e104905. doi:10.1371/journal.pone.0104905
2. Silley P (2003) Campylobacter and fluoroquinolones: a bias data set? Environ Microbiol 5(4):219–230.doi:10.1046/j.1462-2920.2003.00425.x
3. Hofreuter D (2014) Defining the metabolic requirements for the growth and colonization capacity of *Campylobacter jejuni*. Front Cell Infect Nicrobiol 4:137. doi:10.3389/fcimb.2014.00137
4. Ugarte-Ruiz M, Gómez-Barrero S, Porrero MC et al (2012) Evaluation of four protocols for the detection and isolation of thermophilic *Campylobacter* from different matrices. J Appl Microbiol 113(1):200–208. doi:10.1111/j.1365-2672.2012.05323.x
5. Corry JE, Post DE, Colin P et al (1995) Culture media for the isolation of campylobacters. Int J Food Microbiol 26(1):43–76
6. Gharst G, Oyarzabal OA, Hussain SK (2013) Review of current methodologies to isolate and identify *Campylobacter* spp. from foods. J Microbiol Methods 95(1):84–92. doi:10.1016/j.mimet.2013.07.014
7. Duarte A, Botteldoorn N, Coucke W et al (2015) Effect of exposure to stress conditions on propidium monoazide (PMA)-qPCR based *Campylobacter* enumeration in broiler carcass rinses. Food Microbiol 48:182–190. doi:10.1016/j.fm.2014.12.011
8. Kim SA, Lee YM, Hwang IG et al (2009) Eight enrichment broths for the isolation of *Campylobacter jejuni* from inoculated suspensions and ground pork. Lett Appl Microbiol 49(5):620–626. doi:10.1111/j.1472-765X.2009.02714.x
9. Tangvatcharin P, Chanthachum S, Kopaiboon P et al (2005) Comparison of methods for the isolation of thermotolerant *Campylobacter* from poultry. J Food Prot 68(3):616–620
10. Williams LK, Sait LC, Cogan TA et al (2012) Enrichment culture can bias the isolation of *Campylobacter* subtypes. Epidemiol Infect 140(7):1227–1235. doi:10.1017/s0950268811001877
11. de Boer P, Rahaoui H, Leer RJ et al (2015) Real-time PCR detection of *Campylobacter* spp.: a comparison to classic culturing and enrichment. Food Microbiol 51:96–100. doi:10.1016/j.fm.2015.05.006
12. Magajna B, Schraft H (2015) Evaluation of propidium monoazide and quantitative PCR to quantify viable *Campylobacter jejuni* biofilm and planktonic cells in log phase and in a viable but nonculturable state. J Food Prot 78(7):1303–1311. doi:10.4315/0362-028x.jfp-14-583
13. Jokinen CC, Koot JM, Carrillo CD et al (2012) An enhanced technique combining pre-enrichment and passive filtration increases the isolation efficiency of *Campylobacter jejuni* and *Campylobacter coli* from water and animal fecal samples. J Microbiol Methods 91(3):506–513. doi:10.1016/j.mimet.2012.09.005
14. Simmons M, Hiett KL, Stern NJ et al (2008) Comparison of poultry exudate and carcass rinse sampling methods for the recovery of *Campylobacter* spp. subtypes demonstrates unique subtypes recovered from exudate. J Microbiol Methods 74(2-3):89–93. doi:10.1016/j.mimet.2008.03.007
15. Acke E, McGill K, Golden O et al (2009) A comparison of different culture methods for the recovery of *Campylobacter* species from pets. Zoonoses Public Health 56(9-10):490–495. doi:10.1111/j.1863-2378.2008.01205.x
16. Koziel M, Corcoran GD, Sleator RD et al (2014) Detection and molecular analysis of *Campylobacter ureolyticus* in domestic animals. Gut Pathogens 6:9. doi:10.1186/1757-4749-6-9
17. Fitzgerald C, Tu ZC, Patrick M et al (2014) *Campylobacter fetus* subsp. *testudinum* subsp. *nov.* isolated from humans and reptiles. Int J Syst Evol Microbiol 64(Pt 9):2944–2948. doi:10.1099/ijs.0.057778-0
18. Chaban B, Ngeleka M, Hill JE (2010) Detection and quantification of 14 *Campylobacter* species in pet dogs reveals an increase in species richness in feces of diarrheic animals. BMC Microbiol 10:73. doi:10.1186/1471-2180-10-73
19. Kaakoush NO, Mitchell HM, Man SM (2014) Role of emerging *Campylobacter* species in inflammatory bowel diseases. Inflamm Bowel Dis 20(11):2189–2197. doi:10.1097/mib.0000000000000074
20. Casanova C, Schweiger A, von Steiger N et al (2015) *Campylobacter concisus* pseudo-outbreak caused by improved culture conditions. J Clin Microbiol 53(2):660–662. doi:10.1128/jcm.02608-14
21. Lastovica AJ, le Roux E (2000) Efficient isolation of campylobacteria from stools. J Clin Microbiol 38(7):2798–2799

Chapter 2

Recovery of *Campylobacter* spp. from Food and Environmental Sources

Catherine D. Carrillo, Robyn Kenwell, Irene Iugovaz, and Omar A. Oyarzabal

Abstract

The recovery of *Campylobacter* species from food and environmental sources is challenging due to the slow growth of these bacteria and the need to suppress competing organisms during the isolation procedures. The addition of multiple selective antimicrobials to growth media can negatively impact recovery of some *Campylobacter* spp. Here, we describe our current method for the isolation of thermotolerant *Campylobacter* species, mainly *C. jejuni* and *C. coli*, from food and environmental samples. We emphasize the use of membrane filtration during plating for the specific isolation of *Campylobacter* spp. and a reduced use of antimicrobial supplements throughout the whole isolation process.

Key words Campylobacter, Bolton broth, Buffered peptone water, Nitrocellulose membrane filter, Poultry, Produce, Raw milk

1 Introduction

Campylobacter spp., primarily *C. jejuni* and *C. coli*, are commonly found in the intestine of livestock, where these bacteria establish a commensal relationship with the animal host. The main routes for human campylobacteriosis are the handling of raw chicken meat, the consumption of undercooked meats (usually chicken), raw milk, and contaminated water. Procedures for the isolation of *Campylobacter* from food and environmental samples have been adapted from methods originally developed for clinical, mainly fecal, specimens [1, 2]. The first reported method consisted of passive filtration of fecal suspensions through Millipore filters with a pore size of 0.65 μm onto blood agar plates, followed by microaerobic incubation [1]. Due to their small size and motility, *Campylobacter* spp. move through the filters easily, whereas most other bacteria are retained by the filters. Subsequently, the use of membrane filters was largely abandoned in favor of more sensitive methods incorporating selective antimicrobials in the plating media [2, 3].

James Butcher and Alain Stintzi (eds.), *Campylobacter jejuni: Methods and Protocols*, Methods in Molecular Biology, vol. 1512, DOI 10.1007/978-1-4939-6536-6_2, © Springer Science+Business Media New York 2017

For food and environmental samples, the isolation procedure typically requires the enrichment of the sample in liquid media to allow for *Campylobacter* spp. to grow to detectable numbers prior to isolation on plate media. A number of media formulations have been developed to improve the isolation of *Campylobacter* spp. from food samples, such as Campy-Cefex or Campy-Line [4, 5]. Nonetheless, *C. jejuni* and *C. coli* are easily isolated with common microbiology media (e.g., nutrient, tryptic soy, or Brucella broths) supplemented with selective antibiotics, as long as the incubation is done under microaerobic conditions (5% O_2, 10% CO_2, and 85% N_2) that favor the growth of *Campylobacter* spp. (reviewed in [4]). The addition of antibiotics to enhance media selectivity should be done balancing the fact that the growth of *Campylobacter* cells may be inhibited by excessive use of these agents [4–6].

Our current approach for the recovery of *Campylobacter* spp. has been developed to minimize the use of antimicrobial agents. Test samples are added to either Bolton or Park and Sanders enrichment broths supplemented with antimicrobials and are incubated microaerobically for up to 48 h [7–9]. Enrichment broths are plated through membrane filters with pore sizes of 0.45 or 0.65 μm onto antibiotic-free Charcoal Cefoperazone Deoxycholate Agar (CCDA, also known as *Campylobacter* blood-free selective medium) and plates are incubated microaerobically for up to 48 h. While the filters can reduce the sensitivity of *Campylobacter* detection, *Campylobacter* cells are present at high concentrations following the enrichment period, and positive samples can be accurately identified. Resulting colonies are screened using a multiplex PCR (mPCR) to identify the most common species, *C. jejuni* and *C. coli* [10]. The priming oligonucleotides (primers) used in the PCR are highly specific for *C. jejuni* or *C. coli* and do not amplify DNA present in any other *Campylobacter* spp. or non-*Campylobacter* organisms. In the rare cases where other species are recovered, standard approaches can be used for species identification. The method described here has been mainly used for the isolation and identification of *Campylobacter* spp. from raw poultry meat, but has also been tested with water, soil, produce, and chicken feces/cecal contents [7, 8, 11–14].

2 Materials

2.1 Materials for Microbiological Analysis

1. 0.1% Peptone water: Dissolve 15 g of Peptone media (Oxoid) in 1 L of distilled water. Sterilize by autoclaving at 121 °C for 15 min. Store at 4 °C. For Peptone water/1% Tween 80 solutions, add 10 mL of Tween 80 to the solution prior to autoclaving.
2. Park & Sanders broth: Prepare Brucella broth by adding 28 g Brucella broth base to 1 L of deionized water. Sterilize by autoclaving at 121 °C for 15 min and cool to 50 °C. Store at

4 °C until use. Immediately prior to use, add 50 mL/L lysed horse blood and antibiotic supplement. The antibiotic supplement is prepared by combining 10 mg vancomycin, 10 mg trimethoprim lactate, 32 mg cefoperazone, sodium salt in 10 mL sterile water. Heat to 37 °C for 15–30 min with shaking to dissolve then filter sterilize (0.22 μm filter). Add 1 mL to 100 mL of enrichment broth.

3. Bolton broth (standard): Prepare Bolton broth by adding 27.6 g Bolton broth base (Oxoid) to 1 L of deionized water. Sterilize and cool to 50 °C, or store at 4 °C until use. Add 50 mL/L lysed horse blood and four vials per liter Bolton broth-selective supplement (SR0183, Oxoid; 10.0 mg of cefoperazone, 10.0 mg of vancomycin, 10.0 mg of trimethoprim, 25.0 mg of cycloheximide per vial).
4. Bolton broth with reduced supplements: Prepare Bolton broth base as described above, except omitting the addition of supplements. Instead, combine cefoperozone (33 mg) and amphotericin B (0.04 mg) in 10 mL sterile, distilled water. Filter sterilize and add 1 mL to each 100 mL of enrichment broth.
5. Sterile filter bag (e.g., Whirl-Pak® Filter Bag).
6. Charcoal Cefoperazone Deoxycholate Agar (CCDA): Add 45.5 g of the agar base (Oxoid) to 1 L of distilled water. Sterilize by autoclaving at 121 °C for 15 min, cool to 50 °C, and pour into sterile Petri dishes. Store at 4 °C.
7. Filters: 0.45 μm, 47 mm nitrocellulose membrane filters or 0.65 μm, 47 mm nitrocellulose membrane filters.
8. Stomacher® lab paddle blender or equivalent.

2.2 Materials for Multiplex PCR Identification of *C. jejuni* and *C. coli*

1. Block heater capable of accommodating 1.5 mL microfuge tubes.
2. Mini centrifuge capable of centrifuging 1.5 mL microfuge tubes.
3. Adjustable micropipettors to cover a wide range of volumes (1.0–1000 μL) and DNAse-free pipettor tips.
4. Vortex mixer.
5. DNAse-free microcentrifuge tubes (0.2–1.5 mL) or equivalent.
6. Lysis buffer (0.25 % SDS, 0.05 M NaOH). Store at room temperature for up to 1 month.
7. Prepare oligonucleotide primers: Rehydrate primers in appropriate volumes of TE (10 mM Tris, 1 mM EDTA, pH 8.0) to make 100 μM stocks for each of the 6 oligonucleotide primers (Table 1). Prepare a 100 μL, 10 μM primer mix by adding 10 μL of each of the primers (100 μM) to 40 μL of nuclease-free water (*see* **Note 1**).

Table 1
Oligonucleotide primers for identification of *C. jejuni* and *C. coli* [10]

Primer name	Reference	Gene	Sequence (5′-3′)	Target species
16 s-F V2 16 s-R V2	Adapted from [10]	16S rDNA	GGGAGGCAGCAGTRRGGAAT[a] TGACGGGCGGTGRGTACAAG[a]	Universal (eubacteria)
CC18-F CC519-R	Adapted from [15]	Aspartokinase (*asp*)	GGTATGATTTCTACAAGCGAG[a] ATAAAAGACTATCGTCGCGTG	*C. coli*
HipO-F HipO-R	[10]	Hippuricase (*hipO*)	GACTTCGTGCAGATATGGATGCTT GCTATAACTATCCGAAGAAGCCATCA	*C. jejuni*

[a]*see* **Note 2**

8. For each 25 μL PCR reaction, combine 12.5 μL TopTaq Master Mix (2×, Qiagen) or equivalent (*see* **Note 3**), 3 μL primer mix (10 μM) and 7.5 μL PCR grade water (nuclease-free, DNA-free, and sterile) (*see* **Note 4**).
9. Thermocycler (Eppendorf MasterCycler or equivalent).
10. DNA molecular size marker (100–1500 bp).
11. Agarose gel electrophoresis system.

3 Methods

3.1 Sample Preparation and Enrichment

Keep samples refrigerated and analyze as soon as possible after collection (*see* **Note 5**). Unless otherwise indicated, Bolton and Park and Sanders enrichment broths can be used interchangeably in this method (*see* **Note 6**). Include pure cultures of *C. jejuni* and *C. coli* control strains, and an uninoculated broth as positive and negative controls to ensure that the proper incubation conditions are met, and that the broth is not contaminated. Choose the appropriate sample preparation procedure depending on the material to be tested (**steps 1A–1D**).

1A. Raw meat (*see* **Note 7**):

   a. Ground meat or meat samples cut into small (0.3–0.5 cm^3) pieces: Add 25 g into a sterile 250 mL flask/bag containing 100 mL of enrichment broth.

   b. Raw whole chicken carcasses, meat pieces, boned meat pieces: Place the sample (up to 1.5 kg) into a large sterile plastic bag and add 200 mL 0.1% cold (4 °C) peptone water. Shake the contents gently and manually massage for approximately 2 min. Inoculate 25 mL of rinse into 100 mL of enrichment broth.

1B. Fresh produce: Using aseptic technique, place a test portion of 50 g in a sterile filter bag (e.g., Whirl-Pak® Filter Bag, Nasco, Fort Atkinson, WI). Add 100 mL of peptone water containing Tween 80. Manually massage samples for 2 min (*see* **Note 8**). Collect 50 mL from the filtered part of the bag and place in a 50 mL falcon tube. Centrifuge at 15,000 × *g* for 20 min and discard supernatant. Resuspend pellets in 5 mL of enrichment broth and add contents into a sterile 250 mL flask/bag containing 100 mL of enrichment broth.

1C. Water samples: Place 0.45 μm filter into a sterile filter unit using sterile forceps. Apply vacuum and, depending on sampling requirements, filter 10–1000 mL of water through the membrane. Place filter into a sterile 250 mL flask/bag containing 100 mL of enrichment broth (*see* **Note 9**).

1D. Chicken fecal/cecal contents (*see* **Note 10**): Using aseptic technique, place the desired test portion in a sterile filter bag. Weigh sample and add 9× w/v of enrichment broth. Mix for 2 min using a Stomacher® lab paddle blender at low speed (e.g., 200 rpm or equivalent). Plate a portion of the suspension directly as described in **step 3**, and proceed to enrichment for the remaining suspension (*see* **Note 11**).

2. Incubate under microaerobic conditions (10 % CO_2, 5 % O_2, 85 % N_2) at 42 °C for 24–48 h (*see* **Note 12**). For environmental and produce samples, or samples that have been previously frozen, a pre-enrichment step (4 h at 37 °C) in enrichment broth without antimicrobials is recommended (*see* **Note 13**).

3. Plating: Place a mixed cellulose ester filter (0.45 or 0.65 μm, *see* **Note 14**) on the surface of the CCDA plates prepared without selective supplements (*see* **Note 15**, Fig. 1a). Pipette a

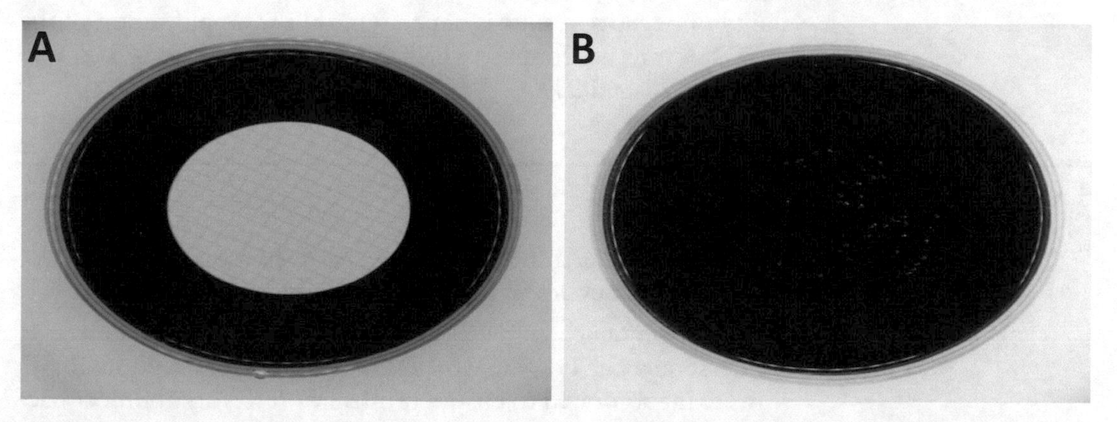

Fig. 1 Recovery of *Campylobacter* spp. from enrichment broths using passive filtration. Mixed cellulose filters are placed on antibiotic-free CCDA (**a**). Approximately 100 μL of enrichment broth is pipetted onto filters in five drops and left to dry for 15 min prior to aseptic removal of the filter. Following overnight microaerobic incubation of plates at 37 °C, isolated *Campylobacter* colonies are observed (**b**)

100 μL volume of enrichment onto the filter in five drops of approximately 20 μL each, and leave to dry for 15–20 min (*see* **Note 16**). Using sterile forceps remove and discard filters and incubate the plates microaerobically at 37 °C for 24–48 h.

4. Visually inspect plates. Typical *Campylobacter* colonies will be small, shiny and grey to translucent (*see* Fig. 1b).

3.2 Confirmation of C. jejuni and C. coli Colonies Using Multiplex PCR (See Note 17)

1. Using a sterile inoculating needle, pick a presumptive positive colony and resuspend in 20 μL of Lysis buffer in a 1.5 mL microfuge tube (*see* **Note 18**).
2. Incubate tubes in the heat block set to 100 °C (±2 °C) for 15 min.
3. Cool the tubes on ice (or equivalent) for 5 min. Add 180 μL of nuclease-free water to each tube.
4. Centrifuge for 5 min at 14,000 × *g*, 4 °C (*see* **Note 19**).
5. Add 2 μL of cell lysate or positive control to 23 μL of PCR reaction mixture in a microfuge tube suitable for thermocycler use. Mix by pipetting. Centrifuge briefly to collect PCR mixture at the bottom of the tube.
6. Place tubes into the thermocycler and start the temperature cycling program as follows: 95 °C (10 min), 35 cycles of 95 °C (30 s); 50 °C (30 s), 72 °C (1 min), with a final 72 °C (10 min) at the end of the program (*see* **Note 20**).
7. PCR products are separated by size by gel electrophoresis on a 1.5% agarose gel or an equivalent system according to the manufacturer's instructions.

Interpretation of results: The colony can be assessed as positive for *C. coli* if a DNA fragment of 500 bp in length is observed or as *C. jejuni* if a DNA fragment of 344 bp is observed (Fig. 2). The validity of negative results is dependent on the amplification of the 16s rDNA internal control, which exhibits a fragment of 1062 bp (*see* **Notes 21** and **22**).

4 Notes

1. Prepare several tubes at a time to decrease the number of freeze/thaw cycles of the 100 μM primer stock solutions.
2. Primer sequence CC18-F was modified to correct an error in the sequence. For the 16S rDNA primers, ambiguous bases were incorporated to ensure primers would work for most bacterial species.
3. This system includes most of the necessary components for PCR (e.g., dNTPs, PCR buffer, Magnesium Chloride, and Taq DNA polymerase) in the master mix. An equivalent system

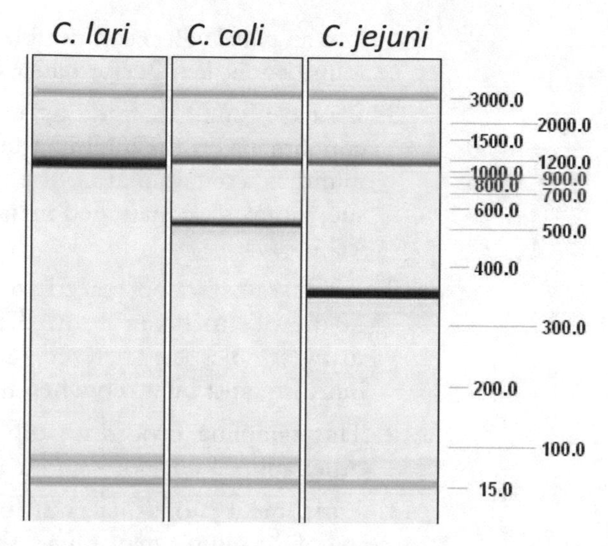

Fig. 2 Multiplex PCR for confirmation of *Campylobacter jejuni* and/or *C. coli*. PCR products were run on a QIAxcel Capillary Electrophoresis system using the DNA screening cartridge. A 500 bp fragment amplified from a portion of the aspartokinase gene identifies the isolate as *C. coli* and a 344 bp fragment of the hippuricase gene (*hipO*) is amplified from *C. jejuni* colonies. In non-target bacterial colonies (e.g., *C. lari*) only the larger (1062 bp) 16 s rRNA fragment is amplified and further testing for identification is conducted

can be used. Preparation of a master mix for other PCR systems may require adaptation of the method.

4. Stock solutions and working solutions should be stored at −20 °C. Thaw stock solutions at room temperature before use. A master mix for multiple reactions can be prepared and 23 μL aliquots dispensed into 0.2 mL microfuge tubes and stored at −20 °C until use.
5. *Campylobacter* spp. are extremely sensitive to oxygen, particularly at room temperature. *Campylobacters* are also sensitive to freezing, thawing, and drying conditions.
6. Park and Sanders and Bolton enrichment broths have been compared and found to have similar performance using standard antimicrobials. A Bolton broth formulation with reduced supplements (cefoperazone and amphotericin B) has been used successfully with poultry products, but not tested with other samples. Other enrichment media formulations may be equally effective. For example, buffered peptone water was found to perform similarly to Bolton broth with raw poultry samples [16]. However, enrichment broth may affect the genetic diversity of species recovered [17].
7. In many methods, enrichment ratios are 1:9 (meat:broth). We have found that a ratio of 1:4 performs similarly for retail broiler meat [18]. We have also tested the addition of rinse

into an equal volume of double strength enrichment broth and found equivalent performance [6].

8. We have found that adding non-intact leafy greens to spiked enrichment broths inhibits the growth of campylobacters. To minimize contamination of enrichment broths with plant tissue, samples are massaged manually rather than using mechanical means.
9. Filters can also be placed in sterile petri dishes containing 20 mL of enrichment broth. This approach may be desirable if many samples are analyzed (e.g., for a quantitative analysis), but care must be taken when handling samples.
10. This sampling procedure can also be used with feces from other animals and soil/sludge samples.
11. Enrichment is optional, but for fecal samples, significant drying of the sample may impact the viability of the *Campylobacter* cultures and enrichment may be required [14].
12. Longer incubations (up to 96 h) may be preferable to improve *Campylobacter* recovery; however, increased time may impact the diversity of *Campylobacter* strains recovered [9].
13. Some studies have shown that lower temperatures improve recovery of *Campylobacter* spp. from food and environmental samples in which *Campylobacter* cells are likely to be in a "stressed" state [19].
14. *Campylobacter* pass through 0.65 μm filters more efficiently; however, enriched cultures typically contain high concentrations of *Campylobacter* cells so sensitivity is not an issue and 0.45 μm filters work well. For direct plating of fecal/cecal samples, the 0.65 μm filters should be used (*see* Ref. [8]).
15. Any plate that is appropriate for growth of campylobacters could be used. We prefer to use the CCDA plates as *Campylobacter* colonies are easy to identify due to their characteristic color and morphology. Addition of standard supplements may be required for some sample types.
16. Filters should be visibly dry when they are removed. If the sample does not appear to have absorbed, check if moisture on plates is too high. Typically, agar plates are dried for 5 h in a laminar flow hood or an atmosphere controlled incubator prior to plating to ensure that moisture on the plates does not interfere with rapid drying of the filters without overdrying which may result in unfavorable growth conditions. Humidity levels and age of the plates could impact drying times and may need to be optimized.
17. Note that this procedure can also be used to screen enrichment broths prior to plating. One mL of enrichment broth is centrifuged (14,000 × *g* for 5 min) and pellets are lysed as described for the colony PCR.

18. If time permits, colonies can be purified by streaking for single colonies on nonselective agar and incubated at 37 °C overnight prior to the PCR-confirmation test. Alternatively, spot colony onto an appropriate agar plate, then lyse cells remaining on the inoculating loop. Typically three colonies are tested for every sample. More may be desirable, as we have found poultry samples to be commonly contaminated with multiple genotypes [6, 17].
19. The lysate can be stored at −20 °C for up to a week and thawed at room temperature immediately before use.
20. PCR products can be stored at 4 °C for up to a month.
21. In some cases the internal control band will not be observed when the smaller amplicons indicative of *C. coli* or *C. jejuni* are present. The amplification of these small products may be favored in the multiplex reaction. We have evaluated this procedure in parallel with biochemical identification methods [9] with over 500 colonies of *C. jejuni/C. coli* and have found greater than 98 % concordance of the two methods. Most of the discrepancies arose due to misidentification of *C. jejuni* as *C. coli* with the biochemical method because of a weak hippuricase reaction.
22. When using the method described herein, we do not commonly find other species; however, in such cases, presence of *Campylobacter* spp. is confirmed using identification methodology described in [9] or equivalent systems. Isolates that are not *C. jejuni* or *C. coli* by PCR are visualized by phase contrast microscopy to confirm typical *Campylobacter* morphology. Young cells of *Campylobacter* are Gram-negative, small (0.2–0.8 μm wide by 1.5–5 μm long) and S-shaped. They possess a characteristic corkscrew-like motility. Old cells (grown for more than 72 h) or cells exposed to air often change to coccoid forms.

Acknowledgements

The authors would like to thank Dr. Albert Lastovica for introducing us to the filtration technique.

References

1. Butzler JP, Dekeyser P, Detrain M et al (1973) Related vibrio in stools. J Pediatr 82(3):493–495
2. Skirrow MB (1977) Campylobacter enteritis: a "new" disease. Br Med J 2(6078):9–11
3. Goossens H, De Boeck M, Coignau H et al (1986) Modified selective medium for isolation of *Campylobacter* spp. from feces: comparison with Preston medium, a blood-free medium, and a filtration system. J Clin Microbiol 24(5):840–843
4. Gharst G, Oyarzabal OA, Hussain SK (2013) Review of current methodologies to isolate and identify *Campylobacter* spp. from foods. J Microbiol Methods 95(1):84–92. doi:10.1016/j.mimet.2013.07.014
5. Oyarzabal OA, Macklin KS, Barbaree JM et al (2005) Evaluation of agar plates for direct

enumeration of *Campylobacter* spp. from poultry carcass rinses. Appl Environ Microbiol 71(6):3351–3354

6. Carrillo CD, Plante D, Iugovaz I et al (2014) Method-dependent variability in determination of prevalence of *Campylobacter jejuni* and *Campylobacter coli* in Canadian retail poultry. J Food Prot 77(10):1682–1688. doi:10.4315/0362-028X.JFP-14-133
7. Williams A, Oyarzabal OA (2012) Prevalence of *Campylobacter* spp. in skinless, boneless retail broiler meat from 2005 through 2011 in Alabama, USA. BMC Microbiol 12:184. doi:10.1186/1471-2180-12-184
8. Speegle L, Miller ME, Backert S et al (2009) Use of cellulose filters to isolate *Campylobacter* spp. from naturally contaminated retail broiler meat. J Food Prot 72(12):2592–2596
9. Carrillo CD, Iugovaz I (2014) Isolation of *Campylobacter* from food (MFLP-46). In: Health Products and Food Branch (HPFB) HC (ed) Compendium of analytical methods, Vol 3. Health Canada, Ottawa, Ontario
10. Persson S, Olsen KE (2005) Multiplex PCR for identification of *Campylobacter coli* and *Campylobacter jejuni* from pure cultures and directly on stool samples. J Med Microbiol 54(Pt 11):1043–1047
11. Jokinen CC, Koot JM, Carrillo CD et al (2012) An enhanced technique combining pre-enrichment and passive filtration increases the isolation efficiency of *Campylobacter jejuni* and *Campylobacter coli* from water and animal fecal samples. J Microbiol Methods 91(3):506–513. doi:10.1016/j.mimet.2012.09.005
12. Carrillo CD, Sproston E, Kenwell R, Ivanov N, Todd BP, Huang H (2013) Prevalence of *Arcobacter* spp. in fresh vegetables from farmers' outdoor markets in Ottawa, Canada. In: Abstracts of posters and oral presentations from the 17th International Workshop on *Campylobacter, Helicobacter* and Related Organisms, 15–19th September, Aberdeen, Scotland
13. Carrillo CD, Kenwell R, Mohajer S, *et al.* (2011) *Campylobacter* spp. in farm environmental water sources in Eastern Ontario. Paper presented at the 16th International Workshop on Campylobacter, Helicobacter and Related Organisms, Vancouver, Canada
14. Potturi-Venkata LP, Backert S, Vieira SL et al (2007) Evaluation of logistic processing to reduce cross-contamination of commercial broiler carcasses with *Campylobacter* spp. J Food Prot 70(11):2549–2554
15. Linton D, Lawson AJ, Owen RJ et al (1997) PCR detection, identification to species level, and fingerprinting of *Campylobacter jejuni* and *Campylobacter coli* direct from diarrheic samples. J Clin Microbiol 35(10):2568–2572
16. Oyarzabal OA, Backert S, Nagaraj M et al (2007) Efficacy of supplemented buffered peptone water for the isolation of *Campylobacter jejuni* and *C. coli* from broiler retail products. J Microbiol Methods 69(1):129–136
17. Williams LK, Sait LC, Cogan TA et al (2012) Enrichment culture can bias the isolation of *Campylobacter* subtypes. Epidemiol Infect 140(7):1227–1235. doi:10.1017/S0950268811001877
18. Oyarzabal OA, Liu L (2010) Significance of sample weight and enrichment ratio on the isolation of naturally occurring *Campylobacter* spp. in commercial retail broiler meat. J Food Prot 73(7):1339–1343
19. Humphrey TJ, Muscat I (1989) Incubation temperature and the isolation of *Campylobacter jejuni* from food, milk or water. Lett Appl Microbiol 9(4):137–139

Chapter 3

Methods for Isolation, Purification, and Propagation of Bacteriophages of *Campylobacter jejuni*

Yilmaz Emre Gencay, Tina Birk, Martine Camilla Holst Sørensen, and Lone Brøndsted

Abstract

Here, we describe the methods for isolation, purification, and propagation of *Campylobacter jejuni* bacteriophages from samples expected to contain high number of phages such as chicken feces. The overall steps are (1) liberation of phages from the sample material; (2) observation of plaque-forming units on *C. jejuni* lawns using a spot assay; (3) isolation of single plaques; (4) consecutive purification procedures; and (5) propagation of purified phages from a plate lysate to prepare master stocks.

Key words *Campylobacter*, *C. jejuni*, Bacteriophage, Isolation, Purification, Propagation, Titration

1 Introduction

Campylobacter jejuni is the major cause of bacterial food borne gastroenteritis in developed countries [1]. *C. jejuni* colonizes the intestinal tract of poultry to high numbers, and regardless of the biosecurity measures aiming at reducing *C. jejuni* in the primary production, poultry meat remains to be the major source of the disease [1]. Since bacteriophages (phages) are often found in niches associated with their hosts, a number of samples have been used for phage isolation such as poultry meat [2], poultry feces [3–5] and intestines [6, 7], pig manure, slaughterhouse wastewater, and sewage [8, 9]. Here, we describe modifications of general techniques, which are needed to successfully isolate *C. jejuni* phages, including standardized steps and tips in the isolation, purification, and propagation of *Campylobacter* phages (Fig. 1). Following liberation of the phages from the sampled material, supernatants are applied on lawns consisting of the chosen isolation strains using a spot assay. Following incubation, single plaques are identified, collected, and purified by consecutive applications

James Butcher and Alain Stintzi (eds.), *Campylobacter jejuni: Methods and Protocols*, Methods in Molecular Biology, vol. 1512,
DOI 10.1007/978-1-4939-6536-6_3, © Springer Science+Business Media New York 2017

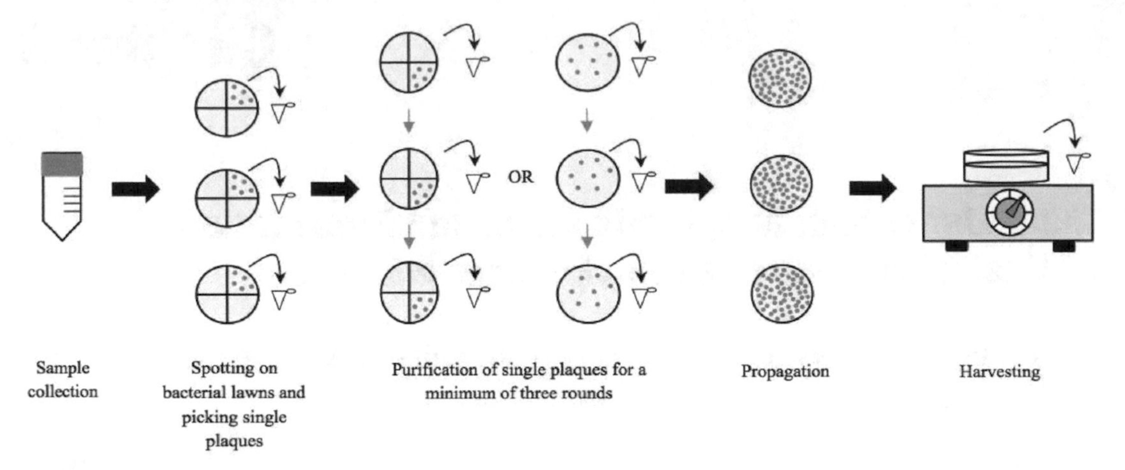

Fig. 1 Brief overview of isolation, purification, and propagation of *C. jejuni* phages

on lawns using spot assays or by co-plating an indicator strain and phage on a full plate using double layer agar. Finally, *Campylobacter* phages are propagated using a plate lysate.

Due to the high genome plasticity of *Campylobacter* and quickness in emergence of phage insensitive mutants in a culture, there are currently no methods that apply an enrichment step for the phages. Therefore, it is important to consider sampling materials that are likely to harbor *Campylobacter* and consequently their phages. Also, the indicator strain used to isolate *Campylobacter* phages is important. *C. jejuni* NCTC12662 (PT14) has been extensively used for this purpose, due to its high sensitivity toward phages [2–4, 7, 10]. However, recently we have showed that the choice of primary isolation strain may result in a strong bias of the isolated phage types either being capsular polysaccharide- (CPS) or flagella-targeting phages [4]. Therefore, for isolating novel and divergent phages against *Campylobacter*, one should consider either using host strains that are much diverse in surface structures (i.e., even different *Campylobacter* species) or using materials from other niches than the poultry environment.

2 Materials

2.1 Equipment and Materials

1. Microaerobic system for *C. jejuni* growth.
2. Microscope, phase contrast, 100× magnification, objective glass and cover slides.
3. Basic laboratory equipment for preparation of broth and solid culture media, and solutions.
4. Analytic balance.
5. Laminar flow hood.
6. Vortex.

7. Pipette controller and variable pipettes, 10–25 mL.
8. Sterile tips and filtertips covering volumes of 10 μL–1 mL.
9. Sterile screw-capped centrifuge tubes, 13 and 50 mL.
10. 1.5 mL sterile capped microcentrifuge tubes.
11. Sterile pasteur pipettes.
12. Sterile syringe filters, 0.2 and 0.45 μm.
13. Sterile syringes, 10 mL.
14. Optical density spectrophotometer.
15. Microwave.
16. Screw-capped autoclavable bottles, 100, 250 and 1 L.
17. Gyratory shaker.
18. Waterbath set at 45 °C.
19. Refrigerator set at 4 °C.
20. Incubator set at 37 °C.

2.2 Culture Media

1. BA (Blood agar): 5% (v/v) calf blood, Blood Agar Base II. Prepare the Blood Agar Base II according to the manufacturer's instructions and sterilize by autoclaving. Add 10 mL calf blood to 200 mL media tempered to 45 °C just before pouring the plates (*see* **Note 1**).
2. cBHI (cation adjusted Brain Heart Infusion): BHI broth with 10 mM $MgSO_4$ and 1 mM $CaCl_2$. Prepare the BHI according to the manufacturer's instructions. Add 2 mL of sterile 1 M $MgSO_4$ and 2 mL of 100 mM $CaCl_2$ to 200 ml (*see* **Note 2**).
3. NZCYM overlay agar: NZCYM broth, 0.6% (w/v) agar. Prepare NZCYM broth as described by the manufacturer. Add 0.6% agar, mix well and sterilize by autoclaving.
4. NZCYM basal agar: NZCYM broth, 1.2% (w/v) agar, 10 μg/mL vancomycin. Prepare NZCYM broth as described by the manufacturer. Add 1.2% agar, mix well and sterilize by autoclaving. Add vancomycin at a final concentration of 10 μg/mL to the media tempered to 45 °C just before the plates are poured (*see* **Note 3**).

2.3 Solutions and Buffers

1. SM Buffer: 5.8 g NaCl, 2 g $MgSO_4{\cdot}7H_2O$, 5 mL 2% (w/v) gelatin, 50 mL of 1 M Tris–HCl. Add water to a volume of 900 mL and mix. Adjust the pH to 7.5 using HCl. Fill up to 1 L with water and autoclave.
2. 1 M $MgSO_4$ solution. Sterilize by autoclaving.
3. 100 mM $CaCl_2$ solution. Sterilize by autoclaving.
4. Vancomycin, 10–25 mg/mL solution in water. Store at −20 °C.
5. 1.5 mL microcentrifuge tubes containing 900 μL of SM buffer.

3 Methods

3.1 Extraction of Phages from the Samples

1. Prepare 10% (w/v) suspension (i.e., 10 g of fecal sample and 90 mL SM buffer) of the sample of interest and SM buffer in a sterile tube or bottle (*see* **Note 4**).
2. Incubate the suspension for 24 h with gentle shaking (~60–80 cycles/min) at 4 °C to dissociate the phages from the sample material.

3.2 Preparation of Bacterial Lawns

1. Streak out *C. jejuni* strains of interest from −80 °C freezer stock cultures on BA plates and incubate under microaerobic conditions at 37 °C for 48 h.
2. Subculture *C. jejuni* strain densely on BA plates and incubate under microaerobic conditions at 37 °C for 20–24 h (*see* **Note 5**).
3. Add 2–3 mL of cBHI to BA plates. Using a bent sterile Pasteur pipette, loosen and harvest bacteria by gently scraping off the colonies.
4. Transfer the dense bacterial harvest to a microcentrifuge tube and vortex thoroughly.
5. To achieve uniform lawns throughout the work, always adjust the final optical density of the initial *C. jejuni* culture to 0.35 at 600 nm (OD_{600}) in fresh cBHI (*see* **Note 6**). This OD_{600} should be achieved regardless of the cBHI volume. Incubate the culture in a sterile petri dish at 37 °C and grow the cells for 4 h.
6. Before the incubation period is over, melt the overlay agar and temper it to 45 °C in a water bath (*see* **Note 7**). Ensure that the basal NZCYM plates are ready (**step 4** in Subheading 2.2 and **Note 3**).
7. Following the incubation period, visualize an aliquot of the freshly grown culture by phase-contrast microscopy at 100× magnification for presence of contaminating bacterial cells (*see* **Note 8**).
8. Mix 1:10 of freshly grown culture with NZCYM overlay agar and quickly pour 5 mL of the mix on the base plate, and gently rotate to produce an even surface with the overlay fresh culture mixture (*see* **Note 9**).
9. Let the overlays settle for 15–20 min and then, dry them in a laminar flow hood for 45 min (*see* **Note 10**).

3.3 Isolation of Campylobacter jejuni Phages

1. Prepare the fresh isolation culture(s), pre-warmed overlay and base agar (*see* **steps 1–5** in Subheading 3.2).
2. Transfer an aliquot of the supernatant of the phage extraction suspension to a sterile tube and centrifuge at 3000 × *g* for 5 min to eliminate the debris that might clog the filter used in the consecutive step. If needed this step can be repeated.
3. Syringe filter (0.2 μm) the supernatant into a sterile tube (*see* **Note 11**).

4. Apply 6 × 10 μL drops of the filtered supernatant to lawns of isolation host culture(s) (*see* **Note 12**).
5. When the spots are totally dried, incubate the plates under microaerobic conditions at 37 °C for 18–24 h.
6. Examine the plates for the presence of plaques (*see* **Note 13**). Collect single plaques with different morphologies from the overlay, using a different sterile 1 mL pipette tip for each of them. Gently dip the pipette tip to where plaque resides, transfer the harvested plaque to 500 μL of SM buffer and thoroughly vortex the suspension (*see* **Note 14**).
7. Incubate the plates for an additional 18–24 h under microaerobic conditions at 37 °C to observe if additional plaques arise.

3.4 Purification of Campylobacter jejuni Phages

Plaque morphology is an important challenge in purification of phages. As the same phage may display several different plaque morphologies, it cannot be concluded that plaques with distinct morphologies are actually distinct phages. Therefore, all plaques with different morphologies in the first round of plaquing must be collected and transferred to SM buffer, and should be treated as a putative individual phage. For concluding that the putative phages are indeed different, further analyses such as host range, genome size, and restriction profile are needed.

1. Prepare the fresh isolation culture(s), pre-warmed overlay and base agar (**steps 1–5** in Subheading 3.2).
2. Dilute the harvested plaques from **step 6** (Subheading 3.3) tenfold in SM buffer (i.e., 100 μL phages: 900 μL SM buffer). Two purification procedures can be performed: plate lysate method or spot test (**steps 1–9** or **steps 1–2** and **10–15**, respectively).

3.4.1 Purification Using Plate Lysate

3. Mix 400 μL of fresh bacterial culture prepared in **steps 1–4** (Subheading 3.2) with 100 μL of the respective phage dilutions in 13 mL tubes.
4. Incubate the phage-host mixture at 37 °C for 15 min under aerobic conditions to ensure proper phage adsorption.
5. Add 5 mL of overlay agar to 13 mL tubes, briefly vortex and pour onto the base plates and gently rotate the plates to produce an even surface.
6. Let the overlay settle and solidify for 15–20 min.
7. Incubate the plates under microaerobic conditions at 37 °C for 18–24 h.
8. After incubation examine the plates for plaques. Transfer an individual plaque to SM buffer and restart the procedure **step 1**.
9. Repeat this purification procedure (**steps 1–9**) at least three times.

3.4.2 Purification Using Spot Test

10. Prepare tenfold serial dilutions of the phage stock in SM buffer up to 10^{-7}.
11. Divide the bottom of two lawn prepared petri dishes into 4 equal segments each, by drawing with a marker, and mark the corresponding dilutions from 0 to –7.
12. Apply the serial tenfold dilutions of phages onto respective lawns of bacteria (Subheading 3.2) by spotting at least 5 × 10 μL to each segment (*see* **Note 12**).
13. When the spots are totally dried, incubate the plates under microaerobic conditions at 37 °C for 18–24 h.
14. After incubation examine the plates for plaques. Transfer an individual plaque to SM buffer and restart this procedure from **step 1**.
15. Repeat this purification procedure (**steps 1–2** and **10–15**) at least three times.

3.5 Phage Propagation and Preparation of Master Stocks

1. Prepare the fresh isolation culture(s), pre-warmed overlay and base agar (**steps 1–5** in Subheading 3.2).
2. Briefly vortex the purified phage suspension (or the plaque that was harvested into 500 μL SM buffer after third purification round in Subheading 3.4) to ensure a homogenous suspension (*see* **Note 15**).
3. In duplicate, mix 200 μL of the phage suspension in SM buffer with 400 μL of fresh propagation culture in a 13 mL sterile tube (*see* **Note 16**).
4. Incubate the phage-host mixture at 37 °C for 15 min under aerobic conditions to ensure proper phage adsorption.
5. Add 5 mL of overlay agar to the 13 mL tubes, briefly vortex and pour on to the base plates. Gently rotate the plates to produce an even surface. Repeat this step with the negative control (*see* **Note 16**) as well.
6. Let the overlays settle for 15 min and then incubate under microaerobic conditions at 37 °C for 18–24 h.
7. Examine the propagation plates for confluent and/or complete lysis in comparison to the phage negative control (*see* **Note 17**).
8. Add 5 mL SM buffer to propagation plates and incubate them at 4 °C on a gyratory shaker (~60–80 cycles/min) overnight.
9. Pour the excessive SM buffer from each of the plates to new petri dish, collect all of the liquid with a syringe and then filter sterilize (0.2 μm) them into a sterile tube.
10. Titrate the harvested phage stock on a lawn of propagation strain, as described below, and calculate the pfu/mL. Remember to write the phage name, date of propagation and the achieved pfu/mL on the master stock tubes.

3.6 Titration

1. Prepare tenfold serial dilutions of the phage stock in SM buffer up to 10^{-7} (*see* **Note 18**).
2. Divide the bottom of the lawn prepared petri dishes into eight equal segments drawing with a marker and mark the corresponding dilutions from 0 to −7 (*see* **Note 19**).
3. Carefully spot 3 × 10 μL of the original phage stock and the phage dilutions to corresponding segments in duplicates on two bacterial lawns (Subheading 3.2, **Note 12**).
4. When the spots are totally dried, incubate the plates under microaerobic conditions at 37 °C for 18–24 h.
5. Following incubation, observe the plaque morphologies, count the formed plaques and calculate the mean plaque-forming units per mL (pfu/mL) (*see* **Note 20**; Fig. 2).

4 Notes

1. The *C. jejuni* strains can also be grown on Mueller-Hinton agar without blood supplement if it is not possible to obtain fresh calf blood.
2. To be sure that the media is sterile, preferably add the divalent solutions to the BHI the day before use and incubate the cBHI overnight at 37 °C. Check for contaminating growth by observing turbidity.
3. NCZYM plates can be made up to 2 weeks before use and stored at 4–5 °C. Storage for more than 2 days should be at 4–5 °C. For optimal plaque formation, pour the plates a day before use and leave them overnight on the bench at room temperature. These plates can then be used without any drying period. If plates were prepared before and stored at 4 °C, they should then be removed from the fridge the day before use and

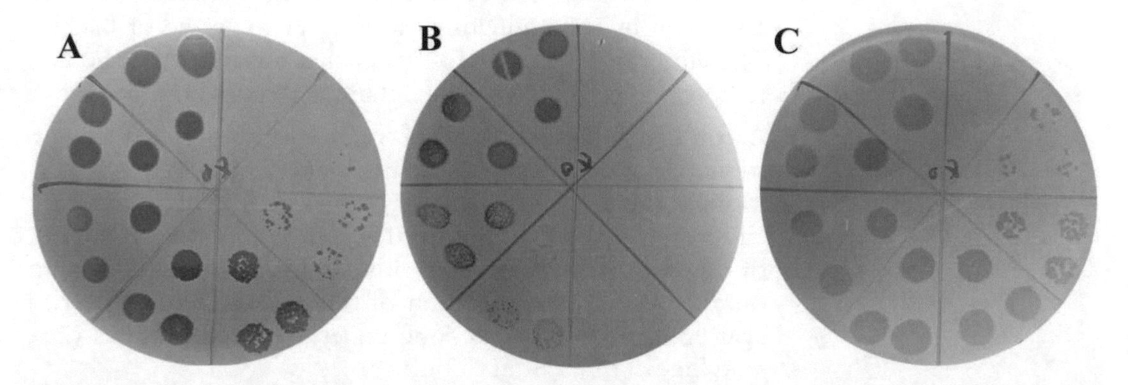

Fig. 2 Examples of plaque morphologies. (**a**) Middle sized (1–2 mm) and clear plaques. (**b**) Pinpoint and clear plaques. (**c**) Large (2–3 mm) and turbid plaques

left overnight at room temperature for standardization. If the NZCYM basal agar plates were poured or taken from the fridge the same day of the experiment, then they should be dried in the laminar flow hood before use. Depending on the moisture content of the plates (i.e., the longer they stay in the fridge, the more humid they will become) the drying period might vary between 10 and 30 min and thus, should be pre-optimized for each laboratory setting.

4. For isolation of *C. jejuni* phages one should collect samples from niches where *C. jejuni* are prevalent. In our experience, collection of feces from poultry farms has proved highly efficient for this purpose. However, to limit the investigated niche to a particular flock or bird, poultry cecum samples obtained from slaughterhouses may be considered. Prior to sample collection, ensure that permission for collecting these types of samples from farm owners, slaughterhouse boards, etc. has been obtained.

5. In most cases one will need a high density of *C. jejuni* cells for preparation of lawns. Thus, subculture enough to support dense growth of the particular *C. jejuni* strain(s) of interest.

6. In this step it is important to have a standard initial amount of cells in every prepared lawn, which should be between 10^8 and 10^9 cfu/mL. In our experience with a variety of *C. jejuni* strains, an OD_{600} of 0.35 adjusted in cBHI corresponds to this amount. However, as equipment used in measuring optical densities may vary, determining the correlation of OD_{600} and the corresponding cfu/mL for strains to be used beforehand is advisable.

7. To avoid lumps, ensure that the overlay agar is brought to a full boil. Otherwise when the agar is solidified, may cause misleading results due to its particulate structure. Base agar should be in an optimum state; neither too dry nor too moist (*see* **Note 3**).

8. *C. jejuni* growing in a broth can easily get contaminated and outgrown by contaminating bacteria (i.e., coccoid or bacilli). Therefore, it is advisable to always check a control of the culture by microscopy to see if contamination has occurred.

9. According to the need, this procedure can be applied using a sterile 10 or 25 mL pipette either in sterile tubes, briefly vortexing 5 mL pre-warmed overlay agar with 500 μL of the fresh culture, or in screw-capped bottles, thoroughly mixing 100 mL of pre-warmed overlay agar with 10 mL of fresh culture. In order to avoid contamination of the fresh culture of overlay agar ensure that work is done under aseptic conditions (i.e., working next to a Bunsen burner).

10. The appearances of plaques, both size and clarity, on the plates is strongly influenced by the moisture and thicknesses of the

NZCYM basal agar and the NZCYM overlay. Nevertheless, this variation in plaque morphologies can be minimized and controlled to some extent by following a strict agar and overlay production/drying procedure to ensure repeatability. Optimal drying is essential for producing clear plaque-forming units and thus it is advisable that some of the moisture from the agar is removed by drying plates in a laminar flow hood (*see* **Note 3**). If the plates are not adequately dried, applied spots will not be easily absorbed and will hinder the reading of plates. Drying plates for too long time will complicate and limit the plaque formation. Therefore, once the poured overlays are settled, drying the plates in the laminar flow hood for 45 min ensures optimal moisture content. However, if the basal plates are prepared the day of the experiment, then there will be a need for two rounds of drying. These drying periods may be between 30 and 60 min and thus needs to be optimized for each laboratory setting.

11. This either can be accomplished directly by filtering through a 0.2 μm syringe filter or following a primary filtration through 0.45 μm filter depending on whether the sample may clog the filter.
12. Avoid sudden movements of the spotted plates and let the lawn soak the droplets for approx. 30 min.
13. If no plaques are observed, one could either try to use an overlay of NZCYM with 0.5% or 0.4% of agar or alternatively, the drying period for the overlay can be lowered. Flagellotropic *C. jejuni* phages are sensitive both to the moisture and agar content of the overlay NZCYM and thus might need further optimization to be recovered [4].
14. To be confident that the plaque is collected with a pipette tip, one can draw a small spot with a permanent marker behind the plaque to ensure that the center of the pipette tip is in the center of the plaque.
15. Ensure that the phage suspension is not contaminated, by observing a clear, not cloudy suspension. Some of the phages might result in a salt-like precipitation, and therefore this should not be considered as a marker of contamination.
16. Repeat this step with 200 μL SM buffer and 400 μL fresh propagation strain as a negative control to ensure that the SM buffer was not contaminated with any other phage.
17. If no plaques, few plaques or contaminations were observed, then do not proceed with the following steps but start from the beginning (**step 1**, Subheading 3.5).
18. The response of *C. jejuni* strains to phage infection is highly variable and with each phage it is possible to observe a broad diversity in plaque formation. Furthermore, when the applied phage titer is high, growth inhibition can be

observed in *C. jejuni* lawns. Therefore, one should apply serial dilutions of the phage stock as a spot assay to clearly observe single plaques and the efficiency of plaque formation.

19. This can be completed during the 4 h incubation period in **step 4** if numerous plates are being assayed simultaneously (*see* Subheading 3.2).

20. The plaques might differ in size from a pinpoint to 1–3 mm, and show either turbid or clear lysis. With a high titer phage, it is possible to observe a clear or a turbid lysis corresponding to the spot. However, as the titer declines with every tenfold dilution, it is possible to observe either a distinct loss of activity within the first few dilutions or a confluent lysis that dilutes out to singular plaques at the spots.

References

1. European Food Safety Authority and European Centre for Disease Prevention and Control (2015) The European Union Summary Report on trends and sources of zoonoses, zoonotic agents and food-borne outbreaks in 2013. EFSA J 13:3991
2. Atterbury RJ, Connerton PL, Dodd CER et al (2003) Isolation and characterization of *Campylobacter* bacteriophages from retail poultry. Appl Environ Microbiol 69:4511–4518
3. Hansen VM, Rosenquist H, Baggesen DL et al (2007) Characterization of *Campylobacter* phages including analysis of host range by selected *Campylobacter* Penner serotypes. BMC Microbiol 7:90
4. Sørensen MCH, Gencay YE, Birk T et al (2015) Primary isolation strain determines both phage type and receptors recognised by *Campylobacter jejuni* bacteriophages. PLoS One 10:e0116287
5. Grawjewski BA, Kusek JW, Gelfand HM (1985) Development of a bacteriophage typing scheme for *Campylobacter jejuni* and *Campylobacter coli*. Epidemiol Infect 104:403–414
6. Atterbury RJ, Dillon E, Swift C et al (2005) Correlation of *Campylobacter* bacteriophage with reduced presence of hosts in broiler chicken ceca. Appl Environ Microbiol 71:4885–4887
7. Carvalho C, Susano M, Fernandes E et al (2010) Method for bacteriophage isolation against target *Campylobacter* strains. Lett Appl Microbiol 50:192–197
8. Khakhria R, Lior H (1992) Extended phage-typing scheme for *Campylobacter jejuni* and *Campylobacter coli*. Epidemiol Infect 108:403–414
9. Salama SM, Bolton FJ, Hutchinson DN (1989) Improved method for the isolation of *Campylobacter jejuni* and *Campylobacter coli* bacteriophages. Lett Appl Microbiol 8:5–7
10. Loc CC, Atterbury RJ, El-Shibiny A et al (2005) Bacteriophage therapy to reduce *Campylobacter jejuni* colonization of broiler chickens. Appl Environ Microbiol 71:6554–6563

Chapter 4

Methods to Study Antimicrobial Resistance in *Campylobacter jejuni*

Orhan Sahin, Zhangqi Shen, and Qijing Zhang

Abstract

Campylobacter jejuni is a leading bacterial cause of foodborne gastroenteritis worldwide and is increasingly resistant to clinically important antibiotics. Detection of antibiotic resistance in *C. jejuni* can be performed with both phenotypic and genotypic methods. In this chapter, we describe the most commonly used molecular biology methods for detection of resistance to clinically important antibiotics. These methods can be employed in both clinical and research settings to facilitate clinical therapy and to monitor the emergence and dissemination of antibiotic-resistant *C. jejuni*.

Key words *Campylobacter jejuni*, Antimicrobial resistance, Molecular detection, Sequencing, PCR

1 Introduction

Campylobacter jejuni is among the leading causes of bacterial gastroenteritis in humans and also a major etiological agent of ruminant abortion [1–4]. Transmission of *C. jejuni* to humans occurs mainly through the foodborne route such as undercooked poultry, unpasteurized milk and dairy products, and foods that are cross-contaminated by raw poultry during preparation [1, 5, 6]. Most human infections are usually self-limited and typically manifest as high fever, abdominal cramps, and watery/bloody diarrhea that usually resolves within a few days [1]. However, severe and prolonged (greater than a week) cases as well as bacteremia and other extraintestinal infections may occur, primarily in young, elderly, and immunocompromised individuals [1, 7]. In these circumstances, therapeutic intervention is usually warranted [7, 8]. For clinical therapy of human campylobacteriosis, macrolide antibiotics are considered the drug of choice if laboratory confirmation is made, and fluoroquinolone (FQ) antimicrobials are frequently used because of their broad spectrum of activity against enteric pathogens [7–9]. Other alternative drugs include tetracyclines and gentamicin, which are occasionally used in cases of systemic infection

James Butcher and Alain Stintzi (eds.), *Campylobacter jejuni: Methods and Protocols*, Methods in Molecular Biology, vol. 1512, DOI 10.1007/978-1-4939-6536-6_4, © Springer Science+Business Media New York 2017

with *Campylobacter* [7]. However, increasing prevalence of resistance to clinically important antibiotics, in particular to fluoroquinolones and macrolides, in *C. jejuni* isolates has been reported worldwide [10–13], compromising the effectiveness of therapeutic intervention. Detection of antimicrobial resistance is important for surveillance of antibiotic-resistant *C. jejuni* and is also useful for guiding the choice of appropriate antibiotics for clinical therapy.

In *Campylobacter*, FQ-resistance is mainly mediated by point mutations in the quinolone resistance-determining region (QRDR) of DNA gyrase (GyrA) and the function of the multidrug efflux pump CmeABC [10, 13]. Macrolide resistance is mainly conferred by mutations in the V domain of 23S rRNA and the efflux function of CmeABC [10, 13]. In addition, a ribosomal RNA methylase enzyme (ErmB) that confers macrolide resistance has been identified recently in *Campylobacter* isolates [14]. Resistance to tetracycline in *Campylobacter* is mediated by Tet(O) that binds to the 30S subunit of ribosome and inhibits protein synthesis [10]. Aminoglycoside resistance in *Campylobacter* is conferred mainly by aminoglycoside-modifying enzymes encoded by genes *aphA* (most commonly reported), *aadA*, *aadE*, and *aacA* [15].

As with other bacteria, phenotypic detection of antibiotic resistance in *C. jejuni* is performed via in vitro antimicrobial susceptibility testing [16]. Currently, agar dilution and broth microdilution tests are the standardized methods for *C. jejuni* as recommended by the Clinical Laboratory Standards Institute (CLSI) in the United States [17, 18]. In addition, a disk diffusion method was recently standardized by the European Committee on Antimicrobial Susceptibility Testing (EUCAST) [19]. Detailed methods of these antimicrobial susceptibility tests are described in those respective documents and will not be repeated here.

In this chapter, we will describe some of the most commonly used molecular methods (e.g., PCR and DNA sequencing) for detecting antibiotic resistance genes or mutations in *C. jejuni*. In addition, the methods for generating a *cmeABC* knockout mutant and measuring active efflux activity in *C. jejuni* are detailed.

2 Materials

1. Standard microbiological media (e.g., Mueller-Hinton broth and agar) and equipment/supplies to grow *C. jejuni* (e.g., 37–42 °C incubator and microaerobic condition-generating systems).
2. Block-based thermal cycler and standard PCR reagents (polymerase, dNTPs, $MgCl_2$, Master Mix, etc.) to amplify the target, and kits for purifying PCR products directly from reaction mixtures if sequencing is performed.

3. *TaKaRa Ex Taq* polymerase (Clontech).
4. Target-specific primer pairs, which are listed in Table 1. Individual primers are first dissolved in sterile deionized (DI) water to make the stock solution (100 pmol/μL), from which the working concentration (5 pmol/μL) is prepared by dilution in sterile DI water. In a PCR reaction mixture of 50 μL of total volume, 2 μL of working concentration for each primer (forward and reverse) is used.
5. Molecular biology grade agarose.
6. Tris–acetate–EDTA (TAE) buffer pH 8.0 (10× stock solution; 400 mM Tris-acetate and 10 mM EDTA), which is diluted in DI water to 1× working solution).
7. Single-cell lysis buffer (SCLB): 1 mL 1× TE Buffer pH 8.0 (10 mM Tris–HCl and 1 mM disodium EDTA) and 10 μL proteinase K (5 mg/mL) solution.
8. A horizontal electrophoresis system that includes gel tank, casting tray, well comb, and power supply.
9. DNA size marker (e.g., 1 Kb DNA Ladder, 100 bp DNA ladder).
10. 6X DNA loading buffer (e.g., Blue/Orange Loading Dye 6X, Promega). The loading dye contains 0.4% orange G, 0.03% bromophenol blue, 0.03% xylene cyanol FF, 15% Ficoll 400, 10 mM Tris–HCl (pH 7.5) and 50 mM EDTA (pH 8.0).
11. Phosphate-buffered saline (PBS) buffer (1×), pH 7.2 (137 mM NaCl, 2.7 mM KCl, 10 mM Na_2HPO_4, 1.8 mM KH_2PO_4).
12. Ethidium bromide: 4 μg/mL in PBS.
13. Ciprofloxacin: 10 mg/mL in sterile DI water.
14. Carbonyl cyanide m-chlorophenylhydrazone (CCCP): 100 mM in Dimethyl sulfoxide (DMSO).
15. Chloramphenicol: 10 mg/mL in 100% ethanol.
16. Glycine hydrochloride: 0.1 M glycine in sterile DI water, adjust pH to 3.0 using hydrochloride.
17. Nucleic acid-staining dye (e.g., SYBR Green 10,000×, Invitrogen).
18. UV transilluminator.
19. UV spectrophotometer.
20. Fluorospectrometer (e.g., FLUOstar Omega, BMG Labtech).
21. Black polystyrene 96-well microplates.
22. Sterile Cotton Swabs.
23. Gibson Assembly Cloning (NEB): Used for cloning PCR-amplified DNA fragments with overlapping ends in a single-tube reaction employing three enzymatic activities

Table 1
Primers used to detect mutations/genes conferring antibiotic resistance in *C. jejuni*

Target	Primer	Sequence (5′ to 3′)	Size (bp)	*Ta* (°C)	Resistance	Ref.
gyrA	GyrAF GyrAR	CAACTGGTTCTAGCCTTTTG AATTTCACTCATAGCCTCACG	1083	55	FQ	[38]
23S rDNA	F1 R1	GTAAACGGCGGCCGTAACTA GACCGAACTGTCTCACGACG	699	52	Ery	[39]
ermB[a]	ermBF ermBR	GGGCATTTAACGACGAAACTGG CTGTGGTATGGCGGGTAAGT	421	55	Ery	[14]
tetO[b]	tetO-F tetO-R	GGCGTTTTGTTTATGTGCG ATGGACAACCCGACAGAAGC	559	57	Tet	[40]
aphA-3	aphA3F aphA3R	GGGACCACCTATGATGTGGAACG CAGGCTTGATCCCCAGTAAGTC	600	55	AMG[c]	[41]
aphA-1	aphA1F aphA1R	CGTATTTCGTCTCGCTCAG CCGACTCGTCCAACATCA	228	52	AMG	[15]
aphA-7	aphA7F aphA7R	ATCCGATAAACTGAAAGTAC ATAATCCGGTTCAAGTCCC	538	52	AMG	[15]
aadE-sat4-aphA3	clustF clustR	CGAGGATTTGTGGAAGAGGCTT TTCCTTCCAGCCATAGCATCATG	1000	55	AMG	[41]
cmeA	F R	TGTGCATCAGCTCCTGTGTAA ACGGACAAGCTTTGATGGCT	957	55	MDR[d]	[42]
cmeB	BF BR	GGTACAGATCCTGATCAAGCC AGGAATAAGTGTTGCACGGAAATT	820	55	MDR	[36]
cmeC	F1 R1	AGATGAAGCTTTTGTAAATT TATAAGCAATTTTATCATTT	500	48	MDR	[43]
cmeA[e]	CmeA-F CmeA-R	AGTTGTTATCAGGGCTACAAA TTTACTCGTTGTCCGCTCCATTGGCTA ATTATATCTTAATTTTGG	832	55	N/A	This study
cmeC[e]	CmeC-F CmeC-R	AATCCCAGTTTGTCGCACTGATGAGCA AAGTGAAGATACGA TATGATTTGGATCGCTTTCG	875	55	N/A	This study
cat	Cat-F Cat-R	TGGAGCGGACAACGAGTAAA TCAGTGCGACAAACTGGGATT	817	55	N/A	This study

[a]Both for *C. jejuni* and *C. coli*

[b]For general detection of *tet*(O) regardless of location. For specific location (pVir plasmid or the chromosome), please refer to Ref. [44]

[c]Aminoglycoside

[d]Multidrug resistance

[e]Used for generating a *cmeABC* knockout mutant. Underlined regions in CmeA-R and CmeC-F are extra bases added to generate complementary overhangs to Cat-F and Cat-R, respectively, to facilitate the ligation (*see* Fig. 1 and text for details)

(see Subheading 3.3 for details). The reaction tube contains 50 ng of each DNA fragment, 10 μL 2× Gibson Assembly Master Mix (NEB #E2611) and DI water in a final volume of 20 μL total.

3 Methods

3.1 Amplification of Target DNA (Antibiotic Resistance Genes or Mutations)

1. Grow *C. jejuni* isolates on MH agar plates for 24–48 h microaerobically.
2. Genomic DNA can be prepared by using either a commercial kit or by simply boiling a few colonies in 100 μL sterile DI water for 10 min, followed by centrifugation at high speed (13,000 ×*g*) for 3 min to remove cellular debris (*see* **Note 1**). All genomic DNA preparations can be stored at −20 °C or −80 °C and used for multiple times.
3. Use 2–4 μL of the supernatant as DNA template in a PCR mixture of 50 μL of total volume. If genomic DNA purified by a kit is used as template, a 2 μL aliquot (50 ng/μL) can be used as template.
4. Genomic DNA from appropriate control strains should be prepared in the same way as from the test strains and included in every PCR experiment.
5. A typical PCR reaction mixture (50 μL total) is prepared by combining 0.25 μL TaKaRa Ex Taq (5 units/μL), 5 μL 10× Ex Taq Buffer, 4 μL dNTP Mixture, 2 μL working solution of each forward and reverse primers (Table 1), 2–4 μL DNA template (approximately 100 ng total DNA) and appropriate amount of sterile DI water to bring the final reaction volume to 50 μL total. All reagents, in particular the polymerase, should be kept on ice until use. If sequencing of PCR amplicons is needed to determine the presence of resistance-associated mutations as in the case of resistance to FQ and macrolides, it is recommended to scale the total PCR reaction volume up to 100 μL.
6. Transfer tubes/microtiter plates to a thermal cycler and amplify the target DNA.
7. Optimal PCR conditions will vary based on the melting temperature of primers and the size of the amplified target (*see* **Note 2** and Table 1); however, a typical PCR reaction consists of an initial denaturation at 95 °C for 5 min; 30 cycles of 98 °C for 10 s, 55 °C (annealing temperature differs with each primer set) for 30 s, and 72 °C for 1 min (elongation time changes based on the size of amplified region); and a final elongation at 72 °C for 10 min.
8. Once the PCR reaction is completed, load 5 μL of the reaction is mixed with 1 μL of 6× Loading Dye, and the mixture is

loaded into the wells of a 1–2% agarose gel (containing SYBR Green; 5 μL of 10,000× stock in 50 mL total gel volume). The first and the last wells are normally loaded with the appropriate DNA size marker (100 bp or 1 Kb, depending on the expected sizes of the amplified products). Electrophorese for about 30 min at 90 V (or 5 V/cm length of the electrophoresis unit) or until the marker dye has run about 75–90% the length of the gel. Make sure to load the first and/or last well with the appropriate DNA size marker (100 bp or 1 Kb).

9. After the electrophoresis is finished, turn off the power and unplug the gel tank. The gel is unloaded onto the UV transilluminator to visualize the target DNA bands, and photographs are taken to document the result (*see* **Note 3** for interpretation).

3.2 Sequencing PCR Amplicons

1. For determining the mutations responsible for resistance to FQ and macrolides, the amplified DNA targets (*gyrA* and 23S rDNA, respectively; Table 1) are further prepared for sequence determination using the standard Sanger technology on an automated sequencer. The primers used for PCR amplifications are also used for DNA sequencing reactions.
2. Once a positive PCR is confirmed on agarose gel (Subheading 3.1, **step 8**), the rest of the PCR product is purified (removal of primers, nucleotides, enzymes, salt, etc.) using a commercial kit.
3. Measure the concentration of purified DNA using a UV spectrophotometer. This is typically done by reading the absorbance taken at wavelength 260 nm, which is then converted using the formula ($OD_{260} \times 50$ ng/μL × dilution factor) to determine the DNA concentration in ng/μL.
4. Prepare sequencing reactions. Each product needs to be sequenced in both directions (forward and reverse). Aliquot 5–10 μL of appropriate concentration (~2.5 ng/100 bases/μL) of template DNA (PCR product) into a microcentrifuge tube. For example, 25–30 ng/μL of *gyrA* amplicon (1083 bp) and 17.5–20 ng/μL of 23S rDNA amplicon (699 bp) are suitable for Sanger sequencing reactions. In addition, 5–10 μL of forward and reverse primers (from 5 pmol/μL working solution) are aliquoted in different tubes. The tubes containing DNA template and primers are submitted to a DNA sequencing facility for sequencing reaction.
5. Check the quality of the sequence data by visualizing the sequence chromatogram using a DNA sequence viewer software (e.g., Chromas, Technelysium). Align the sequences generated from the forward and reserve primers for each amplicon using a DNA sequence analysis software (e.g., SeqMan of DNASTAR Lasergene).
6. Determine if the DNA sequence has the known nucleotide change(s) associated with resistance to the antibiotic in question

by performing homology alignment to the reference sequence (e.g., complete gene sequence of a drug-susceptible wild-type strain) using a DNA sequence alignment program such as BLAST or MegAlign from DNASTAR Lasergene. For interpretation of sequence results for FQ or macrolide resistance, *see* **Note 4**. Also, *see* **Note 5** for additional information on molecular detection of antibiotic resistance.

3.3 Generation of a *cmeABC* Knockout Mutant

The overall strategy for generating a *cmeABC* mutant is illustrated in Fig. 1.

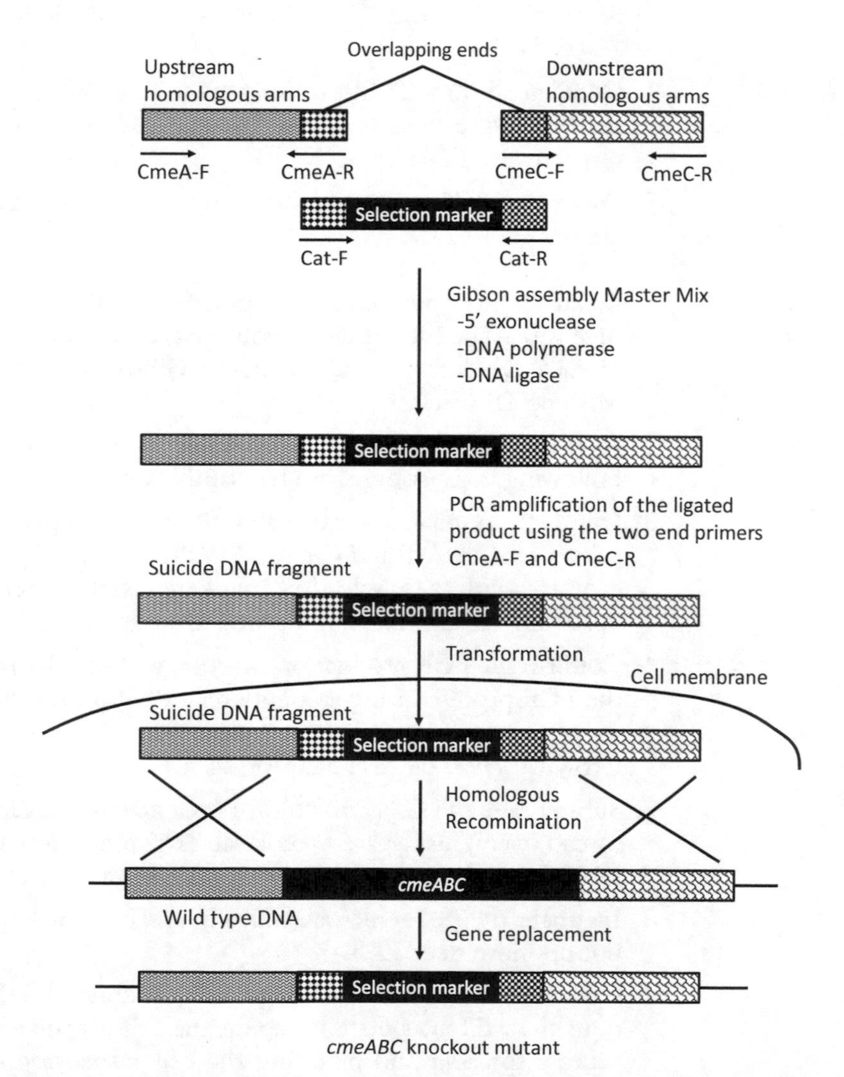

Fig. 1 Strategy for generating a *cmeABC* knockout mutant in *C. jejuni*. Target DNA (*cmeA* and *cmeC*) and selection marker (*cat*) DNA are amplified by PCR, followed by ligation, PCR re-amplification, and natural transformation to replace the wild-type allele of *cmeABC* with the mutated copy. Primers CmeA-R and CmeC-F are purposely designed to have complementary overhangs to Cat-F and Cat-R, respectively, to facilitate the ligation. See Table 1 and the text (Subheading 3.3) for details

3.3.1 Construction of Target DNA Fragments

1. Amplify the *cmeABC* DNA fragments (upstream and downstream homologous arms; Fig. 1) with primer pairs CmeA-F/CmeA-R and CmeC-F/CmeC-R, respectively (Table 1) using a high-fidelity DNA polymerase.
2. Using the same high-fidelity polymerase, amplify the chloramphenicol selection marker (*cat*) from plasmid pUOA18 [21] with primer pairs Cat-F and Cat-R (Table 1). Please note that another selection marker such as kanamycin resistance cassette (*aphA3*) from plasmid pMW10 [22] can also be used instead of *cat*.
3. The PCR conditions for all reactions above consist of an initial denaturation at 98 °C for 30s; 30 cycles of 98 °C for 10 s, 55 °C for 30 s, and 72 °C for 30 s; and a final elongation at 72 °C for 10 min.
4. Confirm all three PCR products by agarose gel electrophoresis. Make sure their sizes are correct (832, 875, and 817 bp, respectively; Table 1).
5. Purify the PCR products using a commercial PCR purification kit.

3.3.2 Construction of Suicide DNA Fragment

1. Ligate all three purified PCR segments in a single microcentrifuge tube. The total reaction volume is 20 μL, including 50 ng of each DNA fragment, 10 μL 2× Gibson Assembly Master Mix and DI water.
2. Incubate samples in a thermocycler at 50 °C for 15 min. Following incubation, store the sample on ice.
3. Dilute the sample for 10–100X times as template. Use the primer set CmeA-F and CmeC-R in PCR to amplify the ligated product with the selection marker inserted between the upstream and downstream homologous arms.
4. Confirm the PCR product on agarose gel (~2.5 Kb) and purify the PCR product using a commercial PCR purification kit.

3.3.3 Natural Transformation

1. Grow *C. jejuni* on MH agar for 24 h.
2. Sub-passage the *C. jejuni* culture on a new MH agar plate and spread evenly using a cotton swab. Drop about 1 μg suicide DNA fragment onto the surface of MH agar plate.
3. Incubate the plate microaerobically at 42 °C at least for 6 h, but no more than 12 h.
4. Collect the cells from the MH plate by adding 2 mL MH broth onto the MH agar plate, scraping the cell off the agar surface using a spreader and pipetting the cell suspension out into a 1.5 mL microcentrifuge tube.
5. Spread the cells onto another MH agar plate containing 4 μg/mL chloramphenicol.

6. Incubate the plate microaerobically at 42 °C for 3–5 days until single colonies are formed.
7. Confirm the insertion of the selection marker in the *cmeABC* gene using PCR. For this purpose, use the primer set CmeA-F and Cat-R (Fig. 1 and Table 1), and DNA template from a well-isolated single colony as detailed in section 3.1 (*see* **Note 6**). The *cmeABC* knockout mutant should yield an amplicon size of ~1.6 Kb, while the wild-type strain should not produce any amplicon.

3.4 Assay for Efflux Activity

3.4.1 Accumulation of Ethidium Bromide

1. Harvest an overnight broth culture of *C. jejuni*, wash it once in 1× PBS, and resuspend it in 1X PBS to an OD_{600} of 0.2 (at least 500 μL) in a 1.5 mL microcentrifuge tube. Prepare triplicate samples for each test strain. To compare a wild-type strain with its *cmeABC* mutant, the cultures should be prepared concurrently.
2. For each strain, incubate the bacterial suspension at 37 °C for 10 min under normal atmosphere.
3. Add 50 μL of cell suspensions into a well of a black polystyrene 96-well microplate. For each test strain, nine wells are needed, including six test wells and three control wells.
4. Add 50 μL ethidium bromide (from 4 μg/mL solution in 1× PBS) to each of the six test wells and 50 μL PBS to each of the three control wells. Make sure to adjust time zero fluorescence prior to addition of ethidium bromide to eliminate the background fluorescence of *C. jejuni* cells.
5. Immediately measure the fluorescence using a fluorospectrometer with excitation and emission wavelengths of 530 and 600 nm, respectively. Set the reading of fluorescence to be taken every minute.
6. Add the efflux pump inhibitor carbonyl cyanide m-chlorophenylhydrazone (CCCP) to three of the test wells at a final concentration of 200 μM at 7 min after the ethidium bromide addition. The other three test wells do not have CCCP added.
7. Measure the fluorescence and express it in arbitrary units (*see* **Note 7** for interpretation).

3.4.2 Accumulation of Ciprofloxacin

1. Harvest an overnight broth culture of *C. jejuni*, wash it once in PBS (pH 7.2), and adjust the concentration to an OD_{600} of 20 (at least 8.5 mL). To compare a wild-type strain with its *cmeABC* mutant, the cultures should be prepared concurrently.
2. For each strain, incubate the cell suspension (8.5 mL) for 10 min at 37 °C under normal atmosphere.

3. Save 0.5 mL of the suspension as control and then add 8 μL of 10 mg/mL ciprofloxacin to the remaining 8.0 mL of cell suspension to a final concentration of 10 μg/mL.
4. Remove 0.5 mL of solutions at 1, 3, 5, and 7 min after the addition of ciprofloxacin. Immediately dilute each of the collected samples in 2.5 mL of ice-cold PBS and then centrifuge for 5 min at 6000 × *g* at 4 °C.
5. Divide the remaining 6 mL solution from step 4 above into two portions (3 mL each). To one 3 mL aliquot, add 6 μL CCCP (from 100 mM stock) to a final concentration of 200 μM, and add 6 μL DMSO as control to the other aliquot (*see* **Note 8**).
6. Continue to collect the samples (0.5 mL each) following the addition of CCCP (e.g., at 3, 8, and 13 min after CCCP addition) from both treated and control (DMSO added) tubes. Immediately dilute each of the collected samples in 2.5 mL of ice-cold PBS and then centrifuge for 5 min at 6000 × *g* at 4 °C.
7. For each of the collected sample from **steps 4** and **6**, wash the pellet again with 2 mL of ice-cold PBS as indicated above in **steps 4** and **6**. Measure the weight of the pellet (weigh the empty centrifuge tube first, and then weigh the tube together with the pellet to determine the weight of the pellet), and resuspend the pellet in 2 mL of 0.1 M glycine hydrochloride (pH 3.0). Shake the samples at 200 rpm at 25 °C under normal atmosphere for 16 h using a tube rotator.
8. Centrifuge the samples at 6000 × *g* for 15 min.
9. Transfer 300 μL of supernatant of each sample into three wells (100 μL/well) of a black polystyrene 96-well microplate.
10. Measure the fluorescence of the supernatant using FLUOstar Omega with excitation and emission wavelengths of 279 nm and 447 nm, respectively.
11. Prepare a standard curve by measuring fluorescence emitting from 0.1 M glycine hydrochloride solution containing serially diluted concentration of ciprofloxacin (0.025, 0.05, 0.1, 0.2, 0.4, 0.8, and 1.6 μg/mL).
12. Calculate the concentrations of ciprofloxacin in the samples by comparing the fluorescence values to those of the standard curve and the corresponding ciprofloxacin amounts.
13. Express the results as ng of ciprofloxacin per mg (wet weight) of bacteria (*see* **Note 7** for interpretation).

4 Notes

1. We commonly employ a single-cell lysis buffer (SCLB) [20] as substitute for DI water to prepare less crude DNA extract, which appears to yield more consistent and reliable PCR results.

As with the boiling method, a few colonies are suspended in 40 μL SCLB, which is then heated to 80 °C for 10 min and cooled down to 55 °C for 10 min in a thermocycler to lyse bacterial cells. The suspension is then diluted 1:2 in DI water and centrifuged at 4500 × *g* for 30 s to remove cellular debris.

2. For PCR, optimal annealing temperature and elongation time can be altered with different target/primer combinations (as mentioned in Subheading **step 5** of Subheading 3.1). The annealing temperature for each primer pair is given in Table 1. Elongation time is chosen based on the size of the target amplicon such that approximately 1 min is used for every 1000 bp.
3. Specificity of PCR amplification is interpreted by visually comparing the size of the amplified product band from the test isolates to that amplified from the control strain. Complete or near-complete match in band size between the control and test isolates is the criterion for a positive result. Sequencing of the amplified product can be performed if a further level of certainty is desired or when the banding pattern is not conclusive.
4. For FQ resistance determination, the nucleotide sequence in the quinolone resistance determining region (QRDR) of *gyrA* of the test isolate is compared to that of the reference strain (**step 5** of Subheading 3.2). The most frequent mutation in FQ-resistant isolates of *C. jejuni* is C257T (change of C in the wild type to T in the resistant strain at nucleotide position 257 of *gyrA*), and followed by G268A, C257A, and G268T [13, 23–25]. The C257T mutation usually confers a high level of FQ resistance (ciprofloxacin MIC ≥ 16 μg/mL) in *C. jejuni*, while other mutations are associated with a moderate level of resistance (MIC = 1–8 μg/mL) [23, 24, 26]. For macrolide resistance, point mutations in domain V of 23S rRNA gene are the most common mechanism in *C. jejuni* and include A2075G (the most frequent one in clinical and field isolates), A2074C, and A2074G [13, 24, 27, 28]. These mutations usually confer a high level of macrolide resistance (MIC > 128 μg/mL) in *C. jejuni*. In most cases, all three copies of the 23S rRNA gene are mutated, but both mutated and wild-type alleles may co-exist in a single macrolide-resistant isolate [13, 24, 28].
5. For molecular detection of resistance in *C. jejuni* to FQ and macrolides, direct sequencing of PCR amplified targets is considered the gold standard; however, there are other molecular methods that have been utilized for identifying the resistance-associated mutations. These include PCR-restriction fragment length polymorphism, PCR-line probe assay, probe-based real-time PCR, mismatch amplification mutation assay (MAMA)-PCR, single-strand conformation polymorphism analysis (SSCPA), and microelectronic chip array [29–35].
6. At least ten colonies should be tested separately.

7. Three independent experiments (biological replicates) are performed to measure the accumulation of ethidium bromide and ciprofloxacin. When ethidium bromide enters the cell and binds to cellular components, fluorescence significantly increases. Thus fluorescence is proportional to the ethidium bromide concentration accumulated in the bacterial cell. Please see references [36, 37] for details.
8. CCCP is added in order to impair the function of efflux pumps; cells will accumulate more ciprofloxacin in the presence of CCCP than without CCCP.

Acknowledgement

This work was supported by the National Institutes of Health grant R56AI118283 and the USDA National Institute of Food and Agriculture grant no. 2012-67005-19614.

References

1. Allos BM (2001) *Campylobacter jejuni* Infections: update on emerging issues and trends. Clin Infect Dis 32:1201–1206
2. Golz G, Rosner B, Hofreuter D et al (2014) Relevance of Campylobacter to public health–the need for a One Health approach. Int J Med Microbiol 304(7):817–23. doi:10.1016/j.ijmm.2014.08.015
3. Scallan E, Hoekstra RM, Angulo FJ et al (2011) Foodborne illness acquired in the United States – major pathogens. Emerg Infect Dis 17:7–15
4. Skirrow MB (1994) Diseases due to *Campylobacter*, *Helicobacter* and related bacteria. J Comp Pathol 111:113–149
5. Altekruse SF, Tollefson LK (2003) Human Campylobacteriosis: a challenge for the veterinary profession. J Am Vet Med Assoc 223:445–452
6. Wagenaar JA, French NP, Havelaar AH (2013) Preventing *Campylobacter* at the source: why is it so difficult? Clin Infect Dis 57:1600–1606
7. Blaser MJ, Engberg J (2008) Clinical aspects of *Campylobacter jejuni* and *Campylobacter coli* infections. In: Nachamkin I, Szymanski C, Blaser M (eds) Campylobacter, 3rd edn. ASM Press, Washington, D.C
8. Frediani-Wolf V, Stephan R (2003) Resistance patterns of *Campylobacter* spp. strains isolated from poultry carcasses in a big Swiss poultry slaughterhouse. Int J Food Microbiol 89:233–240
9. Guerrant RL, Van GT, Steiner TS et al (2001) Practice guidelines for the management of infectious diarrhea. Clin Infect Dis 32:331–351
10. Iovine NM (2013) Resistance mechanisms in *Campylobacter jejuni*. Virulence 4:230–240
11. Andersen SR, Saadbye P, Shukri NM et al (2006) Antimicrobial resistance among *Campylobacter jejuni* isolated from raw poultry meat at retail level in Denmark. Int J Food Microbiol 107:250–255
12. White DG, Zhao S, Simjee S et al (2002) Antimicrobial resistance of foodborne pathogens. Microbes Infect 4(AB):405–412
13. Luangtongkum T, Jeon B, Han J et al (2009) Antibiotic resistance in *Campylobacter*: emergence, transmission and persistence. Future Microbiol 4:189–200
14. Wang Y, Zhang M, Deng F et al (2014) Emergence of multidrug-resistant *Campylobacter* species isolates with a horizontally acquired rRNA methylase. Antimicrob Agents Chemother 58:5405–5412
15. Qin S, Wang Y, Zhang Q et al (2012) Identification of a novel genomic island conferring resistance to multiple aminoglycoside antibiotics in *Campylobacter coli*. Antimicrob Agents Chemother 56:5332–5339
16. Ge B, Wang F, Sjolund-Karlsson M et al (2013) Antimicrobial resistance in *Campylobacter*: susceptibility testing methods and resistance trends. J Microbiol Methods 95:57–67

17. Clinical and Laboratory Standards Institute (2008) Performance standards for antimicrobial disk and dilution susceptibility tests for bacteria isolated from animals-approved standards M31-A3, 3rd edn. CLSI, Wayne, PA
18. Clinical and Laboratory Standards Institute (2010) Methods for antimicrobial dilution and disk susceptibility testing of infrequently isolated or fastidious bacteria: approved guideline–CLSI document M45-A2, 2nd edn. CLSI, Wayne, PA
19. European Committee on Antimicrobial Susceptibility Testing (2012) Clinical breakpoints, epidemiological cut-off (ECOFF) values and EUCAST disk diffusion methodology for *Campylobacter jejuni* and *Campylobacter coli*. EUCAST, Vaxjo, Sweden.
20. Olah PA, Doetkott C, Fakhr MK et al (2006) Prevalence of the *Campylobacter* multi-drug efflux pump (CmeABC) in *Campylobacter* spp. isolated from freshly processed turkeys. Food Microbiol 23:453–460
21. Wang Y, Taylor DE (1990) Chloramphenicol resistance in *Campylobacter coli*: nucleotide sequence, expression, and cloning vector construction. Gene 94:23–28
22. Wosten MM, Boeve M, Koot MG et al (1998) Identification of *Campylobacter jejuni* promoter sequences. J Bacteriol 180:594–599
23. Luo N, Sahin O, Lin J et al (2003) *In vivo* selection of *Campylobacter* isolates with high levels of fluoroquinolone resistance associated with *gyrA* mutations and the function of the CmeABC efflux pump. Antimicrob Agents Chemother 47:390–394
24. Payot S, Bolla JM, Corcoran D et al (2006) Mechanisms of fluoroquinolone and macrolide resistance in *Campylobacter* spp. Microbes Infect 8:1967–1971
25. Yan M, Sahin O, Lin J et al (2006) Role of the CmeABC efflux pump in the emergence of fluoroquinolone-resistant *Campylobacter* under selection pressure. J Antimicrob Chemother 58:1154–1159
26. Piddock LJ, Ricci V, Pumbwe L et al (2003) Fluoroquinolone resistance in *Campylobacter* species from man and animals: detection of mutations in topoisomerase genes. J Antimicrob Chemother 51:19–26
27. Lin J, Yan MG, Sahin O et al (2007) Effect of macrolide usage on emergence of erythromycin-resistant *Campylobacter* isolates in chickens. Antimicrob Agents Chemother 51:1678–1686
28. Gibreel A, Kos VN, Keelan M et al (2005) Macrolide resistance in *Campylobacter jejuni* and *Campylobacter coli*: molecular mechanism and stability of the resistance phenotype. Antimicrob Agents Chemother 49:2753–2759
29. Gibreel A, Taylor DE (2006) Macrolide resistance in *Campylobacter jejuni* and *Campylobacter coli*. J Antimicrob Chemother 58:243–255
30. Beckmann L, Muller M, Luber P et al (2004) Analysis of *gyrA* mutations in quinolone-resistant and -susceptible *Campylobacter jejuni* isolates from retail poultry and human clinical isolates by non-radioactive single-strand conformation polymorphism analysis and DNA sequencing. J Appl Microbiol 96:1040–1047
31. Westin L, Miller C, Vollmer D et al (2001) Antimicrobial resistance and bacterial identification utilizing a microelectronic chip array. J Clin Microbiol 39:1097–1104
32. Payot S, Cloeckaert A, Chaslus-Dancla E (2002) Selection and characterization of fluoroquinolone-resistant mutants of *Campylobacter jejuni* using enrofloxacin. Microb Drug Resist 8:335–343
33. Niwa H, Chuma T, Okamoto K et al (2003) Simultaneous detection of mutations associated with resistance to macrolides and quinolones in *Campylobacter jejuni* and *C. coli* using a PCR-line probe assay. Int J Antimicrob Agents 22:374–379
34. Wilson DL, Abner SR, Newman TC et al (2000) Identification of ciprofloxacin-resistant *Campylobacter jejuni* by use of a fluorogenic PCR assay. J Clin Microbiol 38:3971–3978
35. Hao H, Dai M, Wang Y et al (2010) Quantification of mutated alleles of 23S rRNA in macrolide-resistant *Campylobacter* by TaqMan real-time polymerase chain reaction. Foodborne Pathog Dis 7:43–49
36. Lin J, Michel LO, Zhang Q (2002) CmeABC functions as a multidrug efflux system in *Campylobacter jejuni*. Antimicrob Agents Chemother 46:2124–2131
37. Jeon B, Wang Y, Hao H et al (2011) Contribution of CmeG to antibiotic and oxidative stress resistance in *Campylobacter jejuni*. J Antimicrob Chemother 66:79–85
38. Wang Y, Huang WM, Taylor DE (1993) Cloning and nucleotide sequence of the *Campylobacter jejuni gyrA* gene and characterization of quinolone resistance mutations. Antimicrob Agents Chemother 37:457–463
39. Jensen LB, Aarestrup FM (2001) Macrolide resistance in *Campylobacter coli* of animal origin in Denmark. Antimicrob Agents Chemother 45:371–372
40. Gibreel A, Tracz DM, Nonaka L et al (2004) Incidence of antibiotic resistance in *Campylobacter jejuni* isolated in Alberta, Canada, from 1999 to 2002, with special reference to tet(O)-mediated tetracycline resistance. Antimicrob Agents Chemother 48:3442–3450

41. Gibreel A, Skold O, Taylor DE (2004) Characterization of plasmid-mediated *aphA-3* kanamycin resistance in *Campylobacter jejuni*. Microb Drug Resist 10:98–105
42. Koolman L, Whyte P, Burgess C et al (2015) Distribution of virulence-associated genes in a selection of *Campylobacter* isolates. Foodborne Pathog Dis 12:424–432
43. Fakhr MK, Logue CM (2007) Sequence variation in the outer membrane protein-encoding gene *cmeC*, conferring multidrug resistance among *Campylobacter jejuni* and *Campylobacter coli* strains isolated from different hosts. J Clin Microbiol 45:3381–3383
44. Wu Z, Sippy R, Sahin O et al (2014) Genetic diversity and antimicrobial succeptibility of *Campylobacter jejuni* isolates associated sheep abortion in the United States and Great Britain. J Clin Microbiol 52: 1853–1861

Chapter 5

Method of Peptide Nucleic Acid (PNA)-Mediated Antisense Inhibition of Gene Expression in *Campylobacter jejuni*

Euna Oh and Byeonghwa Jeon

Abstract

Peptide nucleic acid (PNA) is an oligonucleotide mimic that recognizes and binds to nucleic acids. The strong binding affinity of PNA to mRNA coupled with its high sequence specificity enable antisense PNA to selectively inhibit (i.e., knockdown) the protein synthesis of a target gene. This novel technology provides a powerful tool for *Campylobacter* studies because molecular techniques have been relatively less well-developed for this bacterium as compared to other pathogens, such as *Escherichia coli* and *Salmonella*. This chapter describes a protocol for PNA-mediated antisense inhibition of gene expression in *Campylobacter jejuni*.

Key words *Campylobacter*, Antisense, Peptide nucleic acid (PNA), Gene knockdown

1 Introduction

Peptide nucleic acid (PNA) is a DNA mimic where nucleobases are attached to pseudo-peptide backbone, and its N-terminus corresponds to the 5′ end of an oligonucleotide (Fig. 1) [1]. The nucleobases in PNA allow for specific recognition of nucleic acid sequences according to Watson-Crick pairing rules. PNA is not recognized by nucleases or proteases and is highly resistant to enzymatic degradation [2]. Whereas DNA and RNA are joined by negatively charged phosphodiester bonds, PNA is connected by neutral amide linkage and markedly stable in forming hybrid duplexes with DNA or RNA due to the lack of electric repulsion [3]. Although PNA has several advantages as a synthetic antisense agent, the major problem in PNA application results from its deficient membrane penetrability. In order to improve the membrane permeability of PNA, several peptide conjugates have been examined for their effectiveness in improving the permeability of PNA, and the cell-permeabilizing peptide $(KFF)_3K$ is frequently conjugated to PNAs in *E. coli* [4]. We have demonstrated that conjugation of $(KFF)_3K$ to the N-terminus of antisense PNA works successfully to knockdown genes in *C. jejuni* [5, 6]. In addition,

James Butcher and Alain Stintzi (eds.), *Campylobacter jejuni: Methods and Protocols*, Methods in Molecular Biology, vol. 1512, DOI 10.1007/978-1-4939-6536-6_5, © Springer Science+Business Media New York 2017

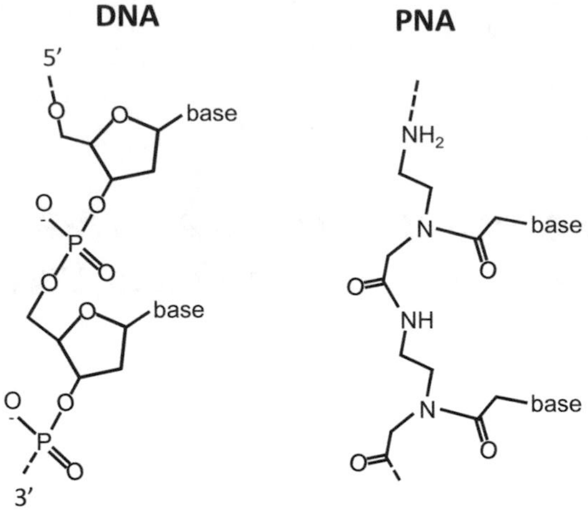

Fig. 1 Chemical structures of DNA and PNA

the limited permeability of PNA also restricts the length of PNA in design. Taken these factors together, 9- to 12-mer PNAs are generally synthesized in conjugation with $(KFF)_3K$ [4].

The efficacy of gene knockdown by antisense PNA critically depends on the target site of inhibition. Antisense PNA binds to regulatory regions for translation in mRNA, and the binding interferes with protein synthesis by steric hindrance. Therefore, ribosomal binding sites (RBS) and start codons are key target sites for antisense inhibition by PNA. Several studies have shown that the start codon region is susceptible to antisense inhibition [7–9], and the RBS is also an useful target in *C. jejuni* [6]. However, a golden rule does not seem to exist yet in PNA design to pinpoint most the effective inhibition site. Based on our previous study to inhibit the expression of CmeABC, a multidrug efflux pump, the choice of targeting flanking regions of the RBS or the start codon also substantially affect the efficacy of antisense inhibition by PNA [6]. For example, a PNA targeting the RBS of *cmeA* plus an RBS upstream region achieved better antisense inhibition than a PNA targeting the RBS plus its downstream region [6]. Although CmeA-PNAs 1, 2, and 3 all bind to the *cmeA* RBS (Fig. 2a), CmeA1-PNA inhibited CmeA expression more significantly than the others (Fig. 2b). Similarly, CmeA-PNAs 4, 5, and 6 all target the start codon of *cmeA* (Fig. 2a), but exhibited different inhibition efficacy (Fig. 2b). The PNA inhibition is also dependent on the concentration (Fig. 2c).

Application of antisense PNA is straightforward and can be done just by adding PNA to *Campylobacter* culture media. Antisense PNA displays effective knockdown activity in *E. coli* at 37 °C [4], and our laboratory often treats *C. jejuni* with antisense PNA at 42 °C [5, 6, 10, 11]. Since antisense PNA inhibits protein synthesis, the efficacy of gene knockdown can be determined by Western blotting to observe changes at the protein level [5, 10]. If a target

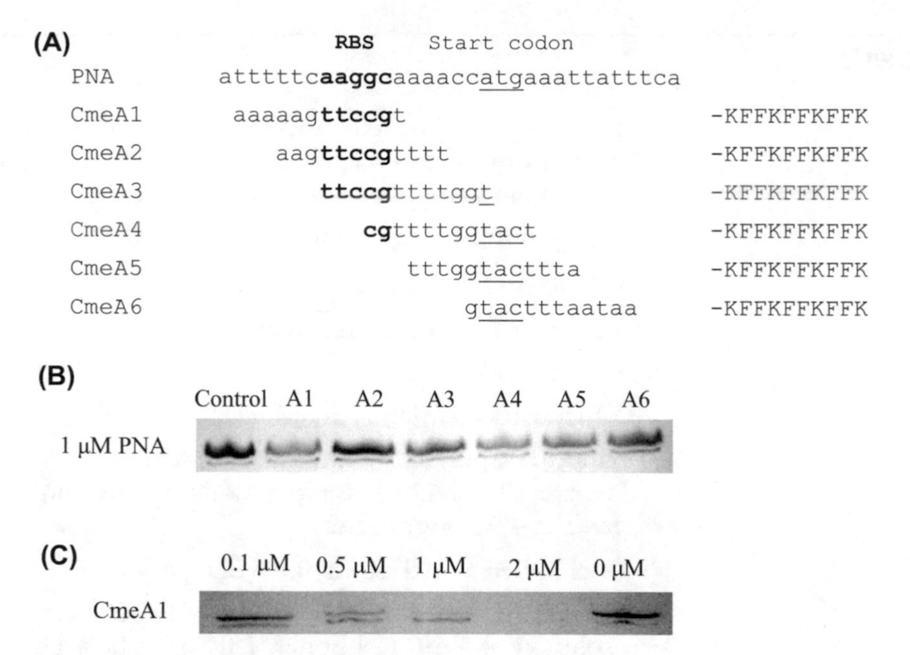

Fig. 2 Several PNA sequences targeting different regulatory regions of *cmeA* (**a**) and their inhibition efficiency visualized by immunoblotting (**b**). Concentration-dependent inhibition of CmeA with CmeA1-PNA (**c**)

gene mediates some unique phenotypes, alternatively, knockdown efficiency can be indirectly determined by measuring phenotype changes for which the target gene is responsible. In this chapter, we describe a method for antisense PNA-mediated gene knockdown by using antisense PNA targeting *cmeA* as an example.

2 PNA Design

Since antisense PNA targets the regulatory region for protein synthesis, only a flanking region containing the RBS and the start codon of *cmeA* was considered in PNA design. Due to permeability, a 9- to 12-mer is recommended for the length of antisense PNA in Gram-negative bacteria [4]. PNAs can be designed to target either the RBS or start codon and include some flanking areas (Fig. 2a). PNA can be designed in the same way as a reverse primer for PCR. The PNA sequence should avoid self-complementary sequences because PNA has stronger interaction with PNA than DNA or RNA. Thus, self-complementarity needs to be examined to avoid formation of hair-pins or homodimers of PNA oligomers. In addition, PNA oligomers rich in purine (>60 %), particularly guanine, may tend to aggregate and reduce solubility according to Panagene Guidelines for PNA oligomers (www.pnabio.com).

3 Materials

1. CmeA PNAs conjugated to $(KFF)_3K$. The synthesized PNAs should be dissolved in water and stored at −80 °C aliquots with small volume to reduce freeze/thaw damage.
2. *C. jejuni* culture components.
   (a) *C. jejuni* NCTC 11168.
   (b) Mueller Hinton broth and agar medium.
   (c) Spectrophotometer.
   (d) Incubators (shaking and cabinet).
3. Tricine–SDS–polyacrylamide gel components

   Prepare all solutions using ultra-pure water and store all reagents at 4 °C prior to use.
   (a) Gel buffer: 3 M Tris–Cl, 0.3% SDS, pH 8.45
   (b) 12.5% Resolving gel buffer: 4.2 mL 30% acrylamide-*bis* solution, 3.3 mL Gel buffer, 1.04 mL 100% glycerol, and 1.5 mL H_2O, Store at 4 °C.
   (c) 5% Stacking gel buffer: 0.4 mL 30% acrylamide-bis solution, 0.75 mL Gel buffer, and 1.8 mL H_2O. Store at 4 °C.
   (d) Tris–Tricine–SDS-PAGE anode buffer – 0.2 M Tris–Cl pH 8.9.
   (e) Tris–Tricine–SDS–PAGE cathode buffer – 0.1 M Tris–Cl, 0.1 M Tricine, and 0.1% (w/v) SDS pH 8.25.
   (f) 2× Tris–Tricine SDS loading buffer: 0.2 M Tris–Cl pH 6.8, 40% Glycerol, 2% (w/v) SDS, 0.04% Commassie Blue G-250 and 2% β-mercaptoethanol. Store at −20 °C and leave one aliquot at 4 °C for current use.
   (g) 10% (w/v) ammonium persulfate. Store at −20 °C.
   (h) *N,N,N, N′*-tetramethyl–ethylenediamine (TEMED).
4. Immunoblotting components.
   (a) PVDF membrane.
   (b) 1× Phosphate-buffered saline (PBS): 137 mM NaCl, 43 mM Na_2HPO_4, 1.47 mM KH_2PO_4, 2.7 mM KCl, adjust to a final pH of 7.2.
   (c) Transfer buffer: 25 mM Tris–base, 192 mM glycine and 20% methanol pH 8.3.
   (d) Blocking buffer: 5% skim milk in 1× PBS pH 7.2.
   (e) 100% Methanol
   (f) Primary antibody: rabbit anti-CmeABC [12] (1:1000 dilution in blocking buffer). Store at 4 °C.
   (g) Washing solution (PBST): 0.05% (v/v) Tween 20 in 1× PBS pH 7.2.

(h) Secondary antibody: goat anti-rabbit immunoglobulin G-horseradish peroxidase (1:5,000 dilution in blocking buffer; KPL). Store at 4 °C.

(i) Detecting solution: 4 CN Membrane Peroxidase Substrate System (KPL). Store at 4 °C.

4 Methods

1. *C. jejuni* culture conditions and PNA treatment.

   (a) Grow overnight culture of *C. jejuni* NCTC 11168 microaerobically (5% O_2, 10% CO_2, 85% N_2) on MH agar at 42 °C.

   (b) Resuspend the overnight bacterial suspension in MH broth to an optical density at 600 nm of 0.05–0.07.

   (c) Grow 1 mL aliquots of these *C. jejuni* cultures in the presence of antisense PNAs for 7 h with shaking at 200 rpm under microaerobic conditions at 42 °C. PNA concentrations and treatment time may vary depending on the PNA and the target gene (*see* **Notes 1** and **2**).

2. Tris–Tricine–SDS-PAGE and immunoblotting.

   (a) Cells are harvested by centrifugation for 5 min at 10,000 × g at 4 °C and washed twice with 200 μL of 1× PBS buffer pH 7.2.

   (b) Resuspend the harvested cells in 50 μL of 1× PBS buffer and boil for 10 min with 2× Tris–Tricine SDS-PAGE loading buffer.

   (c) Run Tris–Tricine–SDS-PAGE until dye reaches the end of the gel; running time depends on the molecular weight of the target protein.

   (d) Separate gel plates and rinse the gel with ultra-pure water.

   (e) Activate PVDF membrane with 100% methanol for 5 min.

   (f) Incubate activated PVDF membrane, rinsed gels, filter papers, and fiber pads with ice-cold transfer buffer at least 10 min.

   (g) Place the PVDF membrane-gel-filter paper sandwich in a holder cassette.

   (h) Transfer proteins to PVDF membrane for 40 min at 80 V at 4 °C.

   (i) Block the transferred membrane with blocking buffer for 1 h.

   (j) Add the primary antibody to the membrane and incubate for 2 h at room temperature or overnight at 4 °C with agitation.

   (k) Wash the membrane with PBST three times for 5 min each time.

(l) Add the secondary antibody and incubate for 1 h at room temperature with agitation.

(m) Wash membrane with PBST three times for 5 min each time.

(n) Add 5 mL of 4 CN Membrane Peroxidase substrate.

(o) After sufficient color develops, rinse the membrane three times with ultra-pure water.

5 Notes

1. PNA concentrations and knockdown efficiency: Since each antisense PNA has different knockdown efficiency (Fig. 2b), treatment concentrations of antisense PNA need to be evaluated with Western blotting. For example, CmeA1 and A4 PNAs substantially reduce CmeA expression, whereas CmeA2 does not exhibit effective knockdown activity at the same concentration (Fig. 2b). Also, antisense inhibition occurs in a concentration-dependent manner (Fig. 2c). As we noticed in a PNA-knockdown of CosR, excessive concentrations of CosR-specific PNA may affect bacterial viability, if the target is an essential gene [10]. Therefore, PNA concentrations need to be determined based on viability changes, if the target gene is essential.
2. PNA treatment time and knockdown efficiency: We have noticed that treatment time affects the efficiency of gene knockdown by antisense PNA. Treatments lasting a few hours with CmeA-PNA does not result in a significant change in CmeA expression under microaerobic conditions with shaking (200 rpm) at 42 °C, whereas PNA treatment for more than 6 h exhibited substantial reduction in CmeA expression (Fig. 3).

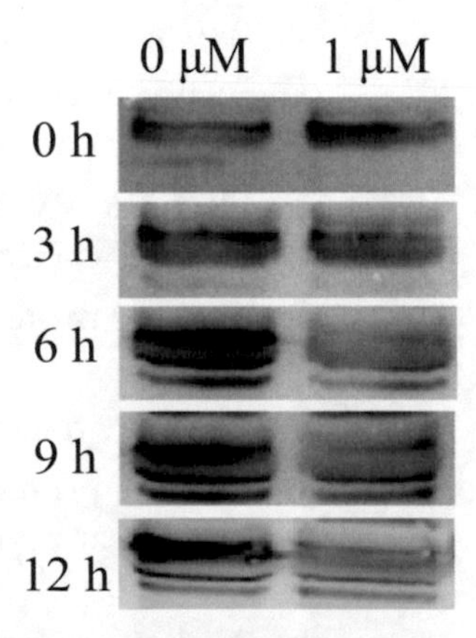

Fig. 3 Effect of treatment time on CmeA inhibition with CmeA1-PNA. The PNA concentrations are indicated on *top*

In the CosR knockdown, we noticed that extended treatment time reduced the inhibition efficiency presumably because increased bacterial populations at a fixed PNA concentration will affect the ratio between the number of PNA molecules and the number of bacterial cells [10]. Therefore, the treatment time for gene knockdown with antisense PNA needs to be optimized for each target gene in each culture conditions.

References

1. Good L, Nielsen PE (1998) Antisense inhibition of gene expression in bacteria by PNA targeted to mRNA. Nat Biotechnol 16(4):355–358
2. Demidov VV, Potaman VN, Frank-Kamenetskii MD et al (1994) Stability of peptide nucleic acids in human serum and cellular extracts. Biochem Pharmacol 48(6):1310–1313
3. Pooga M, Land T, Bartfai T et al (2001) PNA oligomers as tools for specific modulation of gene expression. Biomol Eng 17(6):183–192
4. Good L, Awasthi SK, Dryselius R et al (2001) Bactericidal antisense effects of peptide–PNA conjugates. Nat Biotechnol 19(4):360–364
5. Jeon B, Zhang Q (2009) Sensitization of *Campylobacter jejuni* to fluoroquinolone and macrolide antibiotics by antisense inhibition of the CmeABC multidrug efflux transporter. J Antimicrob Chemother 63(5):946–948
6. Oh E, Zhang Q, Jeon B (2013) Target optimization for peptide nucleic acid (PNA)-mediated antisense inhibition of the CmeABC multidrug efflux pump in *Campylobacter jejuni*. J Antimicrob Chemother dkt381
7. Dryselius R, Aswasti SK, Rajarao GK et al (2003) The translation start codon region is sensitive to antisense PNA inhibition in *Escherichia coli*. Oligonucleotides 13(6):427–433
8. Nekhotiaeva N, Awasthi SK, Nielsen PE et al (2004) Inhibition of *Staphylococcus aureus* gene expression and growth using antisense peptide nucleic acids. Mol Ther 10(4):652–659
9. Mondhe M, Chessher A, Goh S et al (2014) Species-selective killing of bacteria by antimicrobial peptide-PNAs. PLoS One 9(2), e89082
10. Hwang S, Kim M, Ryu S et al (2011) Regulation of oxidative stress response by CosR, an essential response regulator in *Campylobacter jejuni*. PLoS One 6(7):e22300
11. Hwang S, Zhang Q, Ryu S et al (2012) Transcriptional regulation of the CmeABC multidrug efflux pump and the KatA catalase by CosR in *Campylobacter jejuni*. J Bacteriol 1:01636–01612
12. Lin J, Michel LO, Zhang Q (2002) CmeABC functions as a multidrug efflux system in *Campylobacter jejuni*. Antimicrob Agents Chemother 46(7):2124–2131

Chapter 6

Identification of Ligand-Receptor Interactions: Ligand Molecular Arrays, SPR and NMR Methodologies

Christopher J. Day, Lauren E. Hartley-Tassell, and Victoria Korolik

Abstract

Despite many years of research into bacterial chemotaxis, the only well characterized system to date is that of *E. coli*. Even for *E. coli*, the direct ligand binding had been fully characterized only for aspartate and serene receptors Tar and Tsr. In 30 years since, no other direct receptor-ligand interaction had been described for bacteria, until the characterization of the *C. jejuni* aspartate and multiligand receptors (Hartley-Tassell et al. Mol Microbiol 75:710–730, 2010). While signal transduction components of many sensory pathways have now been characterized, ligand-receptor interactions remain elusive due to paucity of high-throughput screening methods. Here, we describe the use of microarray screening we developed to identify ligands, surface plasmon resonance, and saturation transfer difference nuclear magnetic resonance (STD-NMR) we used to verify the hits and to determine the affinity constants of the interactions, allowing for more targeted verification of ligands with traditional chemotaxis and in vivo assays described in Chapter 13.

Key words Chemotaxis receptors, Microarray ligand identification, High-throughput screening, SPR, NMR

1 Introduction

Chemotaxis ligands have traditionally been identified using chemotaxis assays of either targeted mutants or transposon mutant libraries or by selecting mutants that showed no taxis in the presence of media containing specific ligands [1–4], while affinity measurements of interactions were traditionally done using radiometric methodologies [5].

The production of a wider variety of array platforms starting with DNA arrays and now including protein and glycan arrays has increased the availability of chemically reactive substrates including amine, epoxide, NHS, hydroxyl, gold, and avidin. Using epoxide functionalized slides we first used amino acids as an array format with the primary amine groups being reactive with the epoxide surface [6]. This was used to identify the aspartate receptor of

James Butcher and Alain Stintzi (eds.), *Campylobacter jejuni: Methods and Protocols*, Methods in Molecular Biology, vol. 1512, DOI 10.1007/978-1-4939-6536-6_6, © Springer Science+Business Media New York 2017

Campylobacter jejuni, CcaA [6]. With this as a proven technology, extra chemotactic chemicals were added, as epoxide will react from amine groups through to hydroxyl groups. This enabled sugars and salts of organic acids to be added to the array, leading to the discovery of the multiple ligands of CcmL, the multiligand receptor of *C. jejuni* [7].

The advent of higher sensitivity surface plasmon resonance (SPR) with the production of the Biacore T100 and T200 machines has meant measuring the affinity of an interaction between a 25–40 kDa protein and a 100 Da ligand has become possible. These methodologies have been used to determine affinities and verify interactions between CcmL and its multiple ligands [7].

This chapter describes in detail the protocols to screen chemotaxis receptors with high-throughput technologies and verify the interactions with SPR and NMR.

2 Materials

All solutions are prepared using deionized water and analytical grade reagents followed by filtration using a 0.22 μm bottle top or syringe filter. The materials are stored at room temperature unless otherwise stated.

2.1 Buffers, Solutions, and Reagents for Array Analysis

1. Assay buffer: Phosphate buffered saline pH 7.4; 8 g NaCl, 0.2 g KCl, 1.44 g Na_2HPO_4, 0.24 g KH_2PO_4 dissolved in 1 L of MilliQ water and the pH adjusted to 7.4 with HCl.
2. Wash buffer: Phosphate buffered saline pH 7.4 with 0.05% Tween20.
3. Blocking buffer: Phosphate buffered saline pH 7.4 with 1% bovine serum albumin.
4. Amino acids: See Table 1 for full list and source. Prepared as 100 mM stocks in either PBS or MilliQ water. 10 M NaOH or 10 M HCl may be added in incremental steps to improve solubility as indicated.
5. Salts of organic acids: *See* Table 1 for full list and source. Prepared as 100 mM stocks in either PBS or MilliQ water. 10 M NaOH may be added in incremental steps to improve solubility as indicated.
6. Monosaccharides: *See* Table 1 for full list and source. Prepared as 100 mM stocks in either PBS or MilliQ water.
7. Arrayit Superepoxy II glass substrates.
8. Arrayit Spotbot Extreme Printing Robot.
9. 0.22 μm filtered Deionized water
10. 50 mL centrifuge tubes (sterile).

Table 1
Printed chemotaxis ligands

Compounds for printing	Soluble water/PBS	Company and Cat#
Alanine	Yes	Sigma-Aldrich A7469
Arginine	No—Add NaOH	Sigma-Aldrich A8094
Asparagine	No—Add NaOH	Sigma-Aldrich A4159
Aspartic acid	No—Add NaOH	Sigma-Aldrich A7219
Aspartic acid potassium salt	Yes	Sigma-Aldrich 11230
Aspartic acid sodium salt	Yes	Sigma-Aldrich 11195
Cysteine	No—Add HCl	Sigma-Aldrich C7352
Cysteine-HCl	Yes	Sigma-Aldrich C6852
Fucose	Yes	Sigma-Aldrich F2252
Fumaric acid	No—Add NaOH	Sigma-Aldrich F8509
Sodium fumarate	Yes	Sigma-Aldrich F1506
Galactose	Yes	Sigma-Aldrich G5388
Glucosamine	Yes	Sigma-Aldrich G1514
Glucose	Yes	Sigma-Aldrich G7021
Glutamic acid	No—Add NaOH	Sigma-Aldrich G1251
Sodium glutamate	Yes	Sigma-Aldrich 49621
Glutamine	No—Add NaOH	Sigma-Aldrich G3126
Glycine	Yes	Sigma-Aldrich G7126
Histidine	Yes	Sigma-Aldrich H6034
Isoleucine	No—Add NaOH	Sigma-Aldrich I2752
Leucine	No—Add NaOH	Sigma-Aldrich 61819
Lysine	Yes	Sigma-Aldrich 62840
Malic acid	No—Add NaOH	Sigma-Aldrich M7397
Sodium malate	Yes	Sigma-Aldrich M1125
Methionine	Yes	Sigma-Aldrich M5308
Phenylalanine	No—Add NaOH	Sigma-Aldrich P2126
Proline	No—Add NaOH	Sigma-Aldrich 81709
Purine	Yes	Sigma-Aldrich P55805
Ribose	Yes	Sigma-Aldrich R9629
Serine	Yes	Sigma-Aldrich 84959
Succinic acid	No—Add NaOH	Sigma-Aldrich S9512
Sodium succinate	Yes	Sigma-Aldrich 224731

(continued)

Table 1 (continued)

Compounds for printing	Soluble water/PBS	Company and Cat#
Thiamine	Yes	Sigma-Aldrich T4625
Threonine	No—Add NaOH	Sigma-Aldrich T8625
Tryptophan	No—Add NaOH	Sigma-Aldrich T0254
Tyrosine	No—Add NaOH	Sigma-Aldrich 93829
Valine	No—Add NaOH	Sigma-Aldrich 94619
α- Ketoglutaric acid	No—Add NaOH	Sigma-Aldrich K1128
α- Ketoglutaric acid (sodium salt)	Yes	Sigma-Aldrich K2010

11. 1.5 mL centrifuge tubes (autoclaved).
12. Thermo Scientific 65 μL gene frames (AB-0577).

2.2 STD-NMR Reagents

1. D_2O (Store and use as per manufacturers guidelines due to the hygroscopic nature of D_2O).
2. Amicon Ultra-5 K Centrifugal Filter Units for buffer exchange.
3. Bruker optimized Shigemi Tube for D_2O; 5 mm diameter.
4. 10 mM ligand in D_2O.
5. Protein for buffer exchange in to D_2O.
6. Deuterated PBS: 0.4 g NaCl, 0.01 g KCl, 0.072 g Na_2HPO_4, 0.012 g KH_2PO_4 dissolved in 50 mL D_2O at pH 7.4.

2.3 SPR Using Nickel Affinity (NTA) Sensor Chips

1. Series S NTA sensor chip for Biacore T100/T200.
2. Assay Buffer: Phosphate buffered saline pH 7.4; 8 g NaCl, 0.2 g KCl, 1.44 g Na_2HPO_4, 0.24 g KH_2PO_4 dissolved in 1 L of MilliQ water and the pH adjusted to 7.4 with HCl.
3. 100 mM Nickel Chloride.
4. 100 mM Nickel Sulfide.
5. 50 mM NaOH.
6. EDTA regeneration buffer: 1xHBS-EP (1:10 dilution in MilliQ water of 10x HBS-EP purchased from GE) with addition of 300 mM EDTA.
7. Purified His-Tagged Tlp protein at 100 μg/mL minimum concentration.
8. A control His-tagged protein that does not interact with the compounds to be tested at 100 μg/mL minimum concentration (*see* **Note 1**).

9. 1.5 mL, 650 μL tubes, 96-well plates and plate seals.
10. Biacore T100/T200.

2.4 SPR Using Amine Coupling (CM5) Sensor Chips

1. Series S CM5 sensor chip for Biacore T100/T200.
2. 10 mM Sodium acetate solutions pH ranging from 3.0 to 5.5 in 0.5 pH increments.
3. Assay Buffer: Phosphate buffered saline pH 7.4; 8 g NaCl, 0.2 g KCl, 1.44 g Na_2HPO_4, 0.24 g KH_2PO_4 dissolved in 1 L of MilliQ water and the pH adjusted to 7.4 with HCl.
4. 50 mM NaOH.
5. Amine coupling kit (e.g., GE Biacore) containing EDC/NHS and ethanolamine.
6. Purified Tlp protein at 200 μg/mL minimum concentration.
7. 1.5, 650 μL tubes, 96-well plates and plate seals.
8. Biacore T100/T200.

3 Methodologies

3.1 Array Printing Methodology

1. Set for a single 384-well plate (*see* **Note 2**).
2. Set for the use of capillary metal pins 6 (946MP6; *see* **Note 3**) from Arrayit in a 3 column by 2 row format.
3. Set number of slides to be printed.
4. 10 preprint spots with a contact time of 0 s.
5. Between 4 and 10 printed spots with a contact time of 0.5 s.
6. Set up each pin grid as a 16×16 or 20×20 grid depending on print spots per sample.
7. Set a spot to spot distance to 300 μm in each direction.
8. Reload pins after every 50 spots.
9. Use default wash parameters.
10. Use Spocle Generator to generate 3 sub array prints spaced;
    (a) Subarray 1: First spot at 6 mm from top of slide 6 mm from side of slide.
    (b) Subarray 2: First spot at 30 mm from top of slide 6 mm from side of slide.
    (c) Subarray 2: First spot at 54 mm from top of slide 6 mm from side of slide.
11. Save files ensuring you have selected all files to be saved including check boxes to ensure map files are saved.
12. Use methods files, plate file and map file to generate .gal file with the Gal-file generator program.

13. Ensure environment in the printer is between 40 and 60% relative humidity and chilled to at least 10 °C.
14. Lay substrates out as per methodology design with one slide for preprint spots.
15. Insert printing pins capillary metal pins 6 (946MP6) from Arrayit.
16. Insert the source plate.
17. Run the print program within SpotApp.
18. Allow the slides to dry on the printer for at least 2 h and allow the slides to sit dry at 4 °C overnight to complete the sample:substrate chemical interaction.
19. Scan the slide with 488 nm/555 nm/647 nm at 10 μm resolution after overnight at 4 °C to ascertain if all spots printed.. Check with generated .gal file.
20. Neutralize slide by immersing in cold Blocking buffer for 2 h at 4 °C.
21. Rinse in 1× PBS and dry in a 50 mL tube spun at 500*g*.
22. Re-scan the slide with 488 nm/555 nm/647 nm at 10 μm resolution to check if all spots have ceased to auto-fluoresce (*see* **Note 4**).
23. Store slides at 4 °C for a maximum of 3 months.

3.2 Array Assay of Purified Protein (See Note 5)

1. Add 1–2 μg of purified tagged protein (maximum volume of 30 μL) to a 1.5 mL centrifuge tube.
2. Add equal molar concentration of antitag antibody (*see* **Note 6** and 7).
3. Signal amplification can help identify binding in the arrays so use of a labeled secondary and tertiary antibody is recommended with fluorophores such as Alexa488, 555, or 647 (such as Alexa Fluor® 488 Signal-Amplification Kit for Mouse Antibodies A-11054 from Thermo Fisher Scientific). Secondary antibody is added at ½ the concentration (mg/mL) of primary antibody and tertiary antibody (mg/mL) is added at ½ the concentration of secondary antibody. For example if 6 μg of primary antibody is added, 3 μg of secondary and 1.5 μg of tertiary is added. Make up to a final volume of 70 μL with Assay buffer. Incubate the tube at room temperature in the dark for 15 min.
4. Apply gene frames to the slide starting with the first gene frame on the top left hand corner (right top hand corner of slide is notched). Second gene frame lines up (blue to blue) with the bottom of the first gene frame on the very left hand side of the slide. The bottom gene frame lines up on the very left hand edge and with the bottom of the second gene frame (blue to blue) (*see* **Note 8**).

5. If the antibodies have not previously been used on an array of this type, an antibody only control will have to be performed. Set up a second tube containing only assay buffer and the antibodies at the same concentrations used in the described experiment.
6. Prepare 3 of the gene frame cover slips and put to one side.
7. Add 70 μL of the protein:antibody mix to one gene frame area of the slide at a time. Pipette the 70 μL of protein:antibody mix into the bottom corner and quickly put on the cover slip. Use the cover slip to move the full volume across the gene frame area and try to avoid bubbles. A small amount of volume should come out of the gene frame area.
8. Repeat the above until all three areas on slide are filled with protein:antibody under a coverslip and incubate in the dark at room temperature for 15 min.
9. Remove cover slips and gene frames in a pipette box lid under 0.5 cm of Assay buffer to ensure the slide does not dry out.
10. Prepare 50 mL centrifuge tubes containing 45 mL of Wash buffer (×3) and 45 mL of Assay buffer (×1).
11. Transfer array from pipette box lid to the first tube of wash buffer and wash for 2 min with slow inversion of the tube.
12. Repeat the washing step of the array with remaining two tubes of wash buffer (total wash time 6 min).
13. Rinse in tube containing assay buffer and transfer to empty 50 mL centrifuge tube. Dry by centrifuging at 500 ×*g* for 5 min.
14. Scan slide at wave length of antibodies at 10 μm resolution.
15. Analyze using scanners array software or freeware software available from internet and .gal file. For example the ScanArray software from Perkin Elmer or freeware Spotfinder program from TIGR.
16. A positive interaction is identified as a visible group of spots that is significantly more fluorescent than the slide background and was not autofluorescent on the postneutralization scan or did not come up with antibody only control (Fig. 1).

3.3 STD-NMR to Confirm Ligand Binding

1. Prepare a solution of protein:ligand so that the molar ratio is 1:100 in a final volume of 600 μL in D_2O PBS.
2. Prepare a sample of protein at the same concentration as in **step 1** but without the ligand in a final volume of 600 μL of D_2O PBS.
3. Prepare a sample of ligand at the same concentration as in **step 1** but without the protein in a final volume of 600 μL of D_2O PBS.
4. Pippette the solutions from **steps 1–3** into separate Shigemi Tubes.

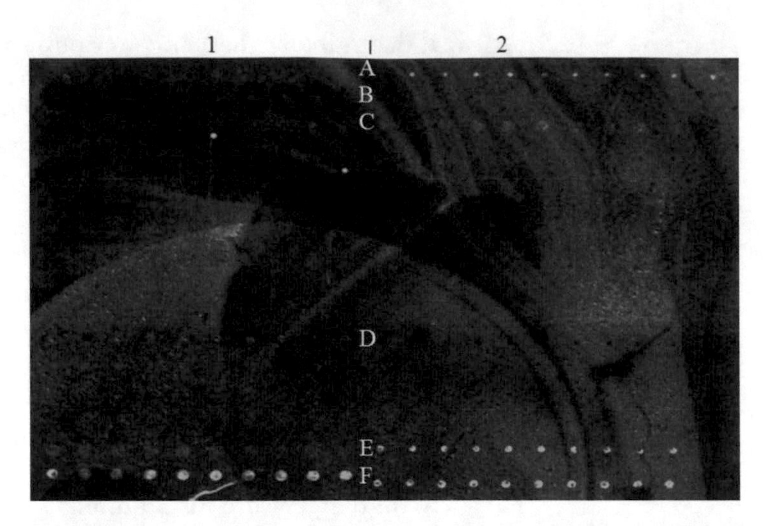

Fig. 1 Chemotaxis Array result. A-F represent the rows of the print with spots present, 1–2 represents the columns with each ligand being printed in 10 replicate spots. *E2* and *F1* and *F2* are printed control spots and do not indicate binding. *A1* represents clear positive binding, while *B1*, *C2*, *D1*, and *E1* represent less clear-cut binding that will need to be subjected to statistical analysis against the average background

5. Acquire H^1 NMR spectra using a 600 MHz NMR spectrometer (e.g., Bruker Avance 600 MHz spectrometer) equipped with a 5 mm TCI cryoprobe.
   (a) Spectra were acquired with 1 k scans and a relaxation delay of at least $>1 \times T_1$ value of the longest T_1 protons of the ligand being studied.
   (b) H^1 T_1 values were determined using the inversion recovery method.
6. For STD experiments protein-ligand interactions
   (a) Protein should be saturated on-resonance at a frequency of −600 Hz in the aliphatic region
   (b) Off resonance at 20,000 Hz
   (c) A cascade of 40 selective Gaussian-shaped pulses of 50 ms with a 100 μs delay between each pulse with a total saturation time of 2 s.
   (d) Control experiments should be performed identical spectra of protein only and ligand only should be measured.
   (e) Protein:Ligand interactions are obtained through subtraction of the on/off resonance spectra. Ligand only should show no signal.

3.4 SPR Using a His-Tagged Protein on NTA Chip

1. Ensure NTA chip is at room temperature and dry before inserting into Biacore.
2. Methodology setup for Biacore:

(a) Buffer A is Assay Buffer (1× PBS).

(b) Temperature is set to 25 °C.

(c) Concentrations of analytes are set to μM.

(d) Use multichannel detection rather than single or dual.

(e) Set for 10 Hz recording rather than 1 Hz.

(f) Set up one cycle with the purpose to be sample and link it to cycle_1.

(g) Cycle_1 contains the following steps.

- Regeneration: Regeneration buffer (EDTA buffer), contact time 30 s, flow rate 30 μL/min. Check extra wash box and use H_2O and check stabilization box with a time of 60 s.
- Enhancement: Nickel solution (1:1 mix of $NiCl_2$ and $NiSO_4$; *See* **Note 9**). Contact time 30 s, flow rate 30 μL/min.
- Capture 1: Flow cell 1 only capture of nonbinding His-tagged protein. Contact time 180 s, flow rate 10 μL/min.
- Capture 2: Flow cell 2 only capture of His-tagged Tlp protein. Contact time 180 s, flow rate 10 μL/min.
- Capture 3: Flow cell 3 only capture of His-tagged tagged protein. Contact time 180 s, flow rate 10 μL/min.
- Capture 4: Flow cell 4 only capture of His-tagged tagged protein. Contact time 180 s, flow rate 10 μL/min. (If only capturing 1 Tlp make 2.7.5 Capture 3 to flow cell 3 and 4 and skip Capture 4).
- Sample: Single cycle kinetics, flow rate of 30 μL/min. Contact time of 60 s and disassociation time of 600 s.
- Regeneration 2: Regeneration buffer (EDTA buffer), contact time 30 s, flow rate 30 μL/min.
- Regeneration 3: 50 mM NaOH, contact time 30 s, flow rate 30 μL/min. Check extra wash box and use H_2O and check stabilization box with a time of 60 s.

3. Set up each analyte with two rows in the table.

(a) Row 1: 5× zero concentration, as nonbinding control. This is just Assay buffer.

(b) Row 2: Dilution series. On first run set up a 1:5 dilution series 0.8, 4, 20, 100, and 500 μM. Once a concentration range is known then subsequent runs can use a 1:2 dilution around the saturation point.

(c) A total of 9 analytes can be tested in tubes or 16 using a 96-well plate.

(d) Set up using the rack positions obtained from the method. Ensure that the same Assay buffer is used to make up dilutions that are being used to run the machine.

(e) Ensure that the amino acids or salts that have been adjusted with HCl or NaOH are pH adjusted to pH 7.4 when diluted into Assay buffer.

(f) Analyze using the Biacore analysis software.

3.5 SPR Using a Purified Protein Immobilized Onto a CM5 Sensor Chip

1. Ensure CM5 chip is at room temperature and dry before inserting into Biacore.

   (a) Methodology setup for Immobilization of protein to Biacore:

   - pH Scout of protein.
     - Calculate pI of your protein.
     - Use pH scout Wizard template.
     - For proteins with pI ranges above pH 6.0 test the pH scout at pH 5.5, 5.0, 4.5, 4.0. For pI below 6.0 adjust the pH range to start at the pI and go down ensuring that the pH scout does not drop below pH 2.5.
     - Make up protein with at least a 1:2 dilution into 10 mM sodium acetate solutions at the pHs to be tested so that the protein is at a final concentration of 100 μg/mL. Ideally, the higher the concentration the better for the pH scout as the dilution into sodium acetate will approach the stated pH.
     - Follow the Wizard template for setup.
     - The best pH to use is the highest pH that provided a strong interaction between the chip surface and the protein.

2. Immobilisation Immobilization of protein to CM5 chip.

   (a) Select amine coupling.

   - Check blank immobilization for flow cell 1.
   - For test protein check the flow cell to be used and select time based immobilization. Immobilize for 600 s at a flow rate of 10 μL/min. This should capture between 3000 and 10,000 RU of protein.

   (b) Set up protein exactly as done in the pH scout with the optimal pH used.

   (c) Follow the wizard set up.

3. Set up a Bicore run to screen for interactions.

   (a) Buffer A is Assay Buffer (1× PBS).

(b) Temperature is set to 25 °C.

(c) Concentrations of analytes is set to μM

(d) Use dual channel detection if analyzing 1 protein or multichannel detection if testing more than 1 Tlp at the same time.

(e) Set for 10 Hz recording rather than 1 Hz.

(f) Set up one cycle with the purpose to be sample and link it to cycle_1.

(g) Cycle_1 contains the following steps

- Sample: Single cycle kinetics, flow rate of 30 μL/min. Contact time of 60 s and disassociation time of 600 s.
- Regeneration 1: use a regeneration buffer that will cause minimum damage to you protein such as the NaAc buffer used for immobilization. Contact time 120 s, flow rate 30 μL/min.

4. Set up each analyte with two rows in the table.

   (a) Row 1: 5× zero concentration, as nonbinding control. This is just Assay buffer.

   (b) Row 2: Dilution series. On first run set up a 1:5 dilution series 0.8, 4, 20, 100, and 500 μM. Once a concentration range is known then subsequent runs can use a 1:2 dilution around the saturation point.

   (c) A total of 9 analytes can be tested in tubes or 16 using a 96-well plate.

5. Set up using the rack positions obtained from the method. Ensure that the same Assay buffer is used to make up dilutions that are being used to run the machine.
6. Ensure that the amino acids and salts that required addition of HCl or NaOH for solubility are readjusted to pH 7.4 when diluted to a 1 mM stock concentration in Assay buffer for Biacore analysis. Highly basic or acidic solutions may damage your protein or remove it from the chip surface.
7. Analyze using the Biacore analysis software.

4 Notes

1. Typically, the best control for an NTA chip is a heat/chemical denatured version of the same His-tagged protein. However, denatured Tlp protein in all cases tested so far is not suitable as a control for Biacore due to loss of solubility in assay buffers when denatured. Use of denatured protein will result in blockage of the flow cells and the need to replace the IFC within the Biacore.
2. Use either Arrayit brand or Griener branded conical/V-bottom polypropylene 384-well plates.

3. 946MP6 were chosen as these provide a 200 μm spot diameter which is preferred when using a low density print. If you wish to improve printing density, such as wishing to print more than 2400 spots/subarray then 946MP4 would be the preferred choice.
4. If slide still contains high levels of autofluorescence block for an additional 1 h in Blocking buffer and 1 h in 1× PBS at 4 °C.
5. This protocol can be applied to glycan arrays with some modifications as per Day et al., 2009 [8]. Glycan arrays can be obtained from several sources including academic sources such as the Institute for Glycomics Glycan Array or the Mammalian Glycan Array for the Consortium for Functional Glycomics from The SCRIPPS Research Institute.
6. Antibody used for His-tagged proteins is the Cell signalling technology His-Tag (27E8) Mouse mAb. For GST-tagged protein the GST (26H1) Mouse mAb is used.
7. IgG antibodies are 150 kDa while Campylobacter Tlp proteins range from 25 to 40 kDa including the tag. Equal molar antibody therefore requires addition of between 3.75 and 6 μg of antibody. Antibody concentrations range from 1 to 10 mg/mL. The above His-tag is typically sold at 10 mg/mL but check each batch and with the manufacturer for concentrations.
8. The gene frames come between 2 pieces of plastic. One side has a hole. The side with the hole in it is the plastic piece you will need to remove to place the gene frame on the slide. You will need to trim the plastic covering to the blue edge before applying to the slide. The frame is wider than it is tall. Ensure that the longest edge is resting across the slide rather than down the slide.
9. In our hands a 1:1 mix of 100 mM $NiCl_2$ and 100 mM $NiSO_4$ is more effective for activation of the NTA chip surface for the capture of proteins. This 1:1 mix allows for greater protein capture levels and for some proteins more stable capture than either solution used separately. This is not the recommended solution supplied by the manufacturer.

References

1. Hartley-Tassell LE, Shewell LK, Day CJ et al (2010) Identification and characterization of the aspartate chemosensory receptor of *Campylobacter jejuni*. Mol Microbiol 75:710–730. doi:10.1111/j.1365-2958.2009.07010.x
2. Melton T, Hartman PE, Stratis JP et al (1978) Chemotaxis of *Salmonella typhimurium* to amino acids and some sugars. J Bacteriol 133(2):708–716
3. Mesibov R, Adler J (1972) Chemotaxis toward amino acids in *Escherichia coli*. J Bacteriol 112(1):315–326
4. Schmidt J, Musken M, Becker T et al (2011) The *Pseudomonas aeruginosa* chemotaxis methyltransferase CheR1 impacts on bacterial surface sampling. PLoS One 6(3):e18184. doi:10.1371/journal.pone.0018184
5. Tareen AM, Dasti JI, Zautner AE et al (2010) *Campylobacter jejuni* proteins Cj0952c and

Cj0951c affect chemotactic behaviour towards formic acid and are important for invasion of host cells. Microbiology 156(Pt 10):3123–3135. doi:10.1099/mic.0.039438-0

6. Mowbray SL, Koshland DE Jr (1987) Additive and independent responses in a single receptor: aspartate and maltose stimuli on the tar protein. Cell 50(2):171–180
7. Rahman H, King RM, Shewell LK et al (2014) Characterisation of a multi-ligand binding chemoreceptor CcmL (Tlp3) of *Campylobacter jejuni*. PLoS Pathog 10(1):e1003822. doi:10.1371/journal.ppat.1003822
8. Day CJ, Tiralongo J, Hartnell RD et al (2009) Differential carbohydrate recognition by *Campylobacter jejuni* strain 11168: influences of temperature and growth conditions. PLoS One 4(3):e4927. doi:10.1371/journal.pone.0004927

Chapter 7

Characterization of High Affinity Iron Acquisition Systems in *Campylobacter jejuni*

Ximin Zeng and Jun Lin

Abstract

Iron acquisition systems are critical for bacterial pathogenesis and thus have been proposed as attractive targets for iron-dependent pathogen control. Of these systems, high-affinity iron acquisition mediated by siderophore, a small iron chelator, is the most efficient iron-scavenging mechanism in gram-negative bacteria. *Campylobacter* does not produce any siderophores but has the ability to utilize exogenous siderophores. In particular, the enterobactin (Ent)-mediated iron scavenging is tightly linked to *Campylobacter* pathogenesis. To date, Ent, a triscatecholate with the highest known affinity for ferric iron, is a well-characterized siderophore used by *Campylobacter* for iron acquisition during in vivo infection. Here, we describe the key methods used to characterize Ent-mediated high affinity iron acquisition system in *Campylobacter jejuni*.

Key words Enterobactin, Growth promotion assay, Insertional mutagenesis, Conjugation, Site-directed amino acid substitution mutagenesis

1 Introduction

To avoid the generation of toxic oxygen-derived free radicals through the Haber–Weiss–Fenton chemistry [1], animal and human hosts have evolved various mechanisms to lower the concentrations of free iron and the resulting concentrations are far below the iron levels required for the growth of gram-negative bacteria upon infection. For example, iron-binding proteins in the intestine could sequester free ferric iron at level of 10^{-24} M and make it unavailable to most bacteria that require at least 10^{-7} M for normal growth [2].

To counteract the iron-restricted conditions in vivo, enteric bacteria have evolved sophisticated genetic systems for iron acquisition and successful colonization in the host. Four major iron acquisition systems have been identified in gram-negative bacteria, which include (1) high-affinity iron acquisition mediated by siderophores; (2) utilization of naturally occurring organic acids as low-affinity iron carriers (e.g., citrate); (3) direct use of host iron complexes

James Butcher and Alain Stintzi (eds.), *Campylobacter jejuni: Methods and Protocols*, Methods in Molecular Biology, vol. 1512, DOI 10.1007/978-1-4939-6536-6_7, © Springer Science+Business Media New York 2017

(such as lactoferrin) in an siderophore-independent way; and 4) utilization of ferrous iron via FeoB-type transport proteins under anaerobic and reducing environment [2–5]. Of these strategies, the siderophore-mediated iron is the most efficient and common iron-scavenging mechanism in gram-negative bacteria [4, 5]. Enterobactin (**Ent**) is a triscatecholate siderophore produced by Enterobacteriaceae, and displays exceptionally high affinity for ferric iron with a *KD* value of 10^{-35} M at physiological pH [6], which is the highest among any known natural siderophore compounds produced by enteric bacteria [7]. Despite the lack of information concerning the diversity and relative abundance of siderophores in the intestine, it is likely that Ent is produced in significant amounts by gut commensal bacteria and ferric Ent (**FeEnt**) may be a significant iron source to enteric pathogens including *Campylobacter jejuni* during colonization in intestine [8, 9].

Several iron-uptake systems have been identified in *C. jejuni*, such as utilization of the ferric iron bound to siderophores, glycoproteins, and heme (reviewed by [8–10]. Notably, *Campylobacter* does not produce any siderophores but has the ability to utilize exogenous siderophores, such as Ent [11]. The FeEnt acquisition system is an effective and highly conserved iron acquisition strategy used by *Campylobacter* to establish successful infection in the intestine [12–15]. As a prototypical triscatecholate, Ent is a well-characterized, physiologically relevant siderophore utilized by *Campylobacter* for intestinal colonization [16]. Therefore, in this chapter, we will focus on major methods used to examine the FeEnt receptor CfrA, an important gatekeeper for FeEnt acquisition in *C. jejuni*. Specifically, this chapter will include the protocols for functional characterization of CfrA, which include Ent purification, Ent growth promotion assays, insertional mutagenesis, site-directed amino acid substitution mutagenesis, and an optimized conjugation protocol for complementation.

2 Materials

All solutions are prepared using ultrapure water (with a sensitivity of 18 MΩ cm at 25 °C) and analytical grade reagents. All growth media should be autoclaved before usage. All chemicals and reagents are stored at room temperature (unless indicated otherwise).

2.1 Purification of Enterobactin (Ent)

1. *Escherichia coli* AN102, an Ent transport mutant (available from Coli Genetic Stock Center, CGSC#: 5130, http://cgsc.biology.yale.edu/Strain.php?ID=7598).
2. Luria-Bertani (**LB**) broth: 25 g of granulated Luria-Bertani powder dissolved in one liter of water and then autoclaved. Store at room temperature.

3. Modified T medium for *E. coli* AN102: Mix 58 g NaCl, 37 g KCl, 1.33 g $CaCl_2$, 1.0 g $MgCl_2 \cdot 6H_2O$, 11.0 g NH_4Cl, 2.72 g KH_2PO_4, 121.0 g Tris base, 1.42 g Na_2SO_4, 5 g of casamino acids, and 100 μL of 100 mM $MnCl_2$ with approximately 9.7 L of water. After all compounds are completely dissolved, adjust pH to 7.4 with HCl and autoclave. The following sterile supplements (via filtration through 0.22 μM membrane) are subsequently added into the autoclaved T medium: 50 mL of 10 mg/mL leucine, 50 mL of 10 mg/mL proline, 100 mL of 5 mg/mL tryptophan, 5 mL of 10 mg/mL thiamine·HCl, and 100 mL of 20% glucose. Store at room temperature.
4. Ethyl acetate (sequencing grade).
5. 0.1 M citrate buffer, pH 5.5.
6. Anhydrous $MgSO_4$.
7. Hexanes (HPLC grade).
8. Rotovap, such as CentriVap Concentrator (LABCONCO).
9. Vacufuge concentrator.

2.2 Enterobactin Growth Promotion Assay

1. Müller-Hinton (**MH**) medium: For broth medium, 22 g of Müller-Hinton powder is dissolved in one liter of water, and autoclaved. Store at room temperature. To prepare MH soft agar medium, 22 g of Müller-Hinton and 4 g of granulated agar powder are added into 1 L of water, autoclaved and stored at 50°C for growth promotion assay.
2. 20 mM Deferoxamine mesylate.
3. 40 mM Ferric sulfate solution.
4. 10 mM Enterobactin in methanol.
5. Sterile blank test discs, 1/4 in.
6. Tri-gas incubator (85% N_2, 10% CO_2, 5% O_2) for cultivation of *C. jeuni*.

2.3 Insertional Mutagenesis of *cfrA*

1. *C. jejuni* NCTC 11168 strain.
2. Müller-Hinton (**MH**) medium: Broth medium is prepared as described above. To prepare MH agar plates, 22 g of Müller-Hinton and 15 g of granulated agar powder were added into 1 L of water, autoclaved and poured into petri dishes. Store agar plates at 4 °C.
3. MH agar plates supplemented with 6 μg/mL of chloramphenicol. Store agar plates at 4 °C.
4. Luria-Bertani (**LB**) agar plates: 25 g of granulated Luria-Bertani powder and 15 g of granulated agar are added into 1 L of water, autoclaved and poured to petri dishes.
5. LB agar plates supplemented with 100 μg/mL ampicillin and/or 6 μg/mL chloramphenicol. Store agar plates at 4 °C.

6. pGEM-T easy vector (Promega).
7. Genomic DNA purification kit (e.g., Promega Wizard).
8. Taq DNA polymerase.
9. PCR primers CfrAF2, CfrAR2, CHFL1, CHFR1, CfrAF, and CfrAR (Table 1).
10. NheI restriction enzyme.
11. pUOA18 plasmid (Wang and Taylor, 1990) which harbors the chloramphenicol-resistant gene *cat*.
12. MH agar tubes (used for natural transformation). Approximately 1.5 mL of autoclaved MH agar (melted) is added to sterile 17 × 100 mm polystyrene plastic culture tubes. After solidification, the culture tube is capped and stored at 4 °C.
13. *E. coli* JM109 competent cells (Promega JM109 Competent Cells, or homemade competent cells).
14. CfrA-specific antibody (freely available to interested researchers) [14].

2.4 Complementation of cfrA Mutant

1. Isogenic *cfrA* mutant of *C. jejuni* NCTC 11168 (constructed from Subheading 3.3).
2. pRY107 shuttle vector, kanamycin resistant [17].

Table 1
Primers used in characterization of the FeEnt receptor CfrA in *C. jejuni*

Primer	DNA Sequence (5′-3′)[a]	Product size (bp)	Target gene or function
CHFL1	TTT**GCTAGC**TGCTCGGCGGTGTTCCTTT (*Nhe* I)	802	*cat*
CHFR1	TTT**GCTAGC**GCGCCCTTTAGTTCCTAAAG (*Nhe* I)		
CfrAF	GAGATGTTGCAGAGGCTATCG	527	*cfrA*
CfrAR	TGCCTTTGTAGGACTTTGAGC		
CfrAF2	TTTCATTGGGTTGTATGTGTAAAAA	1,631	Part of *cfrA* plus 445 bp upstream region
CfrAR2	TCTGCAAAAATTGCCAATAAA		
K297A_F	GAAGTTGATgcATTTGTGACTTATTTAAGTCATG	9,300	Create K297A mutation in CfrA
K297A_R	GTCACAAATgcATCAACTTCCATAATATCTGC		

[a]Restriction sites are underlined in the primer sequence and the names are identified in parentheses. The nucleotide in *lower case* indicates desired mutation site to replace target lysine (K) with alanine (A) using partial overlapping PCR as described in Subheading 3.5

3. Primers CfrAF2 and CfrAR1 (Table 1).
4. SmaI restriction enzyme.
5. High-fidelity DNA polymerase producing blunt ends (e.g., *Pfu Ultra* DNA polymerase from Stratagene).
6. *Escherichia coli* DH5α containing helper plasmid pRK212.2 (Ampicillin and tetracycline resistant) [18].
7. LB plates containing ampicillin (100 μg/mL), tetracycline (12.5 μg/mL), kanamycin (50 μg/mL), X-gal (80 μg/mL), and IPTG (0.5 mM).
8. LB broth supplemented with ampicillin (100 μg/mL), tetracycline (12.5 μg/mL), kanamycin (50 μg/mL).
9. Conjugation selective plates. MH plates containing *Campylobacter*-specific selective agents (SR0117E; Oxoid) and kanamycin (50 μg/mL).
10. MH broth containing kanamycin (50 μg/mL).
11. X-gal (50 mg/mL).
12. 0.1 M IPTG solution.
13. Water bath.
14. CfrA specific antibody [14].

2.5 Site-Directed Amino Acid Substitution Mutagenesis

1. pCfrA plasmid containing full length of *cfrA* gene (constructed in Subheading 3.4).
2. Primers K297A_F and K297A_R (Table 1). The lower case in each primer indicates desired mutation site to replace target lysine with alanine using partial overlapping PCR as described in Subheading 3.5.
3. High-fidelity DNA polymerase producing blunt ends (e.g., *Pfu Ultra* DNA polymerase from Stratagene).
4. DH5α competent cells (e.g., Invitrogen™ Library Efficiency cells or homemade competent cells), ideally with transformation efficiencies $>1 \times 10^8$ cfu/μg plasmid DNA.
5. *Escherichia coli* DH5α containing helper plasmid pRK212.2 (Ampicillin and tetracycline resistant) [18].
6. Commercial gel purification kit (e.g., QIAquick gel purification kit from Qiagen).

3 Methods

3.1 Purification of Enterobactin

1. Inoculate *E. coli* AN102 mutant in LB broth (4 mL). Grow culture overnight at 37 °C in a rotary shaker (250 rpm).
2. Transfer 2 mL of overnight culture to 200 mL of the modified T medium. Grow at 37 °C with vigorous shaking (250 rpm) for 8–10 h.

3. Transfer all above bacterial culture to 10 L of the same T medium. Grow overnight at 37 °C with vigorous shaking (250 rpm) (*see* **Note 1**).
4. Centrifuge overnight culture at 4000 × *g* for 30 min to remove the cells. Keep the supernatant that contains Ent.
5. Extract the supernatant with ethyl acetate for three times. After each extraction, transfer the top level of solution (ethyl acetate layer) into a glass beaker. For the first extraction, 150 mL of ethyl acetate/L of supernatant is used. For the other two extractions, 50 mL of ethyl acetate/L of supernatant is used (*see* **Note 2**).
6. Add a small volume of 0.1 M citrate buffer (pH 5.5) into the ethyl acetate extraction solution (~30 mL/L of ethyl acetate). Carefully remove the water phase (the bottom layer) with glass pipettes (*see* **Note 3**).
7. Rotovap ethyl acetate at 37 °C to 500–600 mL in a fume hood. Wash again with small amount of the 0.1 M citrate buffer (*see* **Note 4**).
8. Pour the ethyl acetate solution to a glass flask and add anhydrous $MgSO_4$ to the solution (~¼ inch height from the bottom). Swirl the flask for 3 min and incubate for 30 min at room temperature. Then filter the solution through funnel with Whatman #1 filter paper (*see* **Note 5**).
9. Rotovap the ethyl acetate extract to 5 mL or less (*see* **Note 6**).
10. Pour the ethyl acetate extract into glass centrifuge tube. Add hexanes dropwise with constant shaking to precipitate Ent (white color). Add additional hexanes until no more precipitation appears. Then add 3 mL of excess hexanes into the ethyl acetate extract and wait for 30 min to precipitate residual Ent (*see* **Note** 7).
11. Aliquot the Ent suspension into preweighed, clean 1.5 mL microcentrifuge tubes. Pellet the Ent precipitates at maximum speed (e.g., 15,000 × *g*) for 1 min on desktop microcentrifuge. Wash Ent pellet with 0.5 mL hexanes three additional times.
12. Dry Ent precipitates using vacufuge concentrator for at least 15 min.
13. Weigh each Eppendorf tube and calculate the yield. Dissolve Ent in methanol to prepare stock solutions of 10 mM.

3.2 Enterobactin Growth Promotion Assay

1. Inoculate the *C. jejuni* strain in the glass test tube containing 4 mL of MH broth. Grow the culture overnight at 37 °C in trigas incubator (*see* **Note 8**).
2. Prepare autoclaved MH soft agar (0.4%). Keep it warm by incubating at 50 °C water bath. Add deferoxamine mesylate stock solution to a final concentration of 20 μM to make iron-restricted soft agar (*see* **Note 9**).

3. In biosafety hood, transfer 100 μL of *C. jejuni* culture (OD_{600nm} = 0.1) to a 50 mL sterile tube. Then add 50 mL iron-restricted soft agar. Close the lid and invert 5 times to completely mix *C. jejuni* cells with the soft agar.
4. Pour the *C. jejuni*-soft agar mixture to two sterile petri dishes (about 25 mL per petri dish). Wait until the mixture is completely solidified.
5. Gently put a sterile test disk at the center of each petri dish containing the solidified *C. jejuni*-soft agar mixture. The test disk should stick to the agar surface.
6. Add 10 μL of following solutions sterile test disks in different petri dishes: 10 mM Ent, 20 mM ferric sulfate (positive control), and sterile water or methanol (negative controls).
7. Let plates sit on bench for 15–30 min (*see* **Note 10**).
8. Cover plates with lid and incubate the plates up-side-down at 37 °C in a trigas incubator for 2 days. Check for a growth halo around test disc (*see* **Note 11**).

3.3 Insertional Mutagenesis of cfrA

1. Extract genomic DNA from *Campylobacter jejuni* NCTC 11168 using a commercial genomic DNA purification kit by following the manufacturer's instructions.
2. PCR amplify a 1.6-kb fragment of the *cfrA* gene using Taq polymerase and primer pair CfrAF2 and CfrAR2 with 200 ng of *C. jejuni* NCTC 11168 genomic DNA as template. Purify PCR product using a PCR purification kit.
3. Ligate the purified *cfrA* PCR fragment into the pGEM-T easy vector (Promega) by following the manufacturer's instructions (*see* **Note 12**).
4. 2 μL of the above ligation mix is used to transform *E. coli* JM109 by following the manufacturer's instructions. Screen positive clones on LB plates containing 100 μg/mL ampicillin, resulting in plasmid pCA.
5. Digest 4 μg of plasmid pCA with 10 U of restriction enzyme NheI at 37 °C for 2 h. Purify the digested product using a PCR purification kit.
6. PCR amplify the 0.8-kb *cat* gene fragment from pUOA18 using primer pair CHFL1 and CHFL1. Purify the digested product using a PCR purification kit.
7. Digest 1 μg of *cat* fragment with NheI at 37 °C for 2 h and then purify the NheI-digested *cat* fragment using a PCR purification kit.
8. Ligate 100 ng of NheI-digested pCA with 200 ng of NheI-digested *cat* PCR product using 3 U of T4 DNA ligase. Incubate ligation mix overnight at 16 °C.

9. 2 μL of the above ligation solution is used to transform *E. coli* JM109 competent cells (50 μL) by following the manufacturer's instructions. Screen desired positive clones in LB plates containing 20 μg/mL of chloramphenicol. Following overnight incubation at 37 °C, the colonies on the LB plates containing chloramphenicol will be randomly picked, cultured, and plasmids purified using a plasmid purification kit. The recombinant plasmids should be sequenced to confirm the orientation of the *cat* gene is the same as that of *cfrA*. Positive plasmids, named pmCA, will serve as a suicide vector to transform *C. jejuni* NCTC 11168 using natural transformation.
10. Spread 100 μL of fresh *C. jejuni* NCTC 11168 onto MH plates. Grow overnight at 37 °C under microaerophilic conditions.
11. Harvest *C. jejuni* cells from plates using MH broth. Adjust OD_{600nm} to 0.5 and transfer 0.5 mL of the fresh bacterial suspension to MH agar tube.
12. Incubate MH agar tubes under microaerobic conditions for 3 h.
13. Add 1 μg of pmCA to the bacterial suspension. Incubate for another 3–5 h (*see* **Note 13**).
14. Plate 100 μL of cells onto MH agar plates containing 6 μg/mL chloramphenicol.
15. Incubate the plates under microaerobic conditions for 2–4 days.
16. Pick up a single colony and restreak it on MH plates containing 6 μg/mL chloramphenicol.
17. Pick up single colony, a putative isogenic *cfrA* mutant of NCTC 11168, and inoculate into MH broth medium.
18. Confirm the insertional mutation of *cfrA* in the selected mutant using PCR with primer pairs of CfrAF/CfrAR (*see* **Note 14**) as well as immunoblotting with CfrA-specific antibody.

3.4 Complementation of cfrA Mutant Using an Optimized Conjugation Protocol

1. PCR amplify the 2.5-kb full-length *cfrA* fragment containing its promoter using primer pair CfrAF2 and CfrAR2 and using a DNA polymerase that produces blunt ends from *C. jejuni* NCTC 11168 genomic DNA. Purify the PCR product using a PCR purification kit.
2. Digest 4 μg of pRY107 with 10 U of restriction enzyme SmaI at 25 °C for 2 h. Purify the digested product using a PCR purification kit (*see* **Note 15**).
3. Ligate 100 ng of SmaI-digested pRY107 with 150 ng of 2.5-kb *cfrA* PCR fragment using 3 U of T4 DNA ligase at 16 °C overnight.
4. Transform *E. coli* DH5α/pRK212.2 conjugation helper strain competent cells with the ligation mix. Screen positive colonies

containing both the helper plasmid and the complementation plasmid pCfrA on LB plates containing ampicillin (100 μg/mL), tetracycline (12.5 μg/mL), kanamycin (50 μg/mL), X-gal (80 μg/mL), and IPTG (0.5 mM). Pick up white colonies through blue/white screening.

5. Inoculate the confirmed *E. coli* DH5α/pRK212.2 clone-containing pCfrA into 5 mL of LB broth supplemented with ampicillin (100 μg/mL), tetracycline (12.5 μg/mL), kanamycin (50 μg/mL). Incubate at 37 °C overnight with vigorous shaking (250 rpm).
6. During Step 5, also spread 100 μL of fresh liquid culture of isogenic NCTC11168 *cfrA* mutant (the strain generated in Subheading 3.3) onto MH plates. Incubate plates under microaerobic conditions overnight at 37 °C.
7. Transfer the overnight culture of *E. coli* DH5α/pRK212.2 and pCfrA to fresh LB broth containing ampicillin (100 μg/mL), tetracycline (12.5 μg/mL), kanamycin (50 μg/mL) at an inoculation ratio of 1:10. Grow the fresh culture for 3–4 h with shaking (250 rpm) at 37 °C to reach OD_{600nm} value of about 1.2 (*see* **Note 16**).
8. During **step** 7, prepare *C. jejuni* recipient culture for conjugation. Briefly, harvest the overnight grown *C. jejuni* cells from MH agar plates using MH broth. Adjust the OD_{600nm} of the *C. jejuni* suspension to 1–1.5 and transfer 0.5 mL of the suspension to sterile microcentrifuge tubes (*see* **Note 17**).
9. Heat shock the aliquoted *C. jejuni* recipient cells (0.5 mL per tube) by incubating the tube with a lid closed in a 50 °C water bath for 30 min (*see* **Note 18**).
10. Cool down the heat-shocked *C. jejuni* recipient cells at room temperature for >2 min (*see* **Note 19**).
11. When the OD_{600nm} of the *E. coli* DH5α/pRK212.2 and pCfrA donor cells reaches 1.2, transfer 0.5 mL of *E. coli* donor cells to sterile microcentrifuge tubes and pellet by centrifugation at 15,000 × *g* for 1 min. The pelleted *E. coli* donor cells are further washed three times using MH broth to remove any residual antibiotics. Finally, the donor cells in each tube are resuspended in 0.5 mL of MH broth.
12. Gently mix 0.5 mL of *E. coli* DH5α/pRK212.2 and pCfrA with 0.5 mL of the heat shock-treated *C. jejuni* recipient cells. Pellet the cell mixture and resuspend cells in 100 μL of MH broth.
13. Spot the 100 μL of *E. coli*-*C. jejuni* mixture onto one MH agar plate. Wait until the spots are absorbed and incubate the plates upside-down under microaerobic condition for 5–7 h.
14. Harvest the *E. coli*-*C. jeuni* mixture from plates using 700 μL of MH broth per plate. Spread 100 μL aliquots of the cell suspension onto conjugation selective plates (*see* **Note 20**).

15. Incubate plates upside-down under microaerobic conditions at 37°C for 2–4 days.
16. Pick up single colonies and restreak on the conjugation selective plate.
17. Pick up single colonies and inoculate into MH broth containing kanamycin (50 μg/mL).
18. Confirm the expression of CfrA from pCfrA in the *cfrA* mutant using immunoblotting with CfrA-specific antibody (*see* **Note 21**).

3.5 Site-Directed Amino Acid Substitution Mutagenesis

1. Identify functionally important regions and/or amino acids in CfrA using the ClustalW program in MEGA 4.0 [19]. As an example, Q271, K297, R327, and F337 were identified by Zeng et al. (2009) as being likely important for FeEnt acquisition using this approach, and can be replaced by alanine as described later using partial overlapping PCR [20] (*see* **Note 22**).
2. Design mutagenic oligonucleotide primers to replace target amino acids with alanine (A). Here, we use lysine 297 (K297) as an example for amino acid substitution mutagenesis (K297A mutation in CfrA). Figure 1 shows a strategy to design the primers for this purpose (*see* **Note 23**).
3. The designed mutagenic PCR primers are used for overlapping PCR with the 9.3 kb pCfrA plasmid as a template. Use a high-fidelity DNA polymerase that produces blunt ends for overlapping PCR by following the cycling conditions below: 95 °C for 2 min, followed by 18 cycles of 94 °C for 1 min, 56 °C for 1 min, and 68 °C for 9.5 min and then 68 °C for 30 min (*see* **Note 24**).
4. The PCR products, which contain pCfrA plasmids bearing desired site-directed amino acid substitution mutation, are purified using a PCR purification kit. Approximately 1 μg of purified PCR products is digested with 20 U of DpnI at 37 °C for 1 h, run on an agarose gel (0.7%) and subsequently gel purified using a commercial kit (*see* **Note 25**).
5. Transform the putative mutated pCfrA plasmids into *E. coli* DH5α competent cells by following the manufacturer's instructions for nick repair.

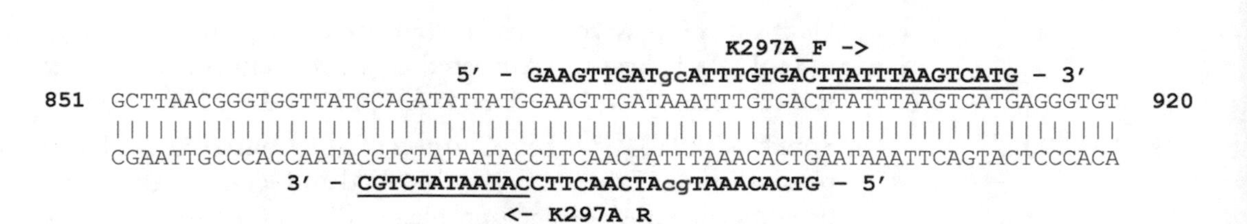

Fig. 1 Design of mutagenic oligonucleotide primers for replacing lysine 297 (K297) with alanine (A). The introduced mutations are *lower case letters in red*

6. Confirm the selected mutated pCfrA plasmids by sequencing, and then transform the plasmid into *E. coli* DH5α/pRK212.2 competent cells.
7. The *E. coli* DH5α/pRK212.2 containing mutated pCfrA plasmids are used as donor strain to conjugally transfer mutated pCfrA into the *cfrA* mutant strain (constructed in Subheading 3.3) using heat shock-enhanced conjugation in *C. jejuni* (please refer to **steps 5–18** of Subheading 3.4).
8. Pick up single transformant colony and restreak on conjugation selective plates.
9. Perform Ent growth promotion assay (Subheading 3.2) to determine if mutated plasmid can rescue the defect of *cfrA* mutant strain for utilizing FeEnt as a sole iron source. If not, the corresponding mutated amino acid is likely functionally important for CfrA-mediated FeEnt acquisition.

4 Notes

1. The OD_{600nm} end-point is approximately 1–1.5. Growing the *E. coli* culture too long (>14 h) may cause the oxidization of Ent, reflected by the red color of final Ent product.
2. Use glass containers and pipettes all the time. Ethyl acetate erodes plastic materials. In this step, most of ethyl acetate goes to water until it is saturated. Maximum Ent yield can be calculated by measuring OD value at 316 nm of the final ethyl acetate extract. The mM extinction coefficient of Ent at 316 nm is 0.7. Normal yield is 100 mg of Ent/5 L culture.
3. The purpose of this step is to remove other compounds, such as 2,3-Dihydroxybenzoic acid and other charged species. Use a minimal amount of citrate buffer because the aqueous residue should be discarded later.
4. If the volume of ethyl acetate extract is reduced very slowly during rotovap, increase temperature to 77 °C. Keeping more ethyl acetate solution before the next $MgSO_4$ dry step will increase yield.
5. Ethyl acetate extract should become very clear after the $MgSO_4$ drying.
6. At this stage ethyl acetate extract may become yellow–orange.
7. After a few drops in the beginning, you may observe brown or orange precipitates, which are polymers and oxidation products. Remove them by centrifuging at maximum speed in a desktop microcentrifuge. Large amounts of white Ent precipitation should form after adding more hexanes dropwise.
8. The overnight cultivated *C. jejuni* should reach mid-log phase ($OD_{600\ nm}$ = 0.1–0.2).

9. MH soft agar should maintain gel-like fluid status at 50 °C. Make sure to mix soft agar with deferoxamine mesylate well.
10. This step will make the test disk stick to the surface of soft agar very well and it will not fall during the next incubation step when the plates are up-side-down.
11. Chemicals in the test disk will diffuse around the test disks. The chemicals (such as enterobactin) that can compete for iron with deferoxamine mesylate would serve as the iron source and promote *C. jejuni* growth.
12. The pGEM-T Easy vector is a vector for easy PCR cloning. Specifically, this vector has a single 3′-T overhang at both ends, which provides a compatible overhang for typical PCR product generated by Taq polymerase that has A-overhang at both ends.
13. Add control MH agar tube (containing *C. jejuni* culture) with equal volume of sterile water.
14. In wild-type *C. jejuni* NCTC 11168, a 527-bp fragment is generated using primer CfrAF and CfrAR, while in *cfrA* mutant, due to the insertion of *cat* fragment (800 bp), a 1327-bp fragment is generated.
15. SmaI digestion generates blunt end product, which can be ligated with the blunt-end PCR product generated by some DNA polymerase, such as *Pfu* polymerase.
16. If chloramphenicol antibiotics are used for growing the *E. coli* donor due to the choice of specific shuttle vector (e.g., pRY111 *C. jejuni*-*E. coli* shuttle vector), *E. coli* may grow very slowly. Addition of 1% sterile glucose can significantly enhance the growth of such donor cells in the presence of chloramphenicol.
17. Generally, *C. jejuni* recipient cells are harvested approximately 1 h before *E. coli* donor cells reach desired growth phase.
18. Heat shock treatment of *C. jejuni* recipient cells can enhance the conjugation efficiency in *C. jejuni* NCTC 11168 100-fold [21], likely by inactivating relevant restriction-modification system in recipient cells.
19. The heat shock-enhanced conjugation efficiency can persist for 6 h without significant change [21].
20. *Campylobacter*-specific selective agents will suppress the growth of *E. coli*, while kanamycin will screen transconjugants with the pRY107-derived shuttle vector, such as pCfrA.
21. To perform immunoblotting, various *C. jejuni* strains are grown under iron-limited condition (MH supplemented with 5 μM deferoxamine mesylate) to late log phase at 42 °C under microaerophilic conditions. *Campylobacter* cells grown in this iron-limited condition are harvested and solubilized by boiling for 5 min in sodium dodecyl sulfate-polyacrylamide gel electrophoresis (SDS-PAGE) sample buffer. The SDS-PAGE and

immunoblotting were performed as detailed in a previous publication [22] except that the dilution of primary (Rabbit anti-CfrA) and secondary (Goat anti-rabbit IgG) antibody was 1:1000 and 1:2000, respectively. The blotted nitrocellulose membrane could be developed with commercial kit such as the 4 CN Peroxidase Substrate System (Kirkegaard & Perry Laboratories, Inc).

22. Cross-species alignment of FepA (CfrA homologs) sequences from four other bacteria (*E. coli*, *S. enterica*, *P. aeruginosa*, and *B. pertussis*) and the CfrA sequences of 15 *Campylobacter* strains were performed to identify conserved and functionally important amino acid residues, which usually are basic or aromatic amino acids [14, 23].
23. The following general principles should be taken into consideration for designing mutagenic primers. (1) At least one G or C is placed at the 3′-end of each primer; (2) the desired mutation should be in the middle of the primer with at least nine bases of correct sequence on both sides; (3) at least eight nonoverlapping bases (underlined) are introduced to the 3′-end of each primer; and (4) the melting temperature (T_m) of each primer should be greater than 69 °C. (The following formula can be used for calculating the T_m of primers: $T_m = 81.5 + 0.41[\%GC] - [675/N]$ -%Mismatch, where N is the primer length in bases and the values for %GC and %Mismatch are whole numbers).
24. The cycle number should be less than 20 to avoid introducing nondesired mutations at other sites.
25. The DpnI endonuclease (target sequence: 5′-Gm^{6}ATC-3′) is specific for methylated and hemimethylated DNA. Therefore, DpnI can digest the original template pCfrA plasmid which originates from *E. coli* and is therefore methylated. However, the newly synthesized pCfrA through overlapping PCR is not methylated and therefore resistant to DpnI digestion. Following DpnI digestion, vectors are subjected to agarose gel electrophoresis (0.7%). Subsequently, the nicked pCfrA DNA band incorporating the desired mutation, reflected by the 9.3 kb-band on the gel, is excised from gel for purification. It is not unusual that multiple bands may be shown on the gel due to various reasons (e.g., inefficient digestion of methylated, supercoiled plasmids by DpnI). In this case, recovery and purification of the specific 9.3 kb-DNA band from gel would improve screening efficiency by lowering false-positive colonies.

Acknowledgments

We are grateful to Sandra K. Armstrong (University of Minnesota) for providing *E. coli* AN102 for purification of enterobactin. This work was supported by NIH grant 1R56AI090095-01A1.

References

1. Koppenol WH (2001) The Haber-Weiss cycle--70 years later. Redox Rep 6(4):229–234
2. Braun V, Hantke K, Koster W (1998) Bacterial iron transport: mechanisms, genetics, and regulation. Met Ions Biol Syst 35:67–145
3. Andrews SC, Robinson AK, Rodriguez-quinones F (2003) Bacterial iron homeostasis. FEMS Microbiol Rev 27(2-3):215–237
4. Miethke M, Marahiel MA (2007) Siderophore-based iron acquisition and pathogen control. Microbiol Mol Biol Rev 71(3):413–451. doi:10.1128/MMBR.00012-07
5. Wandersman C, Delepelaire P (2004) Bacterial iron sources: from siderophores to hemophores. Annu Rev Microbiol 58:611–647. doi:10.1146/annurev.micro.58.030603.123811
6. Harris WR, Carrano CJ, Cooper SR et al (1979) Coordination chemistry of microbial iron transport compounds. 19. Stability constants and electrochemical behavior of ferric enterobactin and model complexes. J Am Chem Soc 101(20):6097–6104. doi:10.1021/ja00514a037
7. Raymond KN, Dertz EA, Kim SS (2003) Enterobactin: an archetype for microbial iron transport. Proc Natl Acad Sci U S A 100(7):3584–3588. doi:10.1073/pnas.0630018100
8. Stintzi A, Vliet AHMV, Ketley JM (2008) Iron metabolism, transport, and regulation. In: Nachamkin I, Szymanski CM, Blaser MJ (eds) Campylobacter, 3rd edn. ASM Press, Washington, DC, pp 591–610
9. Wooldridge KG, van Vliet AHM (2005) Iron transport and regulation. In: Ketley JM, Konkel ME (eds) Campylobacter: molecular and cellular biology. Horizon Bioscience, Norfolk, pp 293–310
10. Miller CE, Williams PH, Ketley JM (2009) Pumping iron: mechanisms for iron uptake by *Campylobacter*. Microbiology 155(Pt 10):3157–3165. doi:10.1099/mic.0.032425-0
11. Field LH, Headley VL, Payne SM et al (1986) Influence of iron on growth, morphology, outer membrane protein composition, and synthesis of siderophores in *Campylobacter jejuni*. Infect Immun 54(1):126–132
12. Palyada K, Threadgill D, Stintzi A (2004) Iron acquisition and regulation in *Campylobacter jejuni*. J Bacteriol 186(14):4714–4729, doi:186/14/4714 [pii]
13. Guerry P, Perez-Casal J, Yao R et al (1997) A genetic locus involved in iron utilization unique to some *Campylobacter* strains. J Bacteriol 179(12):3997–4002
14. Zeng X, Xu F, Lin J (2009) Molecular, antigenic, and functional characteristics of ferric enterobactin receptor CfrA in *Campylobacter jejuni*. Infect Immun 77(12):5437–5448. doi:10.1128/IAI.00666-09
15. Xu F, Zeng X, Haigh RD et al (2010) Identification and characterization of a new ferric enterobactin receptor, CfrB, in *Campylobacter*. J Bacteriol 192(17):4425–4435. doi:10.1128/JB.00478-10
16. Zeng X, Lin J (2014) Siderophore-mediated iron acquisition for *Campylobacter* infection. In: Sheppard SK, Méric G (eds) *Campylobacter* ecology and evolution. Caister Academic, Norfolk, UK, pp 111–124
17. Yao R, Alm RA, Trust TJ et al (1993) Construction of new *Campylobacter* cloning vectors and a new mutational *cat* cassette. Gene 130(1):127–130
18. Figurski DH, Helinski DR (1979) Replication of an origin-containing derivative of plasmid RK2 dependent on a plasmid function provided in *trans*. Proc Natl Acad Sci U S A 76(4):1648–1652
19. Kumar S, Nei M, Dudley J et al (2008) MEGA: a biologist-centric software for evolutionary analysis of DNA and protein sequences. Brief Bioinform 9(4):299–306. doi:10.1093/bib/bbn017
20. Zheng L, Baumann U, Reymond JL (2004) An efficient one-step site-directed and site-saturation mutagenesis protocol. Nucleic Acids Res 32(14):e115. doi:10.1093/nar/gnh110
21. Zeng X, Ardeshna D, Lin J (2015) Heat shock-enhanced conjugation efficiency in standard *Campylobacter jejuni* strains. Appl Environ Microbiol 81(13):4546–4552. doi:10.1128/AEM.00346-15
22. Lin J, Sahin O, Michel LO et al (2003) Critical role of multidrug efflux pump CmeABC in bile resistance and *in vivo* colonization of *Campylobacter jejuni*. Infect Immun 71(8):4250–4259
23. Cao Z, Qi Z, Sprencel C et al (2000) Aromatic components of two ferric enterobactin binding sites in *Escherichia coli* FepA. Mol Microbiol 37(6):1306–1317

Chapter 8

Method for the Successful Crystallization of the Ferric Uptake Regulator from *Campylobacter jejuni*

Sabina Sarvan and Jean-François Couture

Abstract

The Ferric Uptake Regulator (FUR) is a transcription factor (TF) regulating the expression of several genes to control iron levels in prokaryotes. Members of this family of TFs share a common structural scaffold that typically comprises two regions that include a DNA binding and dimerization domains. While this structural organization is conserved, FUR proteins employ different mechanisms to bind divergent DNA binding elements and regulate gene expression in the absence or presence of regulatory metals. These findings, combined with the observations that FUR proteins display different geometries in regard to the relative orientation of the DNA binding and dimerization domains, have highlighted an expanding repertoire of molecular mechanisms controlling the activity of this family of TFs. In this chapter, we present an overview of the methods to purify, crystallize, and solve the structure of *Campylobacter jejuni* FUR.

Key words Transcription factor, Fur, Strep-tag affinity chromatography, Apo-metalloregulator, Bacterial metalloregulator

1 Introduction

Metals are inorganic cofactors required for the growth of nearly all bacteria and among those, iron is essential for bacterial pathogenicity. Accordingly, the control of iron assimilation and metabolism is tightly controlled at the transcriptional level. A prevalent family of transcription factors in prokaryotes that regulate gene expression includes the Ferric Uptake Regulator family (FUR). FUR is a metal sensor that both negatively and positively regulates transcription in the absence (apo) or presence (holo) of regulatory metals. For both modes of regulation, FUR binds DNA and controls the expression of genes important for iron internalization, binding, and elimination [1].

FUR is an important regulator of iron homeostasis in prokaryotes; however, its functions extend beyond iron metabolism. *Campylobacter jejuni* FUR (*Cj*FUR) protein regulates flagellar and capsule biogenesis, energy metabolism, oxidative stress defense,

James Butcher and Alain Stintzi (eds.), *Campylobacter jejuni: Methods and Protocols*, Methods in Molecular Biology, vol. 1512, DOI 10.1007/978-1-4939-6536-6_8, © Springer Science+Business Media New York 2017

uptake of other metal ions such as molybdate, tungsten, zinc, and nickel as well as the regulation of noncoding RNAs [2–5]. *Cj*FUR regulates gene expression by four general mechanisms [3]. The TF can act as a repressor of gene transcription by blocking RNA polymerase access to promoters [5], in the absence (apo-form) or presence (holo-form) of regulatory metal ions. Apo-*Cj*FUR and holo-*Cj*FUR can also act as transcription activator; however, the precise mechanism controlling this activity is currently unknown. One proposed model suggests that *Cj*FUR indirectly activates the expression of genes through the repression of small regulatory RNAs [4, 6].

Since the pioneering work of Kendrew et al., on the elucidation of the crystallographic structure of myoglobin [7, 8], X-ray crystallography has played an instrumental role in detailing the biochemical underpinnings controlling the activity of a wide range of biological macromolecules. Accordingly, X-ray crystallography remains a primary technique used by pharmaceutical companies to develop novel therapeutic molecules to treat pathologies such as cancers, viral and bacterial infections [9]. This chapter describes the optimization of the three steps leading to the structural determination of the apo-form of the ferric uptake regulator of *Campylobacter jejuni*. First, we focus on the importance of the purification scheme for the homogeneous preparation of apo-*Cj*FUR. Second, we present the methods by which *Cj*FUR was crystallized. Finally, we detail the steps leading to the crystal structure determination of the metalloregulator.

2 Materials

2.1 Plasmid Construction

The gene corresponding to *Campylobacter jejuni* FUR was PCR-amplified from *Campylobacter jejuni* genomic DNA and cloned in a homemade vector referred to as pStrepSUMO (pSS). The pSS vector is a derivative of the pHis parallel vector (pHis2) [10] and was constructed by replacing the hexahistidine Tag (6XHis) and the Spacer region of pHis2 by a DNA fragment corresponding to a Strep Tag placed in frame with the sequence encoding the Small Ubiquitin-like Modifier (SUMO) protein SMT3 from budding yeast (Fig. 1). The cDNA of SMT3 was PCR amplified with the following primers (Primer #1: 5′ GGATTAGCATATGTGGAGCCACCCGCAGTTCGAAAAGG GTTCGGACTCAGAAGTCAATCAAGAAGC 3′ and Primer #2: 5′ATCCTATCTCCA TGGCATACGTAGCACCACCAAT 3′) and *S. cerevisiae* genomic DNA as template. The PCR product was subsequently cloned *Nde1-Nco1* in pHis2. Following amplification and purification of the vector, the *CjFUR* gene was cloned, in frame with the open reading frame of the Strep-SUMO tag, using the *BamHI* and *XhoI* restriction sites.

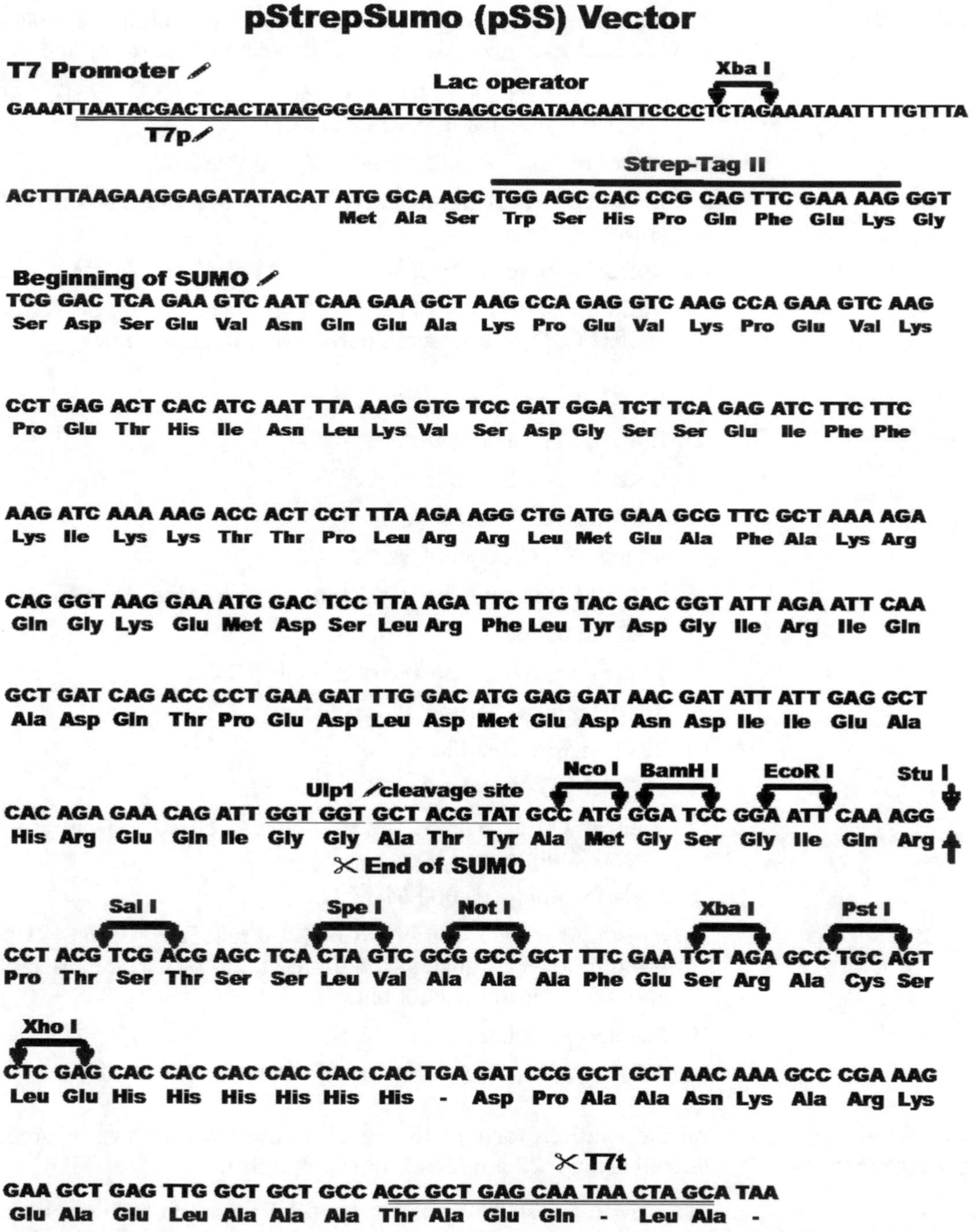

pStrepSumo (pSS) Vector

T7 Promoter

Lac operator

Xba I

GAAATTAATACGACTCACTATAGGGGAATTGTGAGCGGATAACAATTCCCCTCTAGAAATAATTTTGTTTA

T7p

Strep-Tag II

ACTTTAAGAAGGAGATATACAT ATG GCA AGC TGG AGC CAC CCG CAG TTC GAA AAG GGT
Met Ala Ser Trp Ser His Pro Gln Phe Glu Lys Gly

Beginning of SUMO

TCG GAC TCA GAA GTC AAT CAA GAA GCT AAG CCA GAG GTC AAG CCA GAA GTC AAG
Ser Asp Ser Glu Val Asn Gln Glu Ala Lys Pro Glu Val Lys Pro Glu Val Lys

CCT GAG ACT CAC ATC AAT TTA AAG GTG TCC GAT GGA TCT TCA GAG ATC TTC TTC
Pro Glu Thr His Ile Asn Leu Lys Val Ser Asp Gly Ser Ser Glu Ile Phe Phe

AAG ATC AAA AAG ACC ACT CCT TTA AGA AGG CTG ATG GAA GCG TTC GCT AAA AGA
Lys Ile Lys Lys Thr Thr Pro Leu Arg Arg Leu Met Glu Ala Phe Ala Lys Arg

CAG GGT AAG GAA ATG GAC TCC TTA AGA TTC TTG TAC GAC GGT ATT AGA ATT CAA
Gln Gly Lys Glu Met Asp Ser Leu Arg Phe Leu Tyr Asp Gly Ile Arg Ile Gln

GCT GAT CAG ACC CCT GAA GAT TTG GAC ATG GAG GAT AAC GAT ATT ATT GAG GCT
Ala Asp Gln Thr Pro Glu Asp Leu Asp Met Glu Asp Asn Asp Ile Ile Glu Ala

Ulp1 cleavage site

Nco I BamH I EcoR I Stu I

CAC AGA GAA CAG ATT GGT GGT GCT ACG TAT GCC ATG GGA TCC GGA ATT CAA AGG
His Arg Glu Gln Ile Gly Gly Ala Thr Tyr Ala Met Gly Ser Gly Ile Gln Arg

End of SUMO

Sal I Spe I Not I Xba I Pst I

CCT ACG TCG ACG AGC TCA CTA GTC GCG GCC GCT TTC GAA TCT AGA GCC TGC AGT
Pro Thr Ser Thr Ser Ser Leu Val Ala Ala Ala Phe Glu Ser Arg Ala Cys Ser

Xho I

CTC GAG CAC CAC CAC CAC CAC CAC TGA GAT CCG GCT GCT AAC AAA GCC CGA AAG
Leu Glu His His His His His His - Asp Pro Ala Ala Asn Lys Ala Arg Lys

T7t

GAA GCT GAG TTG GCT GCT GCC ACC GCT GAG CAA TAA CTA GCA TAA
Glu Ala Glu Leu Ala Ala Ala Thr Ala Glu Gln - Leu Ala -

Fig. 1 Multicloning site of the pStrepSumo (pSS) vector. The T7 promoter (T7p), T7 terminator (T7t), and the lac operator are underlined. The Strep-Tag II coding sequence is identified. The *horizontal arrows* indicate the 5′ and 3′ regions of the sequence corresponding to the *SMT3* gene. The Ulp1 cleavage site and key restriction sites are indicated

2.2 Buffers

All the buffers and solutions were prepared using ultrapure milliQ H_2O and analytical grade reagents. All the buffers were stored at 4 °C.

1. Buffer A: PBS 1× (140 mM NaCl, 2.7 mM KCl, 10 mM Na_2HPO_4, and 1.8 mM KH_2PO_4, pH 7.4).
2. Buffer B: PBS 1× and 2.5 mM D-desthiobiotin.
3. Buffer C: 20 mM Citrate pH 6.0 and 5 mM β-mercaptoethanol (BME).
4. Buffer D: 20 mM Citrate pH 6.0, 1 M NaCl and 5 mM BME.
5. Buffer E: 20 mM Tris pH 7.0, 250 mM NaCl and 5 mM BME. Buffers C, D, and E were filtered using 0.22 μm filters.

2.3 Protein Purification

1. BL-21 Rosetta competent cells.
2. Luria-Bertani agar plates.
3. Luria-Bertani broth.
4. 100 mg/mL Ampicillin.
5. 50 mg/mL Chloramphenicol.
6. 100 mM Isopropyl β-D-1-thiogalactopyranoside (IPTG).
7. 0.45 μm filters.
8. Shaking incubator capable of reaching 37 °C.
9. Strep-Tactin Superflow Plus (Qiagen).
10. SP Sepharose Fast Flow.
11. HiLoad 16/60 Superdex 75 prep grade.
12. Amicon Ultra-15 and Amicon Ultra-4 10 K concentrators (Merck Millipore Ltd).
13. Dialysis Tubing 10 000 MWCO.
14. 5× sample loading buffer (SLB) (200 mM Tris pH 6.8, 10% sodium dodecyl sulfate, 40% glycerol, 5% β-mercaptoethanol and 0.04% bromophenol blue).
15. Stainless-steel beaker.
16. 2.5 mg/mL ubiquitin-like specific Protease 1 (ULP1).

2.4 *Cj*FUR Crystallization

All the solutions for *Cj*FUR crystallization were freshly prepared, filtered using 0.22 μm filters, and kept at 4 °C.

1. 24-well crystallization plates (Grainer ComboPlate from Hampton Research).
2. High vacuum grease.
3. Crystallization solutions: 10–20% polyethylene glycol (PEG) 3350 and 0.1–0.25 M $MnSO_4$.
4. Plastic cover slips.
5. Cryoloops and vials.
6. Cryocanes.

7. Liquid nitrogen dewar.
8. Micromax 007-HF (Rigaku) and image plate detectors (RAXIX IV^{++}).

3 Methods

In order to purify *Cj*FUR, we screened several constructs containing various tags placed on the N-terminus. First, we cloned *CjFUR* in the parallel vector pGST2 (GST-*Cj*FUR) enabling the expression of the transcription factor in fusion with the Glutathione S-transferase (GST) [10]. While GST-*Cj*FUR expressed relatively well and was amenable to purification, the yield and the purity, as determined by SDS-PAGE gels, were unsatisfactory to carry out crystallization experiments. We then cloned *CjFUR* in a modified version of the pSS vector (Fig. 1), referred to as pSMT3, in which the Strep tag is replaced by a hexahistidine (His) tag coupled to SUMO. Using this approach, we obtained a higher yield of purified protein; however, separation of the eluted proteins using SDS-PAGE followed by Coomassie-staining revealed several contaminating bands after cleavage of the fusion protein suggesting that, similar to GST-*Cj*FUR, the pSMT3-derived *Cj*FUR was unsuitable to produce a homogeneous preparation of recombinant *Cj*FUR. Finally, the cloning of the *CjFUR* gene in pSS enabled us to express and purify *Cj*FUR to a sufficient quantity/purity for crystallography trials.

3.1 Purification of CjFUR

3.1.1 Expression of SS-CjFUR

1. Transform the pSS-*Cj*FUR plasmid into BL-21 Rosetta competent cells.
2. Plate the cells on Luria-Bertani (LB) agar plates supplemented with Ampicillin (0.1 g/mL) and Chloramphenicol (0.05 g/mL).
3. Incubate the LB agar plates overnight at 37 °C.
4. Prepare the preculture by inoculating several colonies of the transformed Rosetta cells into 100 mL LB broth media supplemented with 100 μL of Ampicillin (100 mg/mL) and 100 μL of Chloramphenicol (50 mg/mL).
5. Grow the cells by shaking the 100 mL preculture at 250 RPM at 37 °C during 3–4 h.
6. Once the cells have reached an OD_{600} of 0.8, use 8 mL of the preculture to inoculate 500 mL of LB broth media supplemented with 500 μL Ampicillin (100 mg/mL) and 500 μL of Chloramphenicol (50 mg/mL). 12 flasks, each containing 500 mL of LB broth, are inoculated for large-scale expression of SS-*Cj*FUR.
7. Grow the cells by shaking at 250 RPM at 37 °C until the cultures reach an OD_{600} of 0.5. Induce protein expression by add-

ing Isopropyl β-D-1-thiogalactopyranoside (IPTG) to a final concentration of 0.1 mM and incubate the cells at 37 °C for 3 h while shaking at 250 RPM.

8. Pellet the cells by centrifugation at 3000 RPM at 4 °C for 30 min.
9. Discard the supernatant and resuspend the pellets corresponding to 1 L of cell culture in 25 mL of Buffer A. Keep the harvested cells at −80 °C.

3.1.2 Purification of CjFUR

Keep the protein on ice at all times and perform all the purification procedures (affinity, ion exchange, and size exclusion chromatography steps) at 4 °C. Samples from different steps of the purification steps were kept for analysis by electrophoresis followed by coomassie staining on 12% sodium dodecyl sulfate polyacrylamide gel (SDS-PAGE) gels (Figs. 2 and 3).

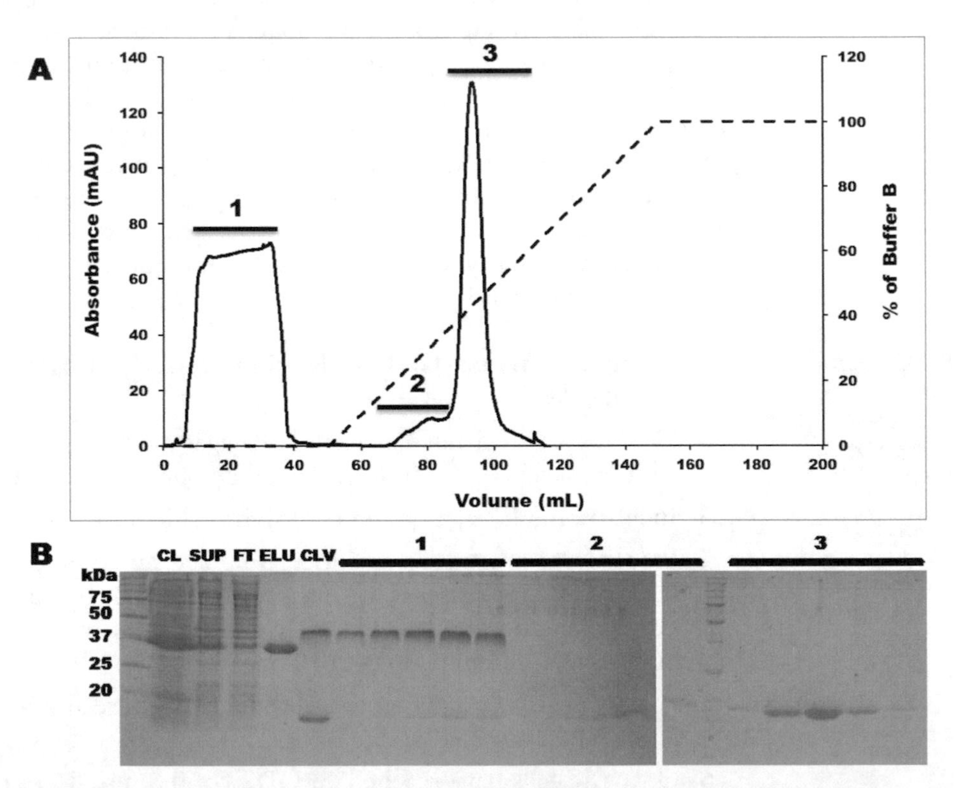

Fig. 2 Purification of *Cj*FUR. (**a**) Ion exchange chromatography. *Dashed* and *solid lines* represent the percentage of Buffer B and the UV absorbance over the volume loaded on the SP sepharose, respectively. Three different peaks are identified and fractions corresponding to these peaks were analyzed on SDS-PAGE gel. (**b**) Coomassie-stained SDS-PAGE gel of the fractions highlighted in panel (**a**). Different steps of *Cj*FUR purification steps were analyzed on SDS-PAGE gel. *CL* cell lysate, *SUP* supernatant, *FT* flow through of the Strep-TACTIN column, *ELU* eluted sample, *CLV* fraction after cleavage with ULP1. 1, 2, and 3 represent the three different peaks obtained during the ion exchange chromatography. Cleaved *Cj*FUR elutes between 30 and 55% of Buffer B, corresponding to peak 3 and to a concentration of NaCl between 300 and 550 mM

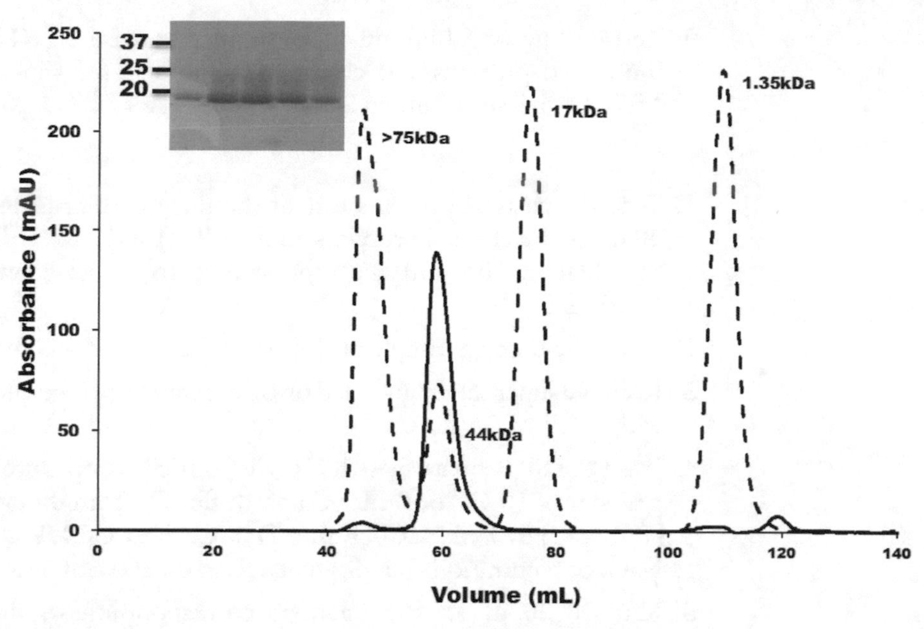

Fig. 3 Size exclusion chromatography of *Cj*FUR. *Solid* and *dashed lines* represent the UV absorbance of *Cj*FUR protein sample and the different molecular weight standards, respectively. *Cj*FUR protein elutes between 50 and 70 mL. The SDS-PAGE of the fractions corresponding to the main peak is presented as inlet in the chromatogram

Day 1

1. Thaw 1 cell pellet corresponding to 1 L of cell culture in cold water.
2. Complete the volume to 100 mL by adding Buffer A to the cell suspension.
3. Transfer the cell suspension into a stainless-steel beaker (*see* **Note 1**) and lyse the cells by sonication. Transfer an aliquot of 20 μL of the cell lysate (CL) to a 1.5 mL eppendorf tube to which 5 μL of 5× sample loading buffer (SLB) was added for subsequent analysis by SDS-PAGE gel.
4. Pellet the remaining cell lysate by centrifugation at 16,000 RPM at 4 °C for 30 min.
5. Filter the supernatant using a 0.45 μm filter. Keep a 20 μL aliquot in a 1.5 mL eppendorf tube corresponding to the supernatant (SUP) and add 5 μL of SLB loading dye and keep the sample to separate on a SDS-PAGE gel.
6. Apply the filtered supernatant on Step-Tactin column pre-equilibrated with 10 column volumes (CV) of Buffer A.
7. Wash the Strep-Tactin column with 10 CV of Buffer A.
8. Elute the protein by adding 100 mL of Buffer B and keep only the first 30 mL of the eluted sample. An aliquot of 20 μL of the eluted sample (ELU) was transferred to a 1.5 mL eppendorf tube to which 5 μL of SLB was added for subsequent analysis by SDS-PAGE gel (*see* **Note 2**).

9. Add 250 μg of Ubiquitin-Like specific Protease 1 (ULP1) to the eluted proteins and cleave overnight (~16 h) in 10,000 MWCO dialysis tubing in 2 L of Buffer A at 4 °C (*see* **Note 3**).

Day 2

1. Take the cleaved proteins out of the dialysis tubing. Remove 20 μL of the cleaved protein sample (CLV) and add 5 μL of 5× SLB loading dye and keep the sample to separate on SDS-PAGE gel.
2. Pellet the protein sample at 3000 RPM at 4 °C for 5 min.
3. Keep the supernatant and load on SP Sepharose pre-equilibrated with 5 CV of Buffer C.
4. Wash the SP Sepharose with 5 CV of Buffer C and elute with a gradient of 10 CV of Buffer C and Buffer D (from 0 mM NaCl [Buffer C] to 1 M NaCl [Buffer D]). Cleaved *Cj*FUR elutes at NaCl concentrations ranging between 300 and 550 mM.
5. Remove 20 μL of the fractions corresponding to the flow through and fractions corresponding to cleaved *Cj*FUR and add 5 μL of 5× SLB loading dye and keep the samples to separate on a SDS-PAGE gel.
6. Pool and concentrate the fractions containing cleaved *Cj*FUR to 2 mL.
7. Centrifuge the solution containing the concentrated protein at 13,000 RPM at 4 °C for 5 min to remove debris.
8. Keep the supernatant and further purify the protein by size exclusion chromatography by applying the concentrated protein on a Superdex S75 column pre-equilibrated with 1.1 CV of Buffer E.
9. Pool and concentrate fractions corresponding to *Cj*FUR to 10 mg/mL. *Cj*FUR elutes between the 50 mL and 70 mL fractions.

3.2 Crystallization of CjFUR

*Cj*FUR crystallization was performed in 24-well plates using the hanging drop vapor diffusion method. Crystallization assays are performed at 4 °C and the protein sample is maintained on ice.

1. Apply high vacuum grease on the edge of each well.
2. Prepare and dispense 24 × 1 mL aliquots of the various crystallography solutions in the 24-well plate (Fig. 4) at 4 °C.
3. On a plastic coverslip, mix 1 μL of protein and 1 μL of reservoir solution. Invert the coverslip over the reservoir and seal the well. *Cj*FUR crystals are grown at 4 °C.
4. *Cj*FUR crystals typically appear after 12 days.
5. After testing different cryoprotectants, we determined that the best cryoprotectant solution for *Cj*FUR crystals was 20% ethylene glycol. *Cj*FUR crystals were then harvested from the

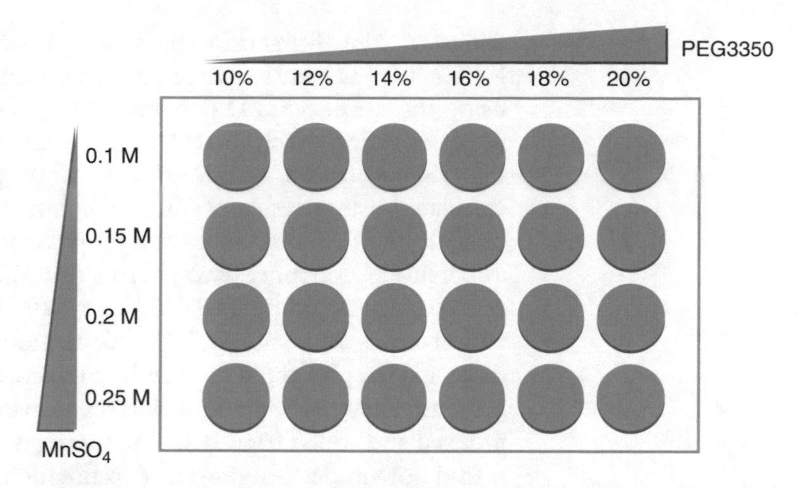

Fig. 4 Crystallization plate setup. Shown is the range of $MnSO_4$ and PEG 3350 concentrations used to obtain diffraction quality crystals of *Cj*FUR

24-well and were subsequently soaked in the crystallization solution supplemented with 5, 10, 15, and 20% of ethylene glycol. Following the last soaking step, the crystal was harvested and flash-frozen in liquid nitrogen.

3.3 Crystal Structure Determination of CjFUR

The conditions listed in Fig. 4 typically yield rod-shape crystals (150 μm × 50 μm × 50 μm) that appear approximately 4 days after the crystallization trials. *Cj*FUR crystals are sensitive to vibration. The trays were therefore not inspected during the 2-week incubation period. Moreover, the crystals tend to disappear if they are not mounted once fully grown. While *Cj*FUR crystals display similar morphologies and dimensions, at least 10 crystals have to be mounted and exposed to X-rays to find one diffraction quality crystal. The screening is typically performed on a Micromax 007-HF (Rigaku) and images are recorded on image plate detectors (RAXIX IV^{++}). Once a well-diffracting crystal is identified, a full-data set is collected and the reflections are indexed using HKL2000 [11]. Initial attempts to solve the structure of *Cj*FUR by molecular replacement, using either the coordinates of the *Helicobacter pylori* or *Pseudomonas aeruginosa* FUR proteins, failed to provide a lead model suggesting that *Cj*FUR structure differed from these two proteins. Sequence alignment of *Cj*FUR with other metalloregulators suggested that *Cj*FUR binds at least one zinc atom per protomer. Zinc is an electron-rich atom when compared to carbon, nitrogen, and oxygen atoms. Consequently, we used the zinc atom to perform single wavelength anomalous diffraction (SAD). To perform SAD, a full dataset was collected at the Life-Science Collaborative Access Team beamline at the Advance Photon Source,

Chicago. The diffraction spots were processed and scaled using HKL2000 [11] and four zinc atoms were identified and refined using the SHELX C/D programs [12]. These initial observations were consistent with our sequence alignment but also revealed that *Cj*FUR harbors an additional zinc-binding site. The phases were calculated using the Arp/Warp program (http://www.embl-hamburg.de/ARP/) and were used to generate an initial model. The initial model contained two polypeptide chains, each traced to near completion, including 140/157 and 120/157 residues for chain A and chain B, respectively. The initial model was used as a search molecule to solve the structure by molecular replacement using the program Phaser [13] and a dataset in which the reflections were indexed and scaled together. The model was completed by iterative rounds of model building and refinement using Coot [14] and Refmac [15], respectively. Following the final round of refinement and model building, the quality of the model was assessed using Molprobity [16].

4 Notes

1. Other beakers can be used (e.g., plastic/glass), but we find that stainless-steel beakers dissipate the heat generated from sonication much more efficiently and lead to improved results.
2. Strep-Tactin resin needs to be regenerated after every usage. First, we apply 100 mL of Strep-Tactin Regeneration buffer (50 mM NaH_2PO_4, 300 mM NaCl, and 1 mM 2-[4-Hydroxyphenylazo] benzoic acid [HABA], pH 8.0) on 10 mL of Strep-Tactin resin. At this step, the color of the resin changes from white to red indicating that the column has been regenerated and the desthiobiotin displaced. Strep-Tactin resin is then washed with 500 mL of Strep-Tactin Lysis buffer (50 mM NaH_2PO_4 and 300 mM NaCl, pH 8.0). At this step, the color of the resin changes to white and the Strep-Tactin resin can be stored in the Lysis buffer at 4 °C.
3. After cleavage with the ULP1 protease, a remnant of the Strep-SUMO tag and residues encoded by the sequence corresponding to the BamH1 site remain on *Cj*FUR which correspond to the following sequence: ATYAMGS.

Acknowledgements

This work was supported by a CIHR grant MIC-274419. S.S. acknowledges a Ph.D. scholarship from Fonds de Recherche en Santé du Quebec.

References

1. Fillat MF (2014) The FUR (ferric uptake regulator) superfamily: diversity and versatility of key transcriptional regulators. Arch Biochem Biophys 546:41–52. doi:10.1016/j.abb.2014.01.029
2. Palyada K, Threadgill D, Stintzi A (2004) Iron acquisition and regulation in *Campylobacter jejuni*. J Bacteriol 186(14):4714–4729, 186/14/4714 [pii]
3. Butcher J, Sarvan S, Brunzelle JS et al (2012) Structure and regulon of *Campylobacter jejuni* ferric uptake regulator Fur define apo-Fur regulation. Proc Natl Acad Sci U S A 109(25):10047–10052. doi:10.1073/pnas.1118321109
4. Butcher J, Stintzi A (2013) The transcriptional landscape of *Campylobacter jejuni* under iron replete and iron limited growth conditions. PLoS One 8(11):e79475. doi:10.1371/journal.pone.0079475
5. Lee JW, Helmann JD (2007) Functional specialization within the Fur family of metalloregulators. Biometals 20(3-4):485–499. doi:10.1007/s10534-006-9070-7
6. Masse E, Salvail H, Desnoyers G et al (2007) Small RNAs controlling iron metabolism. Curr Opin Microbiol 10(2):140–145. doi:10.1016/j.mib.2007.03.013
7. Kendrew JC (1958) Architecture of a protein molecule. Nature 182(4638):764–767
8. Kendrew JC, Bodo G, Dintzis HM et al (1958) A three-dimensional model of the myoglobin molecule obtained by x-ray analysis. Nature 181(4610):662–666
9. Zheng H, Hou J, Zimmerman MD et al (2014) The future of crystallography in drug discovery. Expert Opin Drug Discov 9(2):125–137. doi:10.1517/17460441.2014.872623
10. Sheffield P, Garrard S, Derewenda Z (1999) Overcoming expression and purification problems of RhoGDI using a family of "parallel" expression vectors. Protein Expr Purif 15(1):34–39. doi:10.1006/prep.1998.1003
11. Otwinowski Z, Minor W (1997) Processing of X-ray diffraction data collected in oscillation mode. Method Enzymol 276:307–326. doi:10.1016/S0076-6879(97)76066-X
12. Sheldrick GM (2010) Experimental phasing with SHELXC/D/E: combining chain tracing with density modification. Acta Crystallogr D Biol Crystallogr 66(Pt 4):479–485. doi:10.1107/S0907444909038360
13. McCoy AJ, Grosse-Kunstleve RW, Adams PD et al (2007) Phaser crystallographic software. J Appl Crystallogr 40(Pt 4):658–674. doi:10.1107/S0021889807021206
14. Emsley P, Cowtan K (2004) Coot: model-building tools for molecular graphics. Acta Crystallogr D Biol Crystallogr 60(Pt 12 Pt 1):2126–2132. doi:10.1107/S0907444904019158
15. Winn MD (2003) An overview of the CCP4 project in protein crystallography: an example of a collaborative project. J Synchrotron Radiat 10(Pt 1):23–25
16. Chen VB, Arendall WB 3rd, Headd JJ et al (2010) MolProbity: all-atom structure validation for macromolecular crystallography. Acta Crystallogr D Biol Crystallogr 66(Pt 1):12–21. doi:10.1107/S0907444909042073

Chapter 9

Methods for Initial Characterization of *Campylobacter jejuni* Bacteriophages

Martine Camilla Holst Sørensen, Yilmaz Emre Gencay, and Lone Brøndsted

Abstract

Here we describe an initial characterization of *Campylobacter jejuni* bacteriophages by host range analysis, genome size determination by pulsed-field gel electrophoresis, and receptor-type identification by screening mutants for phage sensitivity.

Key words *Campylobacter*, Bacteriophage, Phage characterization, Phage host range, PFGE, Phage receptor

1 Introduction

Once isolated, *Campylobacter jejuni* bacteriophages may be characterized according to host range, genome size, morphology, and overall receptor dependency, as these have been shown to be important features of phages infecting *C. jejuni*. Knowing these characteristics may help in the classification and grouping of the isolated phage and to study its similarities to previously isolated phages infecting this pathogen. Here we present relevant methods for an initial characterization of *C. jejuni* phages.

1.1 Host Range Analysis

Investigating the host range of an isolated phage is often the first characterization step conducted and is standardly done by performing a plaque assay. This can be done by mixing consecutive phage dilutions with the bacterial strain of interest followed by plating using the double agar layer method or by performing a drop assay where a small volume of diluted phage suspensions are spotted on a bacterial lawn [1]. In some studies, the host range analysis was performed using the drop method where only undiluted high-concentration phage stocks were spotted on lawns of investigated *C. jejuni* strains to determine the host range [2–4].

James Butcher and Alain Stintzi (eds.), *Campylobacter jejuni: Methods and Protocols*, Methods in Molecular Biology, vol. 1512, DOI 10.1007/978-1-4939-6536-6_9, © Springer Science+Business Media New York 2017

However, this procedure is not recommended, as false positives may appear from clearing zones of highly concentrated phage spots due to inhibition of bacterial growth by unspecific phage binding (Fig. 1c). Here we investigate host range using the drop method by spotting consecutive dilutions of the phage of interest on bacterial lawns and only define a sensitive host by the appearance of single plaques in the spotting zones (Fig 1a, b).

1.2 Receptor Type

A large group of isolated *C. jejuni* phages are dependent on either the capsule or motility for successful infection of the bacteria [5–9]. Determining this can be difficult if there is little information available of the *C. jejuni* strain used for phage isolation and will require mutant construction in the capsule and flagella apparatus of the host strain. However, most *C. jejuni* phages are able to infect the *C. jejuni* NCTC12662 strain also known as PT14, which is available from the National Collection of Type Cultures [10]. If the isolated phage can infect NCTC12662, it is possible to study receptor dependency by investigating the sensitivity of the phage on a non-capsulated version of this strain and a nonmotile NCTC12662 mutant already constructed and available [5]. Here we determine receptor dependency by performing a plaque assay by spotting consecutive dilutions of the isolated phage on bacterial lawns of NCTC12662, NCTC12662Δ*kpsM* (no capsule), and NCTC12662Δ*motA* (nonmotile) and define overall receptor dependency as lack of plaque formation on the specific mutant strain compared to the wild-type NCTC12662.

1.3 PFGE

Pulsed-field gel electrophoresis (PFGE) is commonly used to determine the size of viral genomes including *C. jejuni* phages [3, 11]. The following procedure adds to PFGE protocols already

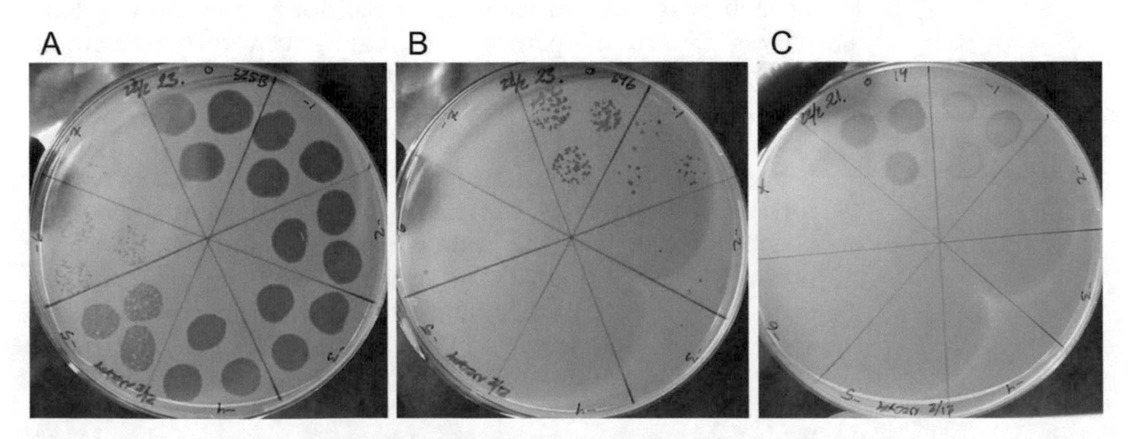

Fig. 1 Plaque assays showing different lysis patterns and plaque formation on *C. jejuni* lawns. (**a**) Clear lysis and plaque formation on a sensitive *C. jejuni* host. (**b**) Clear plaque formation but reduced efficiency of plaquing of a high-titer phage solution. (**c**) Unspecific phage binding showing clearing zones where high-titer phage solutions have been spotted followed by no plaque formation

optimized for phage genome size determination. Phages are embedded in an agarose plug where they are also lysed by proteinase K treatment to access the DNA. Following treatment, the plug is washed and sliced to fit onto an agarose gel for PFGE. If required, the genomic DNA of group III (approx. 140 kb), CP81-type phages can be digested with HhaI for restriction fragment length polymorphism (RFLP) analysis [12].

In addition, phage morphology can be done by transmission electron microscopy (TEM) according to already described protocols [5, 13]. Also, if the phage is to be applied in a therapeutic setting, genome sequencing is important, and guidelines to isolate phage DNA can be found elsewhere [14].

2 Materials

2.1 Equipment

1. Microaerobic incubators or jars for *C. jejuni* growth.
2. Laboratory facilities classified to work with genetically modified *C. jejuni* strains.
3. Microscope, phase contrast, 100× magnification, objective glass, and cover slides.
4. Basic laboratory equipment for preparation of broth and solid culture media and solutions.
5. Laminar flow hood.
6. A vortex mixer.
7. Analytical balance.
8. 10 mL sterile pipettes and pipette controller.
9. Variable pipettes, sterile tips, and filter tips covering volumes of 10 μL to 1 mL.
10. Sterile screw-capped centrifuge tubes, 13 and 50 mL.
11. 1.5 mL sterile-capped microcentrifuge tubes.
12. Sterile Pasteur pipettes.
13. Sterile syringe filters, 0.2 and 0.45 μm.
14. Sterile syringes, 10 mL.
15. Optical density spectrophotometer.
16. Microwave.
17. Screw-capped autoclavable bottles, 100 mL, 250 mL, and 1 L.
18. Gyratory shaker.
19. Water bath set at 45 °C.
20. Refrigerator set at 4 °C.
21. Incubator set at 37 °C.

The following equipment is specific for PFGE analysis:

22. Heating block set at 45 °C.
23. Water bath set at 56 °C.
24. Shaking water bath set at 56 °C.
25. Small flat spatula, sterile.
26. Scalpel blade or gel knife, sterile.
27. PFGE system (Bio-Rad CHEF-DRIII or equivalent) with plug casting molds (Bio-Rad CHEF Mapper XA System, 50-well plug mold or equivalent).
28. Ultraviolet illuminator and a gel documentation system.

2.2 Bacteriophage Host Range Analysis and Receptor-Type Identification

All solutions can be prepared using deionized water and analytical grade reagents. Store all reagents at room temperature unless it is otherwise indicated:

1. BA: Blood agar. 5% calf blood, blood agar base II. Prepare the blood agar base II according to manufacturer's instructions and sterilize by autoclaving. Add 10 mL calf blood to 200 mL media tempered to 45 °C just before the plates are made (*see* **Note 1**).
2. BA + kanamycin: Blood agar, 50 μg/mL kanamycin. Prepare the blood agar Base II according to manufacturer's instructions and sterilize by autoclaving. Add 10 mL calf blood and kanamycin at a final concentration of 50 μg/mL to 200 mL media tempered to 45 °C just before the plates are made (*see* **Note 1**).
3. BA + chloramphenicol: Blood agar, 20 μg/mL chloramphenicol. Prepare the blood agar base II according to manufacturer's instructions and sterilize by autoclaving. Add 10 mL calf blood and chloramphenicol at a final concentration of 20 μg/mL to 200 mL media tempered to 45 °C just before the plates are made (*see* **Note 1**).
4. 1 M $MgSO_4$. Sterilize by autoclaving.
5. 100 mM $CaCl_2$ solution. Sterilize by autoclaving.
6. CBHI: Brain–heart infusion (BHI) broth with 10 mM $MgSO_4$ and 1 mM $CaCl_2$. Add 2 mL of 1 M $MgSO_4$ and 2 mL of 100 mM $CaCl_2$ to 200 mL of BHI (*see* **Note 2**).
7. Phage stock solution of interest. Store at 4 °C (*see* **Note 3**).
8. 1.5 mL microcentrifuge tubes containing 900 μL of SM buffer.
9. SM buffer: 5.8 g NaCl, 2 g $MgSO_4 \cdot 7H_2O$, 5 mL 2% (w/v) gelatin, 50 mL of 1 M Tris–HCl. Add water to a volume of 900 mL and mix. Adjust the pH with HCl to 7.5. Make up to 1 L with water and autoclave.
10. Vancomycin: 10–25 mg/mL solution in water. Store at −20 °C.
11. Chloramphenicol: 50 mg/mL. Store at −20 °C.
12. Kanamycin: 100 mg/mL. Store at −20 °C.

13. NCZYM overlay agar: NZCYM broth, 0.6% (w/v) agar. Prepare NZCYM broth as described by the manufacturer. Add 0.6% agar, mix well, and sterilize by autoclaving.
14. NZCYM basal agar plate: NZCYM broth, 1.2% (w/v) agar, 10 μg/mL vancomycin. Prepare NZCYM broth as described by the manufacturer. Add 1.2% agar, mix well, and sterilize by autoclaving. Add vancomycin at a final concentration of 10 μg/mL to the media tempered to 45 °C just before the plates are poured (*see* **Note 4**).
15. Freezer stock cultures of *C. jejuni* strains of interest. Store at −80 °C.
16. Freezer stock cultures of *C. jejuni* NCTC12662, NCTC12662Δ*kspM* (Kan^R), and NCTC12662Δ*motA* (Cam^R). Store at −80 °C.
17. 1 M Tris–HCl pH 8.0.
18. 0.5 M EDTA pH 8.0.
19. 20 mg/mL proteinase K.

2.3 Bacteriophage Genome Size Determination by Pulsed-Field Gel Electrophoresis (PFGE)

All buffers are freshly prepared using molecular biology grade reagents and ultrapure MilliQ water to minimize run-to-run variation.

1. Purified phages: 10^7–10^9 pfu/mL. Closer to 10^9 pfu/mL is preferred (*see* **Note 5**).
2. TE buffer pH 8.0: 10 mM Tris–HCl, 1 mM EDTA.
3. Plug agarose mixes: 1%, 2%, or 4% (w/v) agarose (e.g., SeaKem Gold agarose) in 10 mL TE buffer. Heat mix until the agarose has dissolved and keep at 50 °C until use.
4. Lysis buffer (TES), pH 6.8: 10 mM Tris–HCl, 100 mM EDTA, 1% sarkosyl, 0.1 mg/mL proteinase K. First, make a fresh 1% sarkosyl solution by weighing out 1 g sarkosyl and dissolve in 10 mL water. Add 1 mL 1 M Tris (pH 8.0) and 20 mL 0.5 M EDTA (pH 8.0). Fill up with water to 70 mL. Adjust pH to 6.8 with HCl, add water to 100 mL, and autoclave. Before use, add proteinase K (20 mg/mL solution in water) to the lysis buffer at a final concentration of 0.1 mg/mL.
5. Washing buffer (TE 20:50): 20 mM Tris, 50 mM EDTA.
6. Restriction nuclease HhaI, 20,000 U/mL.
7. Low range PFGE Marker.
8. 10× TBE buffer pH 8.2: 0.89 M Tris borate, 0.03 M EDTA. Sterilize by autoclaving.
9. Ethidium bromide solution, 2 μg/mL. Preferably make this solution a few hours before use to ensure a saturated evenly dispersed solution.
10. Deionized water.

3 Methods

Always consider a sterile working environment especially when working with bacteriophage stocks. Use filter tips when handling phage stock solutions, wear gloves and work next to a Bunsen burner if possible, and regularly wipe off pipettes with ethanol or other types of disinfectants. Pipettes used for *C. jejuni* and phage work should not be used for other experiments.

3.1 Bacteriophage Host Range Analyses

3.1.1 Preparation of Bacterial Lawns

1. Streak out *C. jejuni* strains of interest from −80 °C freezer stock cultures on BA plates and incubate under microaerobic conditions at 37 °C for 48 h.
2. Subculture *C. jejuni* strains densely on BA plates and incubate under microaerobic conditions at 37 °C for 20–24 h (*see* **Notes 6** and **7**).
3. Add 2–3 mL of CBHI to BA plates. Using a bent sterile Pasteur pipette, loosen and harvest bacteria by gently scraping off the colonies.
4. Transfer the dense bacterial harvest to a centrifuge tube and vortex thoroughly.
5. Measure the optical density at 600 nm (OD_{600}) of the bacterial harvest by making a tenfold dilution in CBHI and measure the OD_{600} on the dilution (*see* **Note 8**).
6. Adjust the OD_{600} to 0.35 in 10 mL of fresh CBHI in a sterile petri dish (*see* **Notes 9** and **10**).
7. Incubate at 37 °C for 4 h (*see* **Note 11**).
8. Before the 4 h incubation period is over, melt the overlay agar and temper it to 45 °C in a water bath (*see* **Note 12**).
9. Following the incubation, visualize the culture by phase-contrast microscopy at 100× magnification for the presence of contaminating bacterial cells (*see* **Note 13**).
10. Mix 1:10 of freshly grown culture with NZCYM overlay agar and quickly pour 5 mL of the overlay mix on the NZCYM base plate, and gently rotate to produce an even surface with the overlay fresh culture mixture (*see* **Note 14**).
11. Let the overlays settle for 15–20 min and then dry them in a laminar flow hood for 45 min for optimal humidity (*see* **Note 15**).

3.1.2 Host Range Analysis

1. Prepare tenfold serial dilutions of the phage stock in SM buffer up to 10^{-7} (*see* **Note 16**).
2. Divide the bottom of the lawn prepared in petri dishes into eight equal segments drawing with a marker and mark the corresponding dilutions from 0 to −7 (*see* **Note 17**).
3. Carefully spot 3 × 10 μL of the original phage stock and the phage dilutions to corresponding segments in duplicates on

two separate lawns containing the same *C. jejuni* strain (*see* **Notes 16** and **18**).

4. When the spots are totally dried, incubate the plates under microaerobic conditions at 37 °C for 18–24 h (*see* **Note 19**).
5. Following incubation, **note** incubation time, observe the plaque morphologies, count formed plaques, and calculate the mean plaque-forming units per mL (pfu/mL) (*see* **Note 20**).
6. Repeat independent host range analyses for a minimum of 2–3 times to ensure reliability of the results.

3.2 Receptor-Type Identification

3.2.1 Preparation of NCTC12662, NCTC12662Δ*kpsM*, and NCTC12662Δ*motA* Bacterial Lawns

1. Streak out *C. jejuni* strains NCTC12662, NCTC12662Δ*kpsM*, and NCTC12662Δ*motA* from −80 °C freezer stock cultures on BA, BA+kanamycin, and BA+chloramphenicol plates, respectively, and incubate under microaerobic conditions at 37 °C for 48 h.
2. Subculture *C. jejuni* strains NCTC12662, NCTC12662Δ*kpsM*, and NCTC12662Δ*motA* densely on BA plates, and incubate under microaerobic conditions at 37 °C for 20–24 h (*see* **Note 6**).
3. Add 2–3 mL of CBHI to BA plates. Using a bent sterile Pasteur pipette, loosen and harvest bacteria by gently scraping off the colonies.
4. Transfer the dense bacterial harvest to a centrifuge tube and vortex thoroughly.
5. Measure the optical density at 600 nm (OD_{600}) of the bacterial harvest by making a tenfold dilution in CBHI and measure the OD_{600} on the dilution (*see* **Note 8**).
6. Adjust the OD_{600} to 0.35 in 10 mL of fresh CBHI in a sterile petri dish (*see* **Notes 9** and **10**).
7. Incubate at 37 °C for 4 h (*see* **Note 11**).
8. Before the 4 h incubation period is over, melt the overlay agar and temper it to 45 °C in a water bath (*see* **Note 12**).
9. Following the incubation, visualize the cultures in a microscope at 100× magnification by phase-contrast microscopy for the presence of contaminating bacterial cells (*see* **Note 13**).
10. Mix 1:10 of freshly grown culture with NZCYM overlay agar and quickly pour 5 mL overly mix on the NZCYM base plate, and gently rotate to produce an even surface with the overlay fresh culture mixture (*see* **Note 14**).
11. Let the overlays settle for 15–20 min and then, dry them in a laminar flow hood for 45 min for optimal humidity (*see* **Note 15**).

3.2.2 Receptor-Type Determination

1. Prepare tenfold serial dilutions of the phage stock in SM buffer up to 10^{-7} (*see* **Note 16**).
2. Divide the bottom of the lawns prepared petri dishes into eight equal segments drawing with a marker and mark the corresponding dilutions from 0 to −7 (*see* **Note 17**).

3. Carefully spot 3 × 10 μL of the original phage stock and the phage dilutions to corresponding segments on the bacterial lawns of all three strains in duplicates on two lawns of the same *C. jejuni* strain (*see* **Notes 16** and **18**).
4. When the spots are totally dried, incubate the plates under microaerobic conditions at 37 °C for 18–24 h (*see* **Note 19**).
5. Following incubation, **note** incubation time, observe lysis and plaque morphologies, count formed plaques, and calculate the mean plaque-forming units per mL (pfu/mL).
6. Repeat independent receptor-type experiments for a minimum of 2–3 times to ensure reliability of the results.

3.3 Bacteriophage Genome Size Determination by Pulsed-Field Gel Electrophoresis (PFGE)

3.3.1 Preparation of Plugs

1. Prepare molding blocks and label the wells.
2. Prepare the plug agarose mix in a volume of 10 mL and keep warm in a water bath at approximately 56 °C (*see* **Note 21**).
3. Transfer 400 μL of the phage stock of interest to a labeled microcentrifuge tube and place in a heating block at 45 °C to warm the phage solution (*see* **Notes 22** and **23**).
4. Make one phage preparation at a time by adding 400 μL of the prepared plug agarose mix to the warm phage solution (*see* **Notes 22** and **23**). Mix carefully by pipetting to avoid air bubbles and then immediately transfer the mixture into wells of the plug casting mold (*see* **Note 24**). Make 4–5 plugs for each phage. Any remaining volume can be cast as extra plugs.
5. Let the plugs solidify at room temperature for 30 min or at 4 °C for 10–15 min.
6. Prepare 13 mL screw-capped tubes for each phage by labeling it and adding 5 mL of TES lysis buffer containing 0.1 mg/mL proteinase K.
7. Gently open the wells of the casting molds and carefully remove the plugs with a small, flat ethanol-sterilized spatula. Place 4–5 plugs from the same phage solution into the corresponding tube containing the TES lysis buffer.
8. Place the tube in a shaking water bath at 56 °C degrees overnight (*see* **Note 25**).
9. Warm the washing buffer (TE 20:50) to 56 °C in a water bath.
10. Take out the tube from the water bath and carefully remove the TES lysis buffer while still saving the plugs in the tube.
11. Wash the plugs by adding at least 5 mL of warm washing buffer to the tube containing the plugs and re-incubate in a shaking water bath at 56 °C for 20 min.
12. Carefully remove the washing buffer while still saving the plugs in the tube.
13. Repeat **steps 11** and **12** an additional five times for a total of six washing steps.

14. Transfer 2–3 plugs to labeled 2 mL sterile screw-capped tubes and add 1 mL TE buffer. Store plugs at 4 °C until ready to load for PFGE:
    (a) Optional: Group III *C. jejuni* phages having a genome size of approximately 140 kb can be further analyzed by restriction fragment length polymorphism (RFLP) by treating the plugs with HhaI restriction enzyme.

3.3.2 Optional HhaI Digestion

Collect the phage plugs one at a time from the TE buffer with a small, flat, ethanol-sterilized spatula and place them on a sterile objective glass.

Cut across the long axis of the plugs with a sterile knife or scalpel blade to make a 3–5 mm thick slice. Store the remaining part of the plugs at 4 °C in TE buffer (*see* **Note 26**).

Place the slice in 200 μL HhaI mix with a final concentration of 40 U HhaI prepared according to manufacturer's instructions (*see* **Note 27**).

Incubate overnight (16–18 h) at 37 °C.

Replace HhaI mix with TE buffer and store at 4 °C until the slices can be analyzed by PFGE.

3.3.3 PFGE

1. Prepare 2 L of 0.5× TBE buffer in ultrapure sterile water, pour it into the PFGE chamber, and allow the buffer to be cooled to 14 °C.
2. Prepare 150 mL of 1% agarose in 0.5× TBE buffer made with ultrapure sterile water in a clean sterile blue cap bottle.
3. Boil until the solution is clear and all the agarose is completely melted. Shake well (*see* **Note 28**).
4. Keep the solution in a water bath at 55–60 °C until the gel is poured.
5. Assemble the gel molding chamber (*see* **Note 29**).
6. Place the comb to make the wells in a horizontal position on top of the molding chamber so that it is only flipped down to fit into the chamber at the correct position.
7. Collect the phage plugs one at a time from the TE buffer with a small, flat, ethanol-sterilized spatula and place them on a sterile objective glass.
8. Cut across the long axis of the plugs with a sterile knife or scalpel blade to make a 3–5 mm thick slice. Store the remaining part of the plugs at 4 °C in TE buffer (*see* **Note 26**).
9. Cut the low-range PFGE marker as thin as possible. Then cut the slice in half, and from the middle cut, create a 1 mm slice.
10. Gently place the PFGE marker slice and the phage plug slices on the bottom of each comb tip avoiding air bubbles (Fig. 2a).

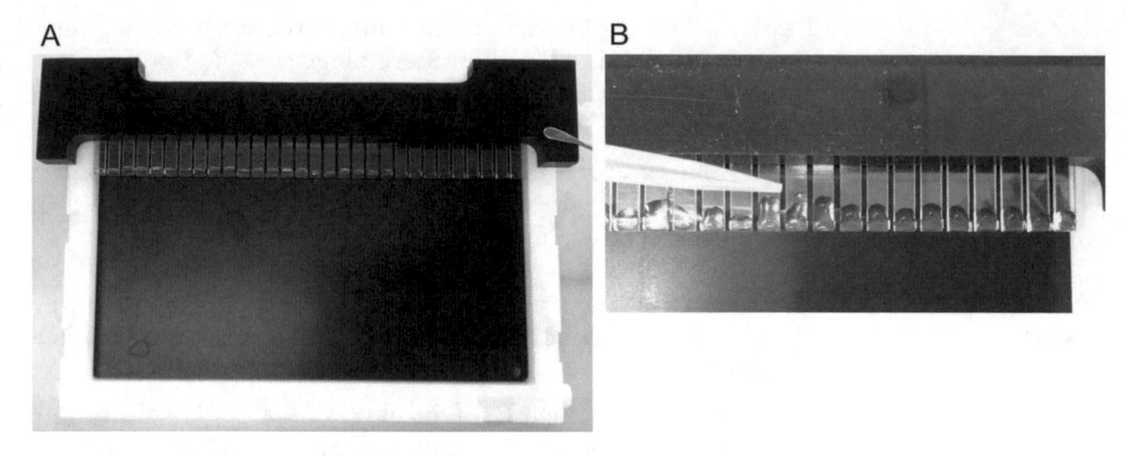

Fig. 2 Positioning (**a**) and fixation (**b**) of plug slices on the pulsed-field gel electrophoresis comb

11. Using a pipette, quickly place a small amount (10–20 μL) of the tempered 1% agarose solution at the top of every slice on the comb tips to seal the slice to the comb (Fig. 2b), again avoiding air bubbles (*see* **Note 30**).
12. Gently place the comb in the correct vertical position on the gel molding chamber making sure the slices do not fall off.
13. Gently pour the gel.
14. Let the cast gel solidify.
15. Carefully remove the comb making sure that the slices are now sealed in the gel (*see* **Note 31**).
16. Place the gel into the PFGE bed with the precooled buffer.
17. Run the gel at 6 V/cm for 14 h using the following conditions. Initial switch time, 2; final switch time, 10; and included angle, 120.
18. Remove the gel and stain it in an ethidium bromide solution (2 μg/mL) for 30 min.
19. Remove the gel into a washing water bath containing fresh deionized water for 1 h.
20. Examine the gel by ultraviolet illumination.
21. Obtain the image of the stained gel using a gel documentation system.
22. Estimate the size of the investigated phage genomes visually by comparison with the DNA marker (Fig. 3a and **Note 32**).

4 Notes

1. It is also possible to grow the *C. jejuni* strains on Mueller-Hinton agar without blood supplement if obtaining fresh calf blood is not possible.

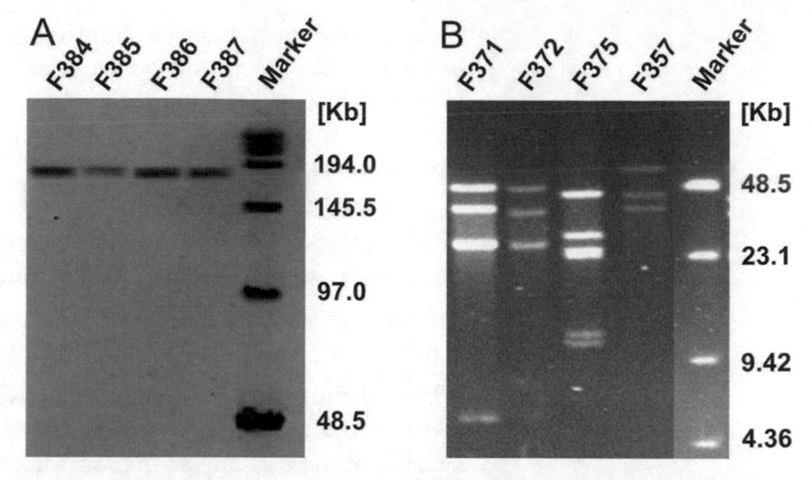

Fig. 3 Pulsed-field gel electrophoresis of *C. jejuni* phages. (**a**) Uncut *C. jejuni* group II CP220-type phages showing genomes of approximately 180 kb. For all phages, the plugs were made by mixing 400 μl 1 % plug agarose mix with 400 μl phage stock, and 3 mm slices were loaded on the gel. *Lane 1*, Phage F384, 6×10^8 pfu/mL; *lane 2*, Phage F385, 4×10^8 pfu/mL; *lane 3*, Phage F386, 6×10^8 pfu/mL; *lane 4*, Phage F387, 9×10^8 pfu/mL; and *lane 5*, Marker (Lambda Ladder PFG marker, New England Biolabs). (**b**) HhaI restriction patterns of *C. jejuni* phages by pulsed-field gel electrophoresis demonstrating band visibility as a result of phage stock concentration. For all phages, the plugs were made by mixing 200 μL 2 % plug agarose mix with 600 μL phage stock, and 5 mm slices were loaded on the gel. *Lane 1*, Phage F371, 1×10^9 pfu/mL; *lane 2*, Phage F372, 5×10^8 pfu/mL; *lane 3*, Phage F375, 3×10^9 pfu/mL, *lane 4*, Phage F357, 2×10^7 pfu/mL; and *lane 5*, Marker (Low Range PFG marker, New England Biolabs)

2. Preferably make the CBHI media the day before use and place it overnight at 37 °C to ensure that it has not been contaminated.
3. The phage stock should be clear and not cloudy; otherwise, contamination might have occurred. A cloudy stock should be discarded. Salty-looking precipitation at the bottom of the phage stock can occur in high-titer stocks over time (approximately 10^9 pfu/mL or above), but this is not contamination, and the stock can still be used. Backup stocks are advisable.
4. NCZYM plates must be stored at 4–5 °C but can be made up to 2 weeks before use. For optimal plaque formation, basal agar plates are preferably made the day before use and left on the lab desk overnight at room temperature (20–25 °C). These plates are then ready for when conducting the experiment and do not need drying. Plates made prior to this and stored at 4 °C should be removed from the cold environment the day before use and left overnight at room temperature to generate the same conditions. Alternately, NZCYM basal agar plates made the same day or taken directly from the fridge the day of

the experiment can be used, but then they need to be pre-dried in the laminar flow hood before use. Dependent on the moisture of the plates, the period in the laminar flow hood can be between 10 and 30 min and should be optimized for each laboratory setting.

5. For the PFGE analysis, phages can be taken from a fresh high-titer stock solution from which bacterial remnants have been removed by sterile filtration using a filter size of 0.2 μm.
6. As a relatively high density of cells is required later on, a dense growth of *C. jejuni* strains of interest is advisable.
7. When performing a host range analysis, it is important both to consider the number and which *Campylobacter* strains to test. Genetic as well as any structural information available for the bacterial strains is important when interpreting host range results. The Penner serotype corresponds to different capsular surface structures expressed by *C. jejuni*, and it is therefore of interest to test strains of different serotypes. Also some *C. jejuni* phages are able to infect other subspecies of *Campylobacter* such as *C. coli*.
8. Measuring the OD_{600} directly on the bacterial harvest is not advisable as the cell density often is so high that it will give an incorrect measurement.
9. *C. jejuni* is a microaerobic bacterium that will often be found just below the surface in liquid medium. Therefore, ensuring a large surface area as can be done in a petri dish compared to tubes provides better growth conditions for the bacteria. Also, *C. jejuni* cultures easily get contaminated by faster-growing bacteria such as *Micrococcus*. Thus, using sterile petri dishes from an unopened batch minimizes possible contamination routes.
10. The desired amount of *C. jejuni* cells should be between 10^8 and 10^9 cfu/mL which in most cases in CBHI is equal to an OD_{600} of 0.35 independently of the strain used. However, as equipment measuring optical densities may vary from lab to lab, an exact correlation of $OD_{600} = 0.35$ and corresponding cfu/mL of the strains used is advisable before making this assumption.
11. The 4 h incubation step is performed to allow the bacteria to adjust to the liquid environment coming from plates. Limited growth is expected during this incubation due to the lag phase and slow growth of the bacterium.
12. Be sure that the overlay agar is brought to a full boil in order to avoid any lumps of undissolved media. This may result in misinterpretation of the plates due to the crystalline structure of the lumps.
13. Due to its slow growth, *C. jejuni* cultures easily get contaminated and overgrown by faster-growing bacteria such as *Micrococcus*. Therefore, it is advisable to always investigate liquid cultures by microscopy to see if contamination has

occurred. *C. jejuni* will show a characteristic spiral-shaped morphology compared to contaminating coccoid or rod-shaped bacteria. A contaminated culture should never be used and should be discarded.

14. Take 500 μL of the adjusted bacterial culture and transfer it into a 13 mL sterile tube followed by the addition of 5 mL pre-warmed overlay agar. The suspension is then briefly mixed by vortexing and poured onto the basal agar plate. If many phages are to be screened on lawns of the same *C. jejuni* strain, the 10 mL adjusted bacterial culture can be added to 100 mL of pre-warmed overlay in a sterile screw-capped bottle followed by thorough mixing. 5 mL overlay mix is then transferred to the basal plates using a sterile 10 mL pipette.
15. The moisture and thickness of the NZCYM basal agar and the NZCYM overlay strongly influence the appearances of plaques, in sizes and clearness, on the bacterial lawns. However, this variation in plaque appearances can be minimized and controlled to some extent by using a strict protocol for the agar and overlay production and can help ensure repeatability between experiments. Optimization of drying is essential for producing clear plaque-forming units, and it is thus recommended to evaporate some of the moisture from the agar by drying plates in a laminar flow hood. If the plates are not dried at all, the moisture in the plates will prevent the liquid spots in adsorbing into the media, and it will not be possible to see any plaques on these lawns. However, drying plates for too long may also result in no plaque formation. Therefore, once the overlay has been poured and settled, dry the plates in a laminar flow hood for 45 min for optimal moisture conditions. If the basal agar plates are dried the day of the experiment (*see* **Note 4**), different drying times will apply and may range between 20 and 60 min but need to be optimized for each laboratory setting.
16. When the applied phage titer is high (10^8–10^9 pfu/mL), inhibition of the growth can be observed in *C. jejuni* lawns as clearing zones in a spot assay. Therefore, one should apply serial dilutions of the phage stock to clearly observe true phage infection by looking at single plaques and determining the efficiency of plaque formation.
17. This may be completed during the 4 h incubation period under **step 6** of Subheading 3.1.1, especially if a high number of plates are used.
18. Duplicate experiments are advisable in case of unevenly distributed bacterial lawns, too warm overlay agar, differences in moisture conditions, etc. that may all influence plaque formation and can be difficult to control completely.
19. Make sure to avoid sudden movements of the spotted plates and let the lawn soak the drops for approximately 20–30 min.

20. Host range patterns are highly variable, and with each phage, it is possible to observe a broad diversity in lysis patterns and plaque formation on different *C. jejuni* strains (Fig. 1). The plaques might differ in size from a pinpoint to 1–3 mm, showing either turbid or clear morphologies. Also it is possible to observe a clear or turbid lysis at the spot where a high-titer phage solution was added, and when the titer declines with every consecutive dilution, it is possible to observe either a loss of activity (no plaque formation) within the first few dilutions (Fig.1c) or a confluent lysis that dilutes out to singular plaques (Fig. 1b). A clearing zone in an undiluted high-titer phage stock and the first couple of consecutive dilutions but no plaque formation should *not* be regarded as a positive result, e.g., that the phage is able to infect that particular *C. jejuni* strain as the clearing most likely is due to a high frequency of unspecific phage binding leading to inhibition of growth of the bacteria.
21. The amount of plug agarose should be 1 mL per phage preparation, but no less than 10 mL should be prepared in order to ensure the correct concentration of the plug agarose.
22. *C. jejuni* phages generally reach between 10^7 and 10^9 pfu/mL and have genomes of approximately 140 kb and 180 kb. However, their DNA can be difficult to visualize on the PFGE gel. Adding a higher amount of a low-titer phage stock and compensating by using a lower volume of a >1% concentration of plug agarose mix can greatly enhance the visibility of the DNA on the gel. The final concentration of agarose in a plug should always be 0.5%, and the total volume of plug agarose and phage mix should be 800 μL. Therefore, 200 μL of a 2% plug agarose mix can be mixed with 600 μL of warmed phage solution, or 100 μL of a 4% plug agarose mix can be mixed with 700 μL warmed phage solution to enhance DNA visibility (Fig. 3b).
23. The phage solution is heated in order to avoid immediate solidification when adding the plug agarose and to ensure an appropriate temperature of the mix for dispensing it into the plug casting mold.
24. The volume of plug casting molds may differ, but it is important for the wells to be filled completely. For the plug casting molds described here, approximately 83 μL can go into a well.
25. Make sure that the water is above the level of the TES lysis buffer in the tube.
26. The plug size is dependent on the pfu/mL concentration of the phage stock. DNA visibility of plugs from low-concentration stocks may be enhanced by increasing the plug size from 3 to 5 mm.
27. It is important that the plug is completely covered by the HhaI mix, so depending on the tube design and the size of slice, it may be necessary to adjust the volume.

28. The smallest amount of dirt, incompletely melted agarose or other contaminates, etc. will interfere with the electrophoresis and give a poor gel.
29. It is important that the gel molding chamber and comb are thoroughly cleaned and wiped with ethanol before use so that there is no leftover agarose clumps or contaminates from previous runs.
30. This is preferably done by taking 1 mL of the tempered 1% agarose solution and placing a small amount droplet wise across the comb.
31. There is no need to seal the holes in the gel generated by removing the comb.
32. If an RFLP analysis with HhaI has been performed, band patterns of the individual phages can be compared.

References

1. Kropinski AM, Mazzocco A, Waddell TE, et al. (2009) Enumeration of bacteriophages by double agar overlay plaque assay. In: Clokie MRJ, Kropinski AM (eds) Bacteriophages: methods and protocols, Volume 1: Isolation, characterization, and interactions. p 69
2. Janež N, Kokošin A, Zaletel E et al (2014) Identification and characterization of new *Campylobacter* group III phages of animal origin. FEMS Microbiol Lett 359:64–71
3. Hansen VM, Rosenquist H, Baggesen DL et al (2007) Characterization of *Campylobacter* phages including analysis of host range by selected *Campylobacter* Penner serotypes. BMC Microbiol 18(7):90
4. El-Shibiny A, Connerton PL, Connerton IF (2005) Enumeration and diversity of *campylobacters* and bacteriophages isolated during the rearing cycles of free-range and organic chickens. Appl Environ Microbiol 71:1259–1266
5. Sørensen MCH, Gencay YE, Birk T et al (2015) Primary isolation strain determines both phage type and receptors recognised by Campylobacter jejuni bacteriophages. PLoS One 10:e0116287
6. Baldvinsson SB, Sørensen MCH, Vegge CS et al (2014) *Campylobacter jejuni* motility is required for infection of the flagellotropic bacteriophage F341. Appl Environ Microbiol 80:7096–7106
7. Holst Sørensen MC, van Alphen LB, Fodor C et al (2012) Phase variable expression of capsular polysaccharide modifications allows *Campylobacter jejuni* to avoid bacteriophage infection in chickens. Front Cell Infect Microbiol 2:11
8. Sørensen MCH, van Alphen LB, Harboe A, Li J et al (2011) Bacteriophage F336 recognizes the capsular phosphoramidate modification of *Campylobacter jejuni* NCTC11168. J Bacteriol 193:6742–6749
9. Coward C, Grant AJ, Swift C et al (2006) Phase-variable surface structures are required for infection of *Campylobacter jejuni* by bacteriophages. Appl Environ Microbiol 72:4638–4647
10. Brathwaite KJ, Siringan P, Moreton J et al (2013) Complete genome sequence of universal bacteriophage host strain *Campylobacter jejuni* subsp. *jejuni* PT14. Genome Announc 1:e00969–13
11. Lingohr E, Frost S, Johnson RP. (2009) Determination of bacteriophage genome size by pulsed-field gel electrophoresis. In: Clokie MRJ, Kropinski AM (eds) Bacteriophages: methods and protocols, Volume 2: Molecular and applied aspects. p 19
12. Javed MA, Ackermann HW, Azeredo J et al (2014) A suggested classification for two groups of *Campylobacter* myoviruses. Arch Virol 159:181–190
13. Ackermann HW (2009) Basic phage electron microscopy. In Clokie MRJ, Kropinski AM (eds) Bacteriophages: methods and protocols, Volume 1: Isolation, characterization, and interactions. p 113
14. Green MR, Sambrook J (2012) Molecular cloning: a laboratory manual, 4th edn. Cold Spring Harbor Laboratory Press, Cold Spring Harbor, NY

Chapter 10

Methods to Assess the Direct Interaction of *C. jejuni* with Mucins

Marguerite Clyne, Gina Duggan, Julie Naughton, and Billy Bourke

Abstract

Studies of the interaction of bacteria with mucus-secreting cells can be complemented at a more mechanistic level by exploring the interaction of bacteria with purified mucins. Here we describe a far Western blotting approach to show how *C. jejuni* proteins separated by SDS PAGE and transferred to a membrane or slot blotted directly onto a membrane can be probed using biotinylated mucin. In addition we describe the use of novel mucin microarrays to assess bacterial interactions with mucins in a high-throughput manner.

Key words *Campylobacter jejuni*, Mucus, Mucin array, SDS PAGE, Slot blot, PVDF

1 Introduction

There is an increasing body of evidence that components of the mucus gel layer play an important role in determining the outcome of *C. jejuni* infection. Whereas human mucus has been shown to promote colonization of cell lines [1] and virulence [2] in *C. jejuni*, chicken mucus has been shown to attenuate these characteristics [3, 4]. Mucins and the biochemical makeup of the mucus layer have also been shown to play an important role in mucosal defense against *C. jejuni* infection in vivo [5–7].

1.1 Probing of C. jejuni Proteins with Biotin-Labeled Mucin

C. jejuni whole cell lysates can be either slot blotted directly to PVDF or separated by polyacrylamide gel electrophoresis and transferred electrophoretically to a polyvinylidene difluoride (PVDF) membrane. Putative mucin-binding proteins can be detected by probing with labeled purified mucins (Fig. 1). Biotin is a suitable label due to its small size, high affinity and resistance to heat, pH, and proteolysis. Slot blots are a simplified version of traditional blotting methods in that biomolecules of interest are not separated by electrophoresis but applied directly to a PVDF membrane. This method allows fast detection of the presence of a particular ligand but not the mass or number of reactive targets in a sample.

James Butcher and Alain Stintzi (eds.), *Campylobacter jejuni: Methods and Protocols*, Methods in Molecular Biology, vol. 1512, DOI 10.1007/978-1-4939-6536-6_10,

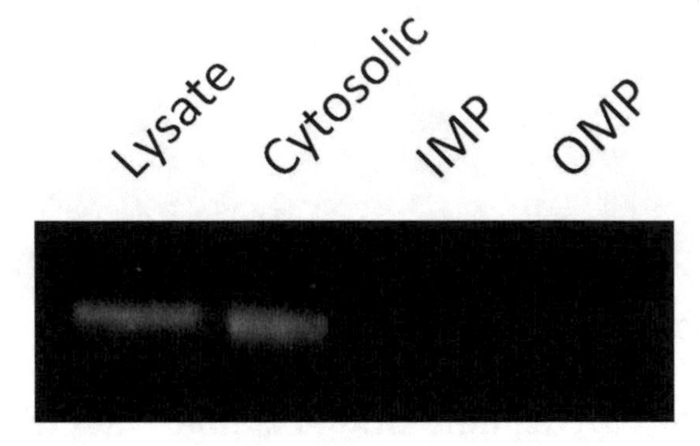

Fig. 1 Mucin overlay of *C. jejuni*. Whole cell lysate, cytosolic fraction (cytosol), inner membrane fraction (IMP), and outer membrane fraction (OMP) showing the presence of a mucin-binding ligand in the whole cell lysate and cytosolic fractions following electrophoretic transfer to a PVDF membrane from an SDS PAGE gel

1.2 Probing of Mucin Printed on an Array Slide with *C. jejuni*

Mucin purification is a tedious process, which often does not yield sufficient amounts of target protein. Differences in glycosylation between individuals present further issues, preventing pooling of samples in order to produce sufficient yields. A novel high-throughput approach to investigating the interaction of bacteria with mucins is the use of mucin microarrays [1]. Mucins isolated from any tissue source can be immobilized onto microarray slides [8]. In the arrays that we use, there are eight identical subarrays per slide, and each mucin is printed in replicates of six per subarray.

2 Materials

2.1 SDS-Polyacrylamide Gel Components

1. Resolving gel buffer: 1.5 M Tris–HCl buffer, pH 8.8.
2. Stacking gel buffer: 0.5 M Tris–HCl buffer, pH 6.8.
3. 30% acrylamide/Bis solution (29.2:0.8 acrylamide/Bis).
4. Ammonium persulfate: 10% (wt/vol) solution in water.
5. N,N,N,N′-tetramethyl-ethylenediamine (TEMED). Store at 4 °C.
6. Distilled H_2O.
7. 10% sodium dodecyl sulfate (SDS) (wt/vol) in water. Store at room temperature.
8. 6× SDS PAGE sample loading buffer (375 mM Tris–HCl pH 6.8, 10% SDS (wt/vol), 50% glycerol (vol/vol),10% β-mercaptoethanol (vol/vol), 0.03% bromophenol blue (wt/vol) in dH_2O.
9. 2× SDS PAGE sample loading buffer: mix 1 part 6× SDS PAGE sample loading buffer with two parts dH_2O.

2.2 Electrophoretic Transfer Components

1. PVDF membrane cut slightly larger than gel size, notched in one corner.
2. Six sheets of Whatman paper per gel to be transferred cut to a size slightly larger than the gel.
3. Galileo mini twin gel unit (catalog #85-1010) or equivalent.
4. Galileo bioscience glass plates (catalog #85-1010-G10) or equivalent.
5. Methanol.
6. Electroblotting tank, cassette, Scotch-Brite pads, voltage power pack.
7. Coomassie Blue Stain R250.
8. 5× transfer buffer: 0.025 M Tris, 0.192 M glycine, and 20% methanol.
9. 1× transfer buffer: 1 part 5× transfer buffer, three parts dH_2O, and one part methanol. Store at 4 °C.
10. 10× Tris-buffered saline (10× TBS): 1.5 M NaCl, 0.1 M Tris–HCl, pH 7.4.
11. 1× TBS: 1 part 10× TBS and 9 parts dH_2O
12. Wash buffer (TBST): 1× TBS containing 0.05% (vol/vol) Tween-20.

2.3 Mucin Overlay and Slot Blot Components

1. PVDF membrane.
2. Methanol.
3. Slot Blot Manifold.
4. 10× Tris-buffered saline (10× TBS): 1.5 M NaCl, 0.1 M Tris–HCl, pH 7.4
5. 1× TBS: 1 part 10× TBS and 9 parts dH2O.
6. TBST: 1× TBS containing 0.05% Tween-20.
7. Sterile Dulbecco "A" phosphate-buffered saline pH 7.3 (PBS).
8. EZ link sulfo-NHS Biotin.
9. Purified mucin resuspended in dH_2O.
10. Dialysis cassette with a molecular weight cutoff of approx. 20,000 Da.
11. 1 L beaker.
12. 3% gelatin in 1× TBS.
13. Streptavidin–peroxidase conjugate (1:50,000) in 3% gelatin in 1× TBS.
14. 22 G needle.
15. Plastic bag.
16. Heat sealer.

17. Cling Film.
18. SuperSignal West Pico Chemiluminescent Substrate (Thermo Scientific).
19. Plastic separator.
20. Heat block.

2.4 Incubation of Mucin Arrays with C. jejuni

1. Log-phase liquid culture of *C. jejuni* grown as described in the chapter entitled "Assays to Study the Interaction of *Campylobacter jejuni* with the Mucosal Surface," Subheading 3.1, **steps 1–3**.
2. 500 μM Syto82 (Life Technologies) (*see* **Note 1**).
3. Sterile Dulbecco "A" phosphate-buffered saline pH 7.3 (PBS).
4. Low-salt Tris-buffered saline (LS-TBS): 20 mM Tris, 100 mM NaCl, 1 mM $CaCl_2$, 1 mM $MgCl_2$, pH 7.2.
5. LS-TBST: LS-TBS supplemented with 0.05 % Tween 20.
6. 8-well gasket slide and incubation cassette system (Agilent Technologies).
7. Stainless steel slide rack.
8. Glass staining dishes.
9. Mucin microarrays.
10. Log-phase culture of *C. jejuni*.
11. TRITC-labeled lectins, e.g., *Ulex europaeus* agglutinin I (UEA-I), *Artocarpus integrifolia* agglutinin (AIA) (EY Laboratories, Inc., San Mateo, CA, USA).

3 Methods

3.1 Sodium Dodecyl Sulfate Polyacrylamide Gel Electrophoresis

1. Mix 2.5 mL of resolving gel buffer, 3.33 mL of 30 % acrylamide, 4 mL of water, and 100 μL of 10 % SDS in a universal tube. Add 100 μL of 10 % ammonium persulfate and 4 μL of TEMED; mix and cast gel immediately within a 10 cm × 10 cm × 1.5 mm gel cassette (*see* **Note 2**). Allow space for the stacking gel and gently add approx. 30 μL isopropanol on top of the poured gel to remove air bubbles. Allow gel to set.
2. Prepare the stacking gel by mixing 0.38 mL of stacking gel buffer, 0.5 mL of 30 % acrylamide, 2.1 mL water, and 30 μL of 10 % SDS in a universal tube. Add 30 μL of 10 % ammonium persulfate and 3 μL of TEMED. Remove the isopropanol from the top of the set resolving gel (*see* **Note** 3) and pour the stacking gel. Insert a 10-well gel comb immediately without introducing air bubbles and allow to set.

3.2 Electrophoretic Transfer

1. Fill the electroblotting tank with 1× transfer buffer (*see* **Note 4**).
2. Wet the PVDF membrane for a few seconds in methanol (*see* **Note 5**).
3. Presoak Whatman paper in some transfer buffer.
4. Using a plastic separator, carefully remove the gel from the electrophoresis plates, cut the gel if necessary, notch the bottom right-hand corner, and place briefly in a tray filled with transfer buffer.
5. Open the cassette containing the gel, and place three presoaked Whatman papers on one side of the apparatus on top of the Scotch-Brite. Place the gel on top of the Whatman paper and cover with the PVDF membrane, lining up the notches. Place the additional three Whatman papers on top, and roll a 50 mL tube over the stack to eliminate any air bubbles.
6. Close the cassette and place it in the tank facing the PVDF membrane toward the positive electrode (red).
7. Transfer at 100 V for 1 h 30 mins or at 15 V overnight at 4 °C.

3.3 Mucin Overlay

All procedures are carried out at room temperature unless otherwise stated.

1. Resuspend the purified mucin (0.1 mg/mL to 0.25 mg/mL) in dH_2O, and biotinylate according to the manufacturer's instructions (EZ-Link sulfo-N-hydroxysuccinimide biotinylation kit (Pierce)) (*see* **Note 6**). To remove excess biotin, transfer the mucin via a 22 G needle to a dialysis cassette with a molecular weight cutoff of approx. 20,000 Da, and dialyze the sample in 1 L of 1× PBS overnight at 4 °C.
2. *C. jejuni* whole cell lysates may be slot blotted directly onto the PVDF membrane or separated by electrophoresis on 10% polyacrylamide minigels and transferred to PVDF membranes for 1 h 30 min in a wet blotter (Bio-Rad, Hercules, CA) at 100 V (*see* **Note 7**).
3. Following transfer, block the PVDF membrane in 3% gelatin in 1× TBS overnight at 30 °C at 50 rpm in a shaking incubator (*see* **Note 8**).
4. Wash the membrane in TBST (*see* **Note 9**) and incubate with 1 mL biotinylated mucin overnight at room temperature. Retain the mucin for reuse (*see* **Note 10**) and wash the membrane three times for 10 min with TBST.
5. Incubate with streptavidin–peroxidase conjugate for 1 h at 37 °C at 50 rpm.
6. Wash the membrane three times for 20 min TBST (*see* **Note 11**).
7. Mucin-binding proteins may be detected by enhanced chemiluminescence. Remove the PVDF membrane from the wash buf-

fer and place in a gel-imaging viewer. Mix equal parts of SuperSignal West Pico Chemiluminescent Substrate (ECL substrate) and pipette directly onto the membrane. A dark exposure time ranging from 30 s to 5 min is recommended (*see* **Note 12**).

3.4 Slot Blot

1. Assemble the slot blot apparatus, removing the upper slot module. Attach a vacuum to the vacuum chamber.
2. Cut a piece of PVDF membrane to the size and shape of the lower slot module.
3. Flood the lower slot module with 1× TBS. Wet the PVDF membrane for a few seconds in methanol, and place in the manifold, ensuring there are no air bubbles present.
4. Place the upper slot module on top of the PVDF membrane and seal tightly.
5. Dilute each sample to be slot blotted in approx. 50 μL 1× TBS and apply to slot. Apply vacuum for approx. 60 s or until the entire sample has been pulled through.
6. Remove the membrane from manifold, and follow instructions from **step 3** of the mucin overlay procedure as above (Subheading 3.3, **steps 3**–**7**).

3.5 Incubation of C. jejuni with Mucin Microarrays

1. Dilute down *C. jejuni* to an OD_{600} of 0.2 from the log-phase culture.
2. Pellet the bacteria by centrifugation at 6500 × *g* for 5 min, wash twice in PBS, and resuspend the pellet in 198 μL LS-TBS to an OD_{600} of 1.0 (*see* **Note 13**).
3. To label bacteria, add 2 μL of the 500 μM Syto82 solution to the bacterial culture in LS-TBS (total volume now 200 μL), vortex to ensure the dye is evenly resuspended in the LS-TBS, and incubate for 30–45 min in the dark at room temperature. From this point onward, all steps must be performed in the dark to ensure that the Syto82 signal does not fade.
4. Following incubation, pellet the bacteria by centrifugation and resuspend the pellet thoroughly in LS-TBS.
5. Repeat **step 4** six more times for a total of seven washes.
6. Resuspend the bacterial culture to a final optical density (OD_{600}) of 0.5 in 400 μL LS-TBST (*see* **Note 14**).
7. Add 70 μL of the fluorescently labeled bacteria to each of the six subarrays on the mucin microarray using an 8-well gasket slide and incubation cassette system (Agilent Technologies). Incubate bacteria on the slide for 1 h at 37 °C in the dark with gentle rotation (*see* **Note 15**). Incubate two subarrays with 70 μL of different TRITC-labeled lectins for quality control (*see* **Note 16**).

8. Wash microarray slides five times in LS-TBST, once in LS-TBS, and once in deionized water (*see* **Note 17**). Dry by centrifugation.
9. Scan slides using a microarray scanner that possesses a 532 nm laser. Syto82 emits a green fluorescence at 560 nm, so binding of fluorescently labeled bacteria is detected using the 532 nm laser (100% laser power, 70% photomultiplier tube [PMT] setting for both) (Molecular Devices). Data generated from scanned slides is extracted and exported to Excel as a text file, where subsequent data analysis is performed. The median feature intensity, with background subtracted, is used for each feature intensity value. The median of six replicate spots per subarray is used as a single data point.

4 Notes

1. Syto82 solution is supplied at a concentration of 5 mM. Make 500 μM aliquots of the dye. To do so, add 10 μL of the 5 mM stock to 90 μL dH_2O. Ensure that the dye is always stored at −20 °C and in the dark as it is light sensitive.
2. Once the TEMED and ammonium persulfate are added, the gel will begin to polymerize so it is important to add these components last; quickly mix the solution using a pipette gun and pipette the solution into the cast.
3. Remove the isopropanol layer by rinsing with dH_2O before adding the stacking gel.
4. Storing 1× transfer buffer at 4 °C before use and placing the electroblotting tank on ice prevent the gel and membrane overheating during transfer.
5. Avoid touching the PVDF membrane with your gloves. Use a tweezers/forceps if necessary. Ensure the membrane is fully submerged in methanol and changes from white to translucent.
6. *N*-Hydroxysuccinimide (NHS) esters of biotin are the most common type of biotinylation reagent. These come with varying spacer arm lengths to prevent steric hindrance of biotin when conjugated to target molecules with complex secondary structures such as mucin. We use a medium-length spacer arm of 22.4 Å for biotinylation of mucin.
7. To make whole cell lysates, resuspend *C. jejuni* to an OD_{600} of 0.4 and pellet by centrifugation. Resuspend pellet in 2× SDS PAGE sample loading buffer and heat at 100 °C for 10 min. Transferring the gel at too high, a voltage will cause the PVDF membrane to overheat. Stain the polyacrylamide gel following transfer in Coomassie Blue R250 to check for transfer efficiency.

8. As the blocking solution is incubated at 30 °C overnight in gelatin, care must be taken in order to avoid any contamination of the solution. For this reason, it is recommended that the container be sterilized and the gelatin heated to 37 °C before use. Transfer the PVDF membrane from the blotter to the container in a laminar flow hood, cover with cling film to avoid contamination, and minimize evaporation within the shaking incubator.
9. Following blocking, ensure the membrane is thoroughly washed to remove excess gelatin before incubating with mucin as the mucin can become gelatinized over time. Due to the small volume of purified mucin, it may be necessary to dilute the sample in TBS in order to cover the membrane. Sealing the membrane and biotinylated mucin in a small plastic bag prevents evaporation of the sample and subsequent drying of the PVDF membrane.
10. Biotinylated mucin may be collected and frozen at −20 °C for reuse.
11. To reduce high background, wash the PVDF membrane overnight in TBST with a final wash in TBS or dH_2O alone as Tween-20 can interfere with the ECL reagent.
12. When removing the membrane, take care to hold it at the corners, and place it in the viewer, adding the ECL substrate promptly to prevent drying of the membrane. The ECL reagent becomes oxidized via the streptavidin–peroxidase resulting in light emission that is detected by the gel-imaging viewer.
13. The optimal OD for staining with Syto82 can vary depending on the bacteria. *C. jejuni* stains best at an OD_{600} of 1.0. To obtain an OD_{600} of 1.0, spin down 1 mL of an OD_{600} of 0.2 and resuspend in 198 μL of LS-TBS. The same amount of bacteria are present, but they have been concentrated five times, giving an OD_{600} of 1.0 in 200 μL of LS-TBS.
14. As the initial 200 μL bacterial culture had an OD_{600} of 1.0, to obtain at a final OD_{600} of 0.5, resuspend the pellet for the final time in 400 μL LS-TBST.
15. Incubating bacteria with the array:
    (a) Ensure that the gasket slide is clean and free of dirt/dust. Also check that the rubber ring around each compartment is not broken to ensure leakage does not occur. Place the gasket slide in the Agilent cassette.
    (b) Take 70 μL of the labeled bacterial culture and pipette into the appropriate compartment on the gasket slide.
    (c) Take mucin array and make note of the number on the bottom to ensure that the correct array is analyzed later. Incubate the array, with the numbered side facing down, with the bacterial culture on the gasket slide. Place the array slide onto the gasket slide at an angle to prevent air bubbles forming.

17. Fluorescently labeled lectins which bind to different glycan structures can be used as a quality control measure to ensure that the mucins are printed on each array. TRITC-labeled lectins are commercially available (EY Laboratories, San Mateo, CA, USA). Generally, two different lectins that bind to different structures are included in each array. Dilute the lectin to the appropriate concentration in LS-TBST and add 70 μL to a compartment on the gasket slide. Lectins are typically used at concentrations ranging from 5 to 15 μg/mL. The optimal concentrations of suitable lectins to use are listed in Ref. [8].
18. Separate the array from the gasket slide in a beaker of LS-TBST to avoid cross contamination between subarrays. Add the array to a stainless steel slide rack, and wash by dipping in and out of glass staining dishes filled with either LS-TBST, LS-TBS, or water. Leave the array in each dish for 30 s at a time.

Acknowledgments

This work was supported by a grant from Science Foundation Ireland (08/SRC/B1393).

References

1. Naughton JA, Marino K, Dolan B et al (2013) Divergent mechanisms of interaction of *Helicobacter pylori* and *Campylobacter jejuni* with mucus and mucins. Infect Immun 81(8):2838–2850. doi:10.1128/IAI.00415-13
2. Tu QV, McGuckin MA, Mendz GL (2008) *Campylobacter jejuni* response to human mucin MUC2: modulation of colonization and pathogenicity determinants. J Med Microbiol 57(Pt 7):795–802. doi:10.1099/jmm.0.47752-0
3. Alemka A, Whelan S, Gough R et al (2010) Purified chicken intestinal mucin attenuates *Campylobacter jejuni* pathogenicity *in vitro*. J Med Microbiol 59(Pt 8):898–903. doi:10.1099/jmm.0.019315-0
4. Byrne CM, Clyne M, Bourke B (2007) *Campylobacter jejuni* adhere to and invade chicken intestinal epithelial cells *in vitro*. Microbiology 153(Pt 2):561–569. doi:10.1099/mic.0.2006/000711-0
5. McAuley JL, Linden SK, Png CW et al (2007) MUC1 cell surface mucin is a critical element of the mucosal barrier to infection. J Clin Invest 117(8):2313–2324. doi:10.1172/JCI26705
6. Linden SK, Florin TH, McGuckin MA (2008) Mucin dynamics in intestinal bacterial infection. PLoS One 3(12), e3952. doi:10.1371/journal.pone.0003952
7. Dawson PA, Huxley S, Gardiner B et al (2009) Reduced mucin sulfonation and impaired intestinal barrier function in the hyposulfataemic NaS1 null mouse. Gut 58(7):910–919. doi:10.1136/gut.2007.147595
8. Kilcoyne M, Gerlach JQ, Gough R et al (2012) Construction of a natural mucin microarray and interrogation for biologically relevant glycoepitopes. Anal Chem 84(7):3330–3338. doi:10.1021/ac203404n

Chapter 11

Methods to Study *Campylobacter jejuni* Adherence to and Invasion of Host Epithelial Cells

Nicholas M. Negretti and Michael E. Konkel

Abstract

Measuring bacterial adherence and invasion of cells in vitro has enabled researchers to dissect the interactions of *Campylobacter jejuni* with eukaryotic cells. Numerous *C. jejuni* virulence determinants and host cell factors that contribute to the process of adherence, invasion, and immune modulation have been identified utilizing in vitro adherence and invasion assays. In this chapter, we describe the evaluation of *C. jejuni* adherence to and invasion of HeLa cells using the gentamicin-protection assay.

Key words Adherence, Attachment, Binding, Invasion, Bacteria-host cell interactions

1 Introduction

Attachment to and invasion of host epithelial cells lining the intestinal tract is required for *Campylobacter jejuni*-mediated enteritis [1]. Attachment of *C. jejuni* to the cells lining the intestinal tract is promoted by the binding of bacterial adhesins to host cell ligands and extracellular matrix components [2–5]. Constitutively synthesized surface proteins mediate *C. jejuni* adherence to host cells [6, 7]. After binding, *C. jejuni* invade host epithelial cells [8], which provides a niche for replication and protection from the immune system [9]. Unlike adhesion that requires the constitutively synthesized adhesins, cellular invasion requires *de novo* protein synthesis that occurs in response to a stimulatory signal (i.e., contact with host cells) [6, 10]. Chloramphenicol-treated *C. jejuni* adhere to epithelial cells equally as well as synthetically active *C. jejuni* [6]; however, chloramphenicol-treated *C. jejuni* demonstrate a significant reduction in host cell invasion.

In vitro tissue culture models have been used to investigate *C. jejuni* adherence to and invasion of host epithelial cells. Investigation of bacterial adherence involves incubation of *C. jejuni* with cultured host epithelial cells, rinses to remove nonadherent bacteria,

James Butcher and Alain Stintzi (eds.), *Campylobacter jejuni: Methods and Protocols*, Methods in Molecular Biology, vol. 1512, DOI 10.1007/978-1-4939-6536-6_11, © Springer Science+Business Media New York 2017

selective lysis of the epithelial cells with a detergent (i.e., 0.1% Triton X-100), and enumeration of the adherent bacteria by spreading serially diluted suspensions on agar plates. The adherence assay will quantify the number of bacteria bound to the surface of host cells and internalized within host cells at a given time. In addition to the cell adherence assay, internalized bacteria can be quantified by the gentamicin-protection assay. The gentamicin-protection assay takes advantage of the fact that the aminoglycoside antibiotic gentamicin does not readily penetrate the membrane of eukaryotic cells [11]. Thus, the extracellular bacteria are killed while the intracellular *C. jejuni* remain protected from the antibiotic. Due to the slow growth rate of *C. jejuni* (doubling time of approximately 90 min [12]), a long gentamicin treatment (3 h) is required to kill the extracellular bacteria. Following incubation with gentamicin, the epithelial cells are lysed and the internalized bacteria quantified by spreading serially diluted suspensions on agar plates.

Compared to other methods of determining cellular adherence and invasion of *C. jejuni*, such as differential staining and microscopy, the gentamicin-protection assay is a relatively rapid method to assess important *C. jejuni* virulence attributes and the conditions that affect these attributes. These assays are amenable to a wide variety of manipulations to both the bacterial inoculum and the host cells. A variety of *C. jejuni* deletion mutants have been tested using this assay, including adhesin mutants (e.g., CadF, FlpA, and JplA) [2, 3, 10, 13] and protein secretion mutants (e.g., CiaB, CiaC, CiaD, and CiaI) [8, 14–16]. Here, we describe a basic assay testing *C. jejuni* adherence to and invasion of HeLa cells as outlined in Fig. 1.

2 Materials

Prepare all solutions in tissue culture grade water that is endotoxin free. Follow appropriate precautions when disposing of biohazardous materials. All protocols, except final dilution and plating of bacteria, take place in a class II biosafety cabinet.

2.1 Mammalian Tissue Culture Components

1. A humidified 5% CO_2 incubator at 37 °C.
2. HeLa cells (ATCC CCL-2) (*see* **Note 1**).
3. Minimal Essential Medium with 10% Fetal Bovine Serum (MEM 10% FBS): Gibco MEM with Earle's salts or equivalent (*see* **Note 2**), supplemented with 1 mM sodium pyruvate, and 10% (v/v) heat inactivated Fetal Bovine Serum (FBS). Filter sterilize through a 0.22 μm filter (*see* **Note 3**).
4. Sterile 24-well plastic tissue culture plates.

2.2 Bacterial Culture Components

1. A humidified microaerobic (5% O_2, 10% CO_2, and 85% N_2) incubator at 37 °C.

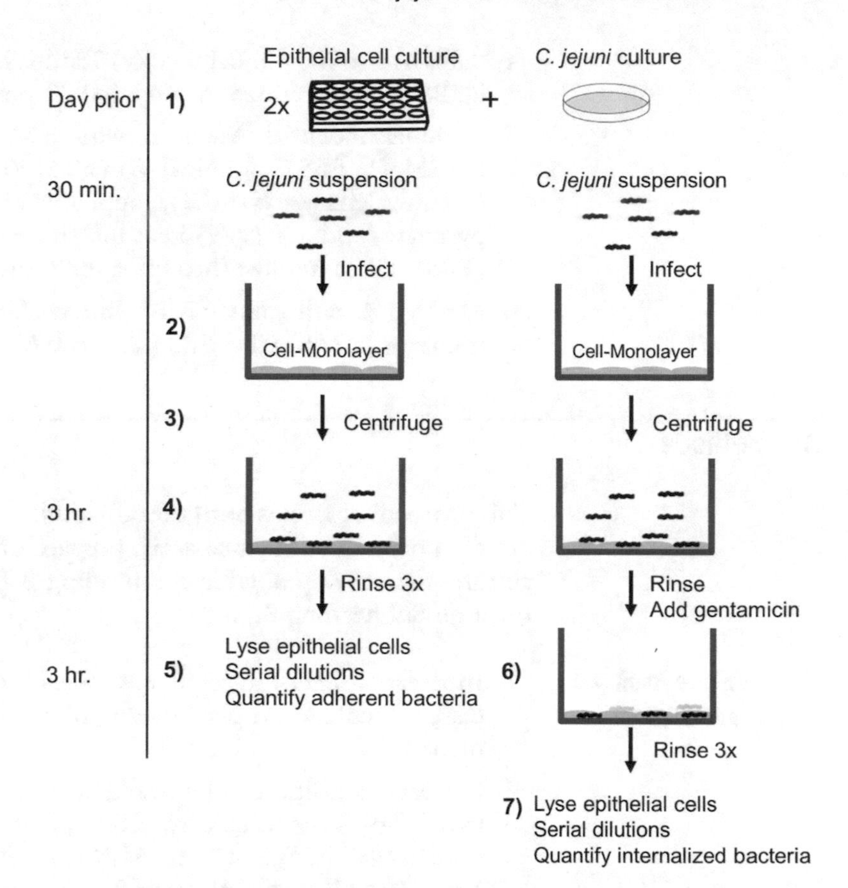

Fig. 1 Overview of adherence and invasion assays. Adherence and invasions assays take a total of approximately six and a half hours and involve: (*1*) Preparation of epithelial cell and *C. jejuni* cultures followed by (*2*) infection epithelial cells with *C. jejuni*, (*3*) centrifugation to promote *C. jejuni*-epithelial cell attachment, (*4*) incubation to permit adherence and invasion, (*5*) selective lysis of the epithelial cells to quantify adherent bacteria, (*6*) treatment with gentamicin, and (*7*) selective lysis of the epithelial cells to quantify internalized bacteria. *Black spirals* indicate live bacteria and *light gray spirals* indicate dead bacteria

2. *Campylobacter jejuni* culture (*see* **Note 4**).
3. Mueller-Hinton agar plates supplemented with 5% citrated bovine blood (MHB): 2.0 g/L beef extract powder, 17.5 g/L acid digest of casein, 1.5 g/L starch, 17.0 g/L agar, 5% (v/v) citrated bovine blood.
4. Mueller-Hinton (MH) broth: 2.0 g/L beef extract powder, 17.5 g/L acid digest of casein, 1.5 g/L starch.

2.3 Adherence and Invasion Assay Components

1. Spectrophotometer set at a wavelength of 540 nm.
2. Centrifuge for tissue culture plates.
3. Sterile PBS: 137 mM NaCl, 2.7 mM KCl, 10 mM Na_2HPO_4, 1.8 mM KH_2PO_4, pH 7.2. Sterilize by autoclave.

4. 0.1% Triton X-100: 0.1% (v/v) Triton X-100 prepared in sterile PBS, filter sterilize through a 0.22 μm filter.
5. Minimal Essential Medium with 1% Fetal Bovine Serum (MEM 1% FBS): (*see* **Note 5**) Gibco MEM with Earle's salts or equivalent (*see* **Note 2**), supplemented with 1 mM sodium pyruvate, and 1% (v/v) heat-inactivated Fetal Bovine Serum (FBS). Filter sterilize through a 0.22 μm filter (*see* **Note 3**).
6. MEM 1% with gentamicin: 250 μg/mL gentamicin sulfate prepared in MEM 1% FBS (*see* **Note 6**).

3 Methods

This protocol describes both the adherence and invasion assays. To perform only an adherence assay, prepare only one 24-well tissue culture plate as described in Subheading 3.1 **step 2**, and proceed through Subheading 3.2.

3.1 Prepare Bacterial and Tissue Cultures

1. Inoculate a MHB agar plate with each *C. jejuni* isolate to be tested. Incubate the plate overnight in a microaerobic environment at 37 °C (*see* **Note 7**).
2. For each condition to be tested, seed HeLa cells into wells of two 24-well tissue culture plates (four wells per plate, eight total wells) in 0.5 mL of MEM 10% FBS at a density of 1.5×10^5 cells/well (*see* **Note 8**). Incubate plates in a 5% CO_2 incubator at 37 °C overnight.

3.2 Adherence Assay

1. Prewarm MEM 1% FBS and sterile PBS in a 37 °C water bath.
2. Suspend a portion of an overnight culture of *C. jejuni* in 10 mL of MEM 1% FBS (*see* **Note 9**) to an approximated OD_{540} of 0.3. Determine the OD_{540} of the culture using a spectrophotometer.
3. Inoculate 10 mL of MEM 1% FBS with *C. jejuni* to an OD_{540} of 0.03 [equivalent to approximately 3×10^7 colony-forming units per mL (CFU/mL)] with the suspended culture prepared in **step 2** (Subheading 3.2).
4. Visualize the HeLa cell culture in the two 24-well tissue culture plates by microscopy to ensure that the HeLa cells are intact (*see* **Note 10**). Aspirate the medium and rinse the cells twice with approximately 0.5 mL MEM 1% FBS (*see* **Note 11**).
5. Inoculate two 24-well tissue culture plates (four wells per plate, eight wells total) with 0.5 mL of the 0.03 OD_{540} *C. jejuni* suspension prepared in **step 3** (Subheading 3.2) (*see* **Note 12**). Centrifuge the plates at $800 \times g$ for 5 min to synchronize the infection (*see* **Note 13**). Label one plate "Adherence" and the other "Invasion." Incubate the tissue culture plates in a 5% CO_2 incubator at 37 °C for 3 h (*see* **Note 14**).

6. After incubation, rinse each well of the "Adherence" plate three times with approximately 1 mL of sterile PBS (*see* **Note 15**).
7. To quantitate the number of adherent (cell-associated) bacteria, add 200 μL of 0.1% Triton X-100 to lyse the HeLa cells and incubate the plate at 37 °C for 5 min (*see* **Note 16**). Add 800 μL of sterile PBS (*see* **Note 17**). Perform tenfold serial dilutions with an aliquot of each well of the "Adherence" plate (*see* **Note 18**) in sterile PBS, and plate the suspensions (with final dilutions of 10^3 through 10^6) in quadrants on a MHB agar plate (*see* **Note 19**). Incubate the plates at 37 °C in a microaerobic environment for two days.

3.3 Invasion Assay

1. After the initial 3 h incubation period (Subheading 3.2, **step 5**), rinse each well of the "Invasion" plate twice with approximately 1 mL of MEM 1% FBS, add 1 mL of MEM 1% FBS with gentamicin (*see* **Note 20**) and return the tissue culture plate to the 5% CO_2 incubator (*see* **Note 21**).
2. After a 3 h incubation period, wash each well of the plate three times with approximately 1 mL of sterile PBS (*see* **Note 11**).
3. To quantitate the number of internalized bacteria, add 200 μL of 0.1% Triton X-100 to each well and incubate at 37 °C for 5 min (*see* **Note 16**). Following incubation, add 800 μL of sterile PBS to each well (*see* **Note 17**). Perform tenfold serial dilutions with an aliquot from each well of the "Invasion" plate in sterile PBS. Plate the suspensions (with final dilutions of 10^2–10^5) in quadrants on a MHB agar plate (*see* **Note 19**), and incubate the plates at 37 °C in a microaerobic environment for 2 days.
4. Once colonies are visible on the plates (*see* **Note 22**), select a dilution that has easily distinguishable colonies, count the colonies in that dilution (*see* **Note 23**), and calculate the CFU/mL per well as outlined in Fig. 2 (*see* **Note 24**).

3.4 Additional Controls

1. When determining the number of internalized bacteria, it is always necessary to determine the number of adherent bacteria (*see* **Note 25**). We also recommend including a *C. jejuni* strain that is noninvasive as a control in every assay (*see* **Note 26**).
2. The viability of nontreated and drug-treated HeLa cells should be assessed by trypan blue staining. To quantitate the number of adherent HeLa cells, cells can be trypsinized prior to trypan blue staining and counted with a hemocytometer. An alternative to assess damaged cells (cellular cytotoxicity and cytolysis) is to measure the amount of lactate dehydrogenase (LDH) released into the medium.

3.5 Data Presentation and Statistical Analysis

Results from these assays are usually presented as a mean ± standard deviation of adherent and internalized bacteria, either plotted on a graph or presented in a table. One alternative is to calculate the percent of inoculated bacteria that adhere to the cells and the

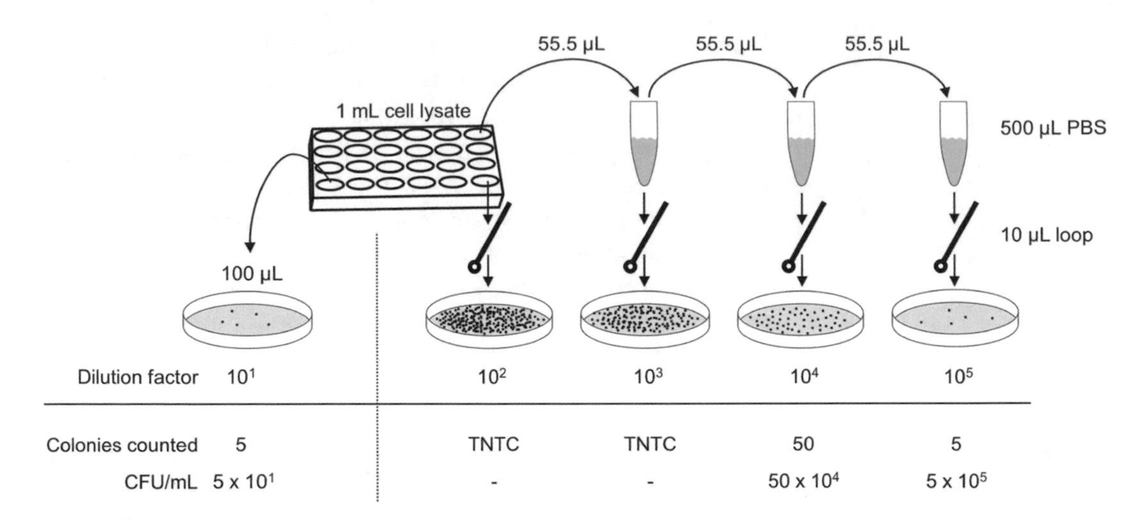

Fig. 2 Serial dilutions to quantitate *C. jejuni*. After adherence and invasion assays, there is 1 mL of cell lysate in the 24-well tissue culture plate. To enumerate the viable *C. jejuni* in the lysate serial dilutions are performed by mixing 55.5 μL of sample with 500 μL of sterile PBS. Calibrated 10 μL inoculating loops are used to plate 10 μL of each dilution on quadrants of a Mueller-Hinton blood agar plate. Example colony counts and calculated colony forming units per mL (CFU/mL) are given

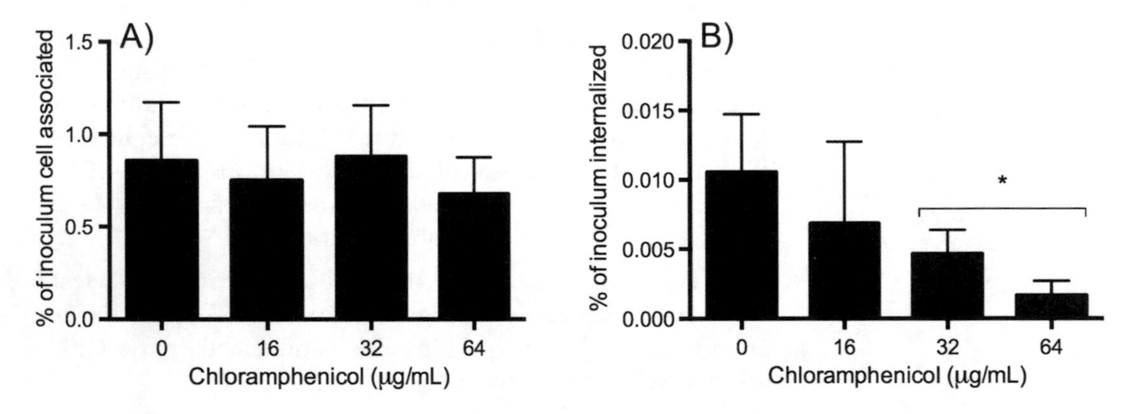

Fig. 3 Maximal invasion of *C. jejuni* requires de novo protein synthesis. INT 407 cells were inoculated with *C. jejuni* strain 81–176 for 3 h (Panel (**a**), cell associated bacteria) and treated with 250 μg/mL of gentamicin for an additional 3 h (Panel (**b**), internalized bacteria). The bacteria were treated with 0, 16, 32, and 64 μg/mL of chloramphenicol for 15 min prior to the inoculation of the cells and during the first 3 h incubation period. Values are plotted as the mean ± the standard deviation of quadruplicate measurements. The *asterisk* indicates a significant difference ($P < 0.05$) compared with the non-treated sample, as judged by one-way ANOVA followed by post-hoc Dunnett's analysis. The concentrations of chloramphenicol used in this assay had no effect on INT 407 cell viability

percent of adherent bacteria that are internalized (Fig. 3). Another alternative is to calculate the ratio of internalized bacteria to adherent bacteria.

Significant changes between two conditions can be determined by a Student's *t*-test. If more than two conditions are being evaluated, then it is necessary to first perform an analysis of variance (ANOVA) followed by a multiple comparison test (*see* **Note 27**).

4 Notes

1. Epithelial or fibroblast cells of any type can be used. The efficiencies of *C. jejuni* adherence and invasion vary with cell type and host species. The major considerations when changing cell type are attaining similar confluence (70–80%) at the start of the assay.
2. Use the tissue culture medium that is appropriate for the cell type chosen. Assays need to be done to ensure that *C. jejuni* can survive in the medium for the duration of the assay (approximately six hours). Medium containing phenol red provides a simple method to monitor the pH over the course of the assay.
3. This protocol uses MEM supplemented with both 10% (v/v) and 1% (v/v) FBS. It is convenient to make 945 mL of MEM and then split it into two aliquots prior to adding 10% (v/v) or 1% (v/v) FBS (450 mL and 495 mL, respectively). Bacterial contamination of MEM supplemented with either 10% or 1% FBS can occasionally be visualized by a cloudy appearance and change in the pH of the medium (it usually becomes yellow).
4. Routine laboratory culture of *C. jejuni* requires passage every 2 days to an MHB agar plate; otherwise, there is a risk of losing motility. Before performing the adherence and invasion assays, test the motility of the *C. jejuni* by inoculating a semisolid agar plate as described by Neal-McKinney et al. [17]. A defect in the flagellum (i.e., the export apparatus and/or motility) will influence the results of assays.
5. MEM with 1% FBS is generally used throughout the assay to reduce the amount of fibronectin in the medium available to bind the *C. jejuni* fibronectin binding proteins. These fibronectin binding proteins are required for adherence to host epithelial cells [13, 18]. While HeLa cells will survive for the duration of the assay in medium with minimal FBS, higher concentrations of FBS can be used if the particular cell line necessitates it. However, there will be a significant reduction in the observed adherent bacteria. Any change in HeLa cell morphology may indicate insufficient FBS concentration.
6. To minimize the time that the cells are in MEM 1% FBS, the gentamicin solution can be prepared in MEM 10% FBS without impact on the assay. This may be necessary for some cell types.
7. We get the most consistent results using biphasic *C. jejuni* cultures for the inoculum. These cultures can be prepared by overlaying a MHB agar plate with a 10 mL suspension of *C. jejuni* in MH broth at an OD_{540} of 0.2. Each plate is incubated in a microaerobic environment at 37 °C for 18–20 h, taking care to avoid contact of the culture medium to the lid of the

plate. Adherence and invasion assays are commonly used to assess differences in *C. jejuni* culture conditions, *C. jejuni* mutants, and *C. jejuni* strains.

8. The ideal HeLa cell confluence is 70–80 %. Due to variations in the growth rate of cell cultures, it may be necessary to adjust the starting density of cells. Prior to setting up the experiment, a range of seeding densities (e.g., 1×10^5 to 5×10^5 cells/well) should be tested.
9. *C. jejuni* is sensitive to both acidic and basic pH. Ensure that the MEM is approximately neutral (pH 7), adjust the MEM to an orange color with filter sterilized 1 M HCl. The optical density measurement will be influenced by the pH (color) of the medium and the pH (color) of the reference sample.
10. Inhibitors to specific cellular processes (e.g., cytochalasin D, an inhibitor of actin polymerization) can be used to investigate the role that host cell processes play in bacterial adherence and invasion. Generally, a pretreatment with inhibitor for 30 min prior to the infection of the cells with *C. jejuni* is sufficient. The inhibitor needs to be maintained in the medium for the duration of the adherence assay. It does not need to be maintained after the addition of gentamicin for the invasion assay.
11. Aspirating the medium with a sterile disposable glass pipette attached to a vacuum is an easy way to quickly remove the medium. Using a 25 mL serological pipette to gently drip the medium onto the cells reduces the risk of dislodging the adherent epithelial cells. Inspect the cells after washes or the addition of reagents to evaluate the integrity of the cells before proceeding with the assay.
12. We typically use a multiplicity of infection (MOI) of approximately 100 bacteria to 1 cell. MOI can be adjusted; however, we do not recommend increasing the MOI because the efficiency of invasion decreases [19]. To determine the actual number of bacteria used for inoculation of the HeLa cells, spread serially diluted suspensions (with final dilutions of 10^5–10^8) of the bacterial inoculum, prepared in **step 3** (Subheading 3.2), in quadrants of a MHB agar plate.
13. Performing the adherence assay with and without the centrifugation step may be useful when analyzing motile and nonmotile bacteria. The centrifugation promotes contact of the bacteria with the cells, and will reduce the differences in the total number of adherent bacteria between motile and nonmotile isolates.
14. The kinetics of *C. jejuni* adherence and invasion can be evaluated by changing the duration of the incubation prior to cellular lysis or the addition of gentamicin.
15. As an alternative to cellular lysis and plating, adherent bacteria can be stained for examination by immunofluorescence

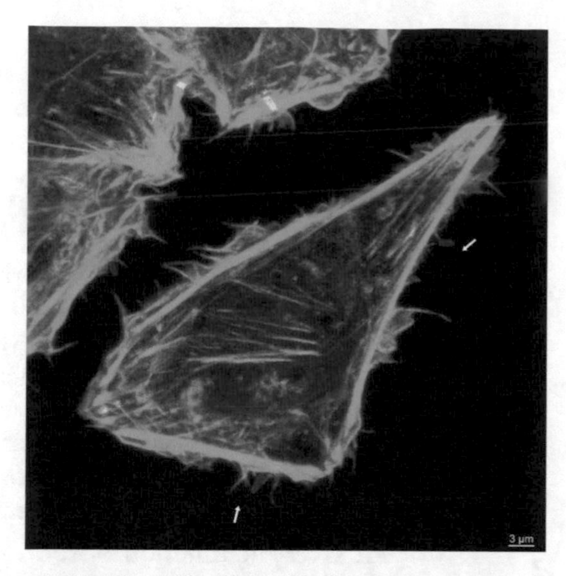

Fig. 4 *C. jejuni* interaction with cells. Epithelial cells were seeded on coverslips and infected with *C. jejuni* strain 81–176. Immunofluorescence confocal microscopy was performed. *C. jejuni* is labeled in *red*, actin is labeled in *green*, and DNA is labeled in *blue*. *Arrows* indicate flagella

microscopy (Fig. 4). The procedure used for immunofluorescence microscopy is described elsewhere [20]. This requires seeding the HeLa cells on coverslips at the beginning of the assay.

16. The time needed to lyse the epithelial cells in 0.1% Triton X-100 will vary between cell types. Observe the cells for lysis by microscopy and increase the incubation time if necessary.
17. Add the sterile PBS to the wells vigorously to begin suspending the cell lysate.
18. Mix the sample by pipetting up and down several times when removing the aliquot from the 24-well tissue culture tray. This will break up cell-debris aggregates and resuspend the *C. jejuni*. Any strategy to perform tenfold serial dilutions can be used; however, we find it convenient to use a repeater pipette to aliquot 500 μL of sterile PBS into sterile 1.5 mL tubes, and then perform a tenfold dilution series by pipetting 55.5 μL of sample into the sterile PBS aliquots. Additionally, disposable 10 μL calibrated loops make plating the dilutions simpler than other methods. The 10 μL loop adds a 10^2 dilution to the final dilution factor and allows for easily plating four dilutions in quadrants on a single MHB agar plate. It is possible to plate the 10^2 and 10^1 dilutions by plating 10 μL or 100 μL directly from the well of the tissue culture plate, respectively.
19. The dilutions will vary depending on the specific conditions being tested; however, a general guideline is that approximately 1% of the *C. jejuni* added to the well will adhere to the cells and approximately 1% of the adherent *C. jejuni* will invade the cells.

20. The concentration and duration of gentamicin treatment are dependent on the *C. jejuni* strain being tested and the specific conditions of the assay. Before performing an invasion assay, be sure to evaluate the effectiveness of the gentamicin treatment as follows. Inoculate 1 mL of MEM 1% FBS containing gentamicin with *C. jejuni* at an OD_{540} of 0.03 and incubate for three hours in a humidified 5% CO_2 incubator at 37 °C. Following incubation, rinse the bacteria with sterile PBS, resuspend the final bacterial pellet in 100 μL of PBS, and plate the entire bacterial suspension on an MHB agar plate. Incubate the plate in a microaerobic environment at 37 °C for 3 days and evaluate for growth. If the conditions are insufficient to kill all the *C. jejuni* (no growth of bacteria on the plate), increase the gentamicin concentration and/or incubation duration until there are no viable bacteria remaining.
21. Over the duration of the assay, there is not enough time for significant growth of *C. jejuni*.
22. Duration of incubation will depend on the strain of *C. jejuni* being tested. Generally, colonies will be visible in 2–3 days.
23. When counting the dilutions, make sure they follow a general tenfold dilution trend. For each test condition, when possible, count the same dilution for each replicate. This will improve consistency.
24. Because the wells were suspended in a total of 1 mL, the number of colonies counted is equal to CFU/mL.
25. *C. jejuni* must adhere to cells prior to invasion. *C. jejuni* strains defective in cellular adherence will have reduced invasion.
26. *C. jejuni* require de novo protein synthesis to invade cells. A widely applicable invasion-negative control is to pretreat the *C. jejuni* for 30 min with a concentration of chloramphenicol (e.g., 32–1024 μg/mL depending on the strain) that prevents protein synthesis (Subheading 3.2 **step 3**). A stock of chloramphenicol is prepared by dissolving the antibiotic in methanol to a concentration of 32 mg/mL.
27. Several multiple comparison tests can be used. When comparing all conditions to one control condition, use Dunnett's test, and when comparing all conditions to each other, use Tukey's test.

Acknowledgements

We thank Christopher R. Gourley for critical review of this manuscript and Joanna Fragoso for technical assistance.

Research in the Konkel Laboratory is supported by funds from the United States Department of Agriculture, National Institute of Food and Agriculture (Award Number 2011-67015-30772).

References

1. Allos BM (2001) *Campylobacter jejuni* infections: update on emerging issues and trends. Clin Infect Dis 32(8):1201–1206
2. Jin S, Joe A, Lynett J et al (2001) JlpA, a novel surface-exposed lipoprotein specific to *Campylobacter jejuni*, mediates adherence to host epithelial cells. Mol Microbiol 39(5):1225–1236
3. Pei Z, Burucoa C, Grignon B et al (1998) Mutation in the peb1A locus of *Campylobacter jejuni* reduces interactions with epithelial cells and intestinal colonization of mice. Infect Immun 66(3):938–943
4. Flanagan RC, Neal-McKinney JM, Dhillon AS et al (2009) Examination of *Campylobacter jejuni* putative adhesins leads to the identification of a new protein, designated FlpA, required for chicken colonization. Infect Immun 77(6):2399–2407. doi:10.1128/IAI.01266-08
5. Moser I, Schroeder W, Salnikow J (1997) *Campylobacter jejuni* major outer membrane protein and a 59-kDa protein are involved in binding to fibronectin and INT 407 cell membranes. FEMS Microbiol Lett 157(2):233–238
6. Konkel ME, Cieplak W Jr (1992) Altered synthetic response of *Campylobacter jejuni* to cocultivation with human epithelial cells is associated with enhanced internalization. Infect Immun 60(11):4945–4949
7. Konkel ME, Christensen JE, Dhillon AS et al (2007) *Campylobacter jejuni* strains compete for colonization in broiler chicks. Appl Environ Microbiol 73(7):2297–2305. doi:10.1128/AEM.02193-06
8. Konkel ME, Kim BJ, Rivera-Amill V et al (1999) Bacterial secreted proteins are required for the internaliztion of *Campylobacter jejuni* into cultured mammalian cells. Mol Microbiol 32(4):691–701
9. Hornef MW, Wick MJ, Rhen M et al (2002) Bacterial strategies for overcoming host innate and adaptive immune responses. Nat Immunol 3(11):1033–1040. doi:10.1038/ni1102-1033
10. Neal-McKinney JM, Konkel ME (2012) The *Campylobacter jejuni* CiaC virulence protein is secreted from the flagellum and delivered to the cytosol of host cells. Front Cell Infect Microbiol 2:31. doi:10.3389/fcimb.2012.00031
11. Vaudaux P, Waldvogel FA (1979) Gentamicin antibacterial activity in the presence of human polymorphonuclear leukocytes. Antimicrob Agents Chemother 16(6):743–749
12. Weingarten RA, Grimes JL, Olson JW (2008) Role of *Campylobacter jejuni* respiratory oxidases and reductases in host colonization. Appl Environ Microbiol 74(5):1367–1375. doi:10.1128/AEM.02261-07
13. Konkel ME, Larson CL, Flanagan RC (2010) *Campylobacter jejuni* FlpA binds fibronectin and is required for maximal host cell adherence. J Bacteriol 192(1):68–76. doi:10.1128/JB.00969-09
14. Christensen JE, Pacheco SA, Konkel ME (2009) Identification of a *Campylobacter jejuni*-secreted protein required for maximal invasion of host cells. Mol Microbiol 73(4):650–662. doi:10.1111/j.1365-2958.2009.06797.x
15. Samuelson DR, Eucker TP, Bell JA et al (2013) The *Campylobacter jejuni* CiaD effector protein activates MAP kinase signaling pathways and is required for the development of disease. Cell Commun Signal 11:79. doi:10.1186/1478-811X-11-79
16. Buelow DR, Christensen JE, Neal-McKinney JM et al (2011) *Campylobacter jejuni* survival within human epithelial cells is enhanced by the secreted protein CiaI. Mol Microbiol 80(5):1296–1312. doi:10.1111/j.1365-2958.2011.07645.x
17. Neal-McKinney JM, Christensen JE, Konkel ME (2010) Amino-terminal residues dictate the export efficiency of the *Campylobacter jejuni* filament proteins via the flagellum. Mol Microbiol 76(4):918–931. doi:10.1111/j.1365-2958.2010.07144.x
18. Konkel ME, Garvis SG, Tipton SL et al (1997) Identification and molecular cloning of a gene encoding a fibronectin-binding protein (CadF) from *Campylobacter jejuni*. Mol Microbiol 24(5):953–963
19. Hu L, Kopecko DJ (1999) *Campylobacter jejuni* 81-176 associates with microtubules and dynein during invasion of human intestinal cells. Infect Immun 67(8):4171–4182
20. Konkel ME, Samuelson DR, Eucker TP et al (2013) Invasion of epithelial cells by *Campylobacter jejuni* is independent of caveolae. Cell Commun Signal 11:100. doi:10.1186/1478-811X-11-100

Chapter 12

Assays to Study the Interaction of *Campylobacter jejuni* with the Mucosal Surface

Marguerite Clyne, Gina Duggan, Ciara Dunne, Brendan Dolan, Luis Alvarez, and Billy Bourke

Key words *Campylobacter jejuni*, Tissue culture, pIVOC, HCT-8, HT-29MTX

1 Introduction

Mucosal colonization and overcoming the mucosal barrier are essential steps in the establishment of infection by *Campylobacter jejuni*. The interaction between *C. jejuni* and host cells, including binding and invasion, is thought to be the key virulence factor important for pathogenesis of *C. jejuni* infections in animals or humans. The intestinal mucosal barrier is composed of a polarized epithelium covered by a thick adherent mucus gel layer. There is a requirement for cell culture assays of infection to accurately represent the in vivo mucosal surface. In this chapter, we describe the use of a number of cell culture models and the use of polarized in vitro organ culture to examine the interaction of *C. jejuni* with mucosal surfaces.

1.1 Tissue Culture Models

Well-characterized cell culture-based models of infection play a key role in identifying and investigating host-microbe interactions, and a wide variety are in use. HCT-8 cells, a human ileocecal cell line, have been used to study the interactions of enterovirulent bacteria with human intestinal epithelial cells (reviewed by Lieven le Moal and Servin, 2013) [1]. We have used these cells to show that chicken intestinal mucin can attenuate *C. jejuni* virulence [2] and also to investigate the effects of mucosal reactive oxygen species on *C. jejuni* pathogenicity [3].

The original version of this chapter was revised. An erratum to this chapter can be found at DOI 10.1007/978-1-4939-6536-6_22.

James Butcher and Alain Stintzi (eds.), *Campylobacter jejuni: Methods and Protocols*, Methods in Molecular Biology, vol. 1512, DOI 10.1007/978-1-4939-6536-6_12, © Springer Science+Business Media New York 2017

HT29 cells, a colonic adenocarcinoma cell line, are composed of columnar absorptive cells with a small proportion of goblet cells [4, 5]. A number of mucus-secreting subclones have been generated from HT29 cells, and these cells provide models to investigate the role of mucus in *C. jejuni* infection. The HT29-MTX mucus-secreting subclone of the HT29 cell line was isolated following treatment with methotrexate [6]. These cells differentiate into a mixed population of enterocytes (50%) and goblet cells (50%) [6] and when grown on Transwell filters over a period of 21 days form a polarized monolayer which secretes mucins and other mucus gel layer components into the culture medium [7]. HT29-MTX-E12 cells are a subclone isolated from the heterogeneous HT29-MTX population. These cells were selected on the basis of development of a confluent monolayer, formation of tight junctions, and the production of an adherent mucus gel layer [8]. HT29-MTX-E12 cells form a polarized monolayer when grown on Transwell filters consisting of mainly mature goblet cells which secrete a continuous mucus layer [9].

1.2 Culture of *C. jejuni*

Prior to use in total association or invasion assays, *C. jejuni* may be cultured on solid, in liquid, or biphasic media. For cultivation of fresh isolates, optimal growth is obtained using biphasic medium supplemented with fetal bovine serum (FBS) [10]. A low passage number is essential to retaining the spiral morphology and high motility of *C. jejuni* isolates. *C. jejuni* may be cultured at either 37 °C or the avian body temperature of 42 °C.

1.3 Adherence and Invasion of Epithelial Cells by *C. jejuni*

To examine the adherence to and invasion of epithelial cells by *C. jejuni*, total association and gentamicin protection assays are used. Following infection of model epithelial cells, gentamicin is used to kill extracellular bacteria. After gentamicin treatment, cells are lysed to expose intracellular bacteria. These bacteria, protected from gentamicin, are enumerated by plating serial dilutions of the lysate on agar plates to determine the numbers of bacteria internalized by the cells. Total association assays determine the number of both adherent and intracellular bacteria associated with the cells (i.e., bacterial counts following cell lysis without the use of gentamicin).

1.4 Measurement of Transepithelial Electrical Resistance

The integrity of polarized cell cultures before and after infection with *C. jejuni* can be determined by measuring the transepithelial electrical resistance (TER). TER measurements are performed throughout the period of cell culture, e.g., days 7, 14, and 21, to monitor the development of the polarized monolayer. The TER should increase over time; high TER values indicate the development of tight junctions and cell polarization.

1.5 Preservation of Mucus Layer in the Liver for Microscopy

Certain fixation methods such as formalin fixation can result in destruction of the mucus layer. The HT29-MTX-E12 cells produce an adherent mucus layer after 21 days growth on Transwell filters (Fig. 1). Snap freezing the cells on a Transwell filter in

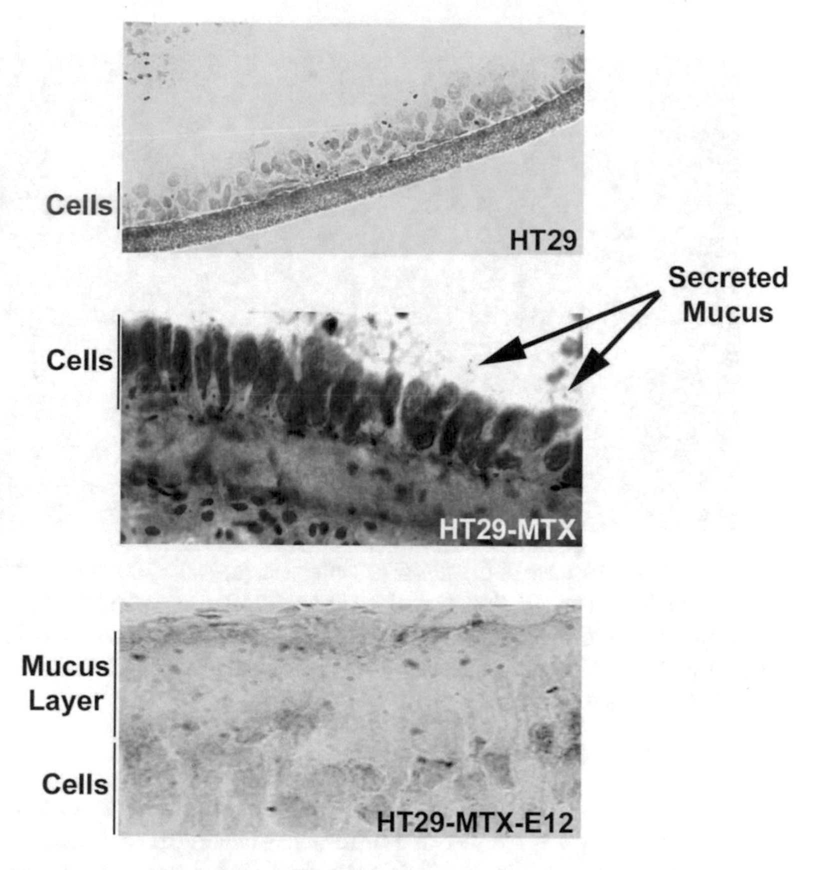

Fig. 1 Mucus secretion and mucus layer formation in HT29 cells and their mucus-producing subclones. HT29, HT29-MTX, and HT29-MTX-E12 cells were grown on Transwell filters, wrapped in chicken liver, snap frozen and mounted in OCT medium. Sections were stained with Alcian blue and counterstained with neutral red. Micrographs show the absence of mucus secretion in HT29 cells, the presence of secreted mucus in HT29-MTX cells, and the formation of a thick adherent mucus gel layer in HT29-MTX-E12 cells

chicken liver allows for the mucus layer to remain intact and subsequently be embedded in OCT compound and sectioned using a cryostat. Cells can then be stained using standard methods for frozen tissue sections.

1.6 Differential Staining of Adherent and Intracellular C. jejuni Bacteria

Differential immunofluorescent microscopy is another method for examining *C. jejuni* interaction with epithelial cells and distinguishing between extracellular and invasive bacteria. The method we use is to infect cells with fluorescently labeled bacteria and subsequently probe the infected cells with a *C. jejuni*-specific antibody that fluoresces at a different wavelength from the dye used to label the infecting bacteria [11] allowing discrimination between the adherent and intracellular bacterial populations. Carboxytetramethylrhodamine (TAMRA) is the fluorescent dye we use to label the bacteria. Following infection with TAMRA-labeled *C. jejuni*, cells are probed

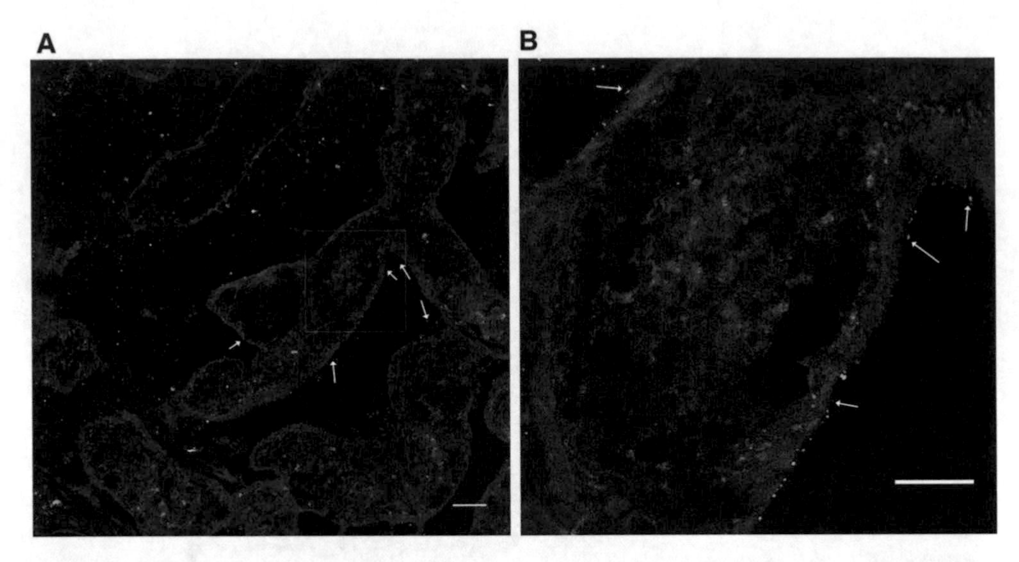

Fig. 2 Immunofluorescence of *C. jejuni* pIVOC infection. (**a**) 700 × 700 μm Overview of a stained pIVOC following *C. jejuni* infection for 3 h. Nucleus stained in *blue* (DAPI), Duox1/2 stain in green, and *C. jejuni* in *red*. *White arrows* indicate the *C. jejuni* adhering to the epithelial cells. Scale bar 50 μm. Zoomed section indicated with a square. (**b**) 150 × 150 μm zoomed view of the biopsy, showing the *C. jejuni* in *red* (*white arrows*) adhering to the epithelium. Scale bar 25 μm

with an anti-*C. jejuni* antibody and a secondary antibody that fluoresces at a different wavelength to TAMRA and cannot penetrate the eukaryotic cells. In this manner, adherent extracellular bacteria that are labeled with the red TAMRA stain and a FITC or Alexa Fluor 488-labeled antibody appear yellow (red from TAMRA, green from FITC-labeled or Alexa Fluor 488-labeled antibody), while intracellular bacteria appear red only.

1.7 Polarized In Vitro Organ Culture (pIVOC) with Intestinal Biopsies

pIVOC allows examination of biopsy intestinal mucosal tissue ex vivo as a proxy for pathogenesis studies in vivo. By challenging only the mucosal side of the biopsy, this methodology, originally described by Schuller et al. for the use with *E. coli* [12], aims to more accurately mimic the intestinal niche (Fig. 2). In our hands, even the mucous layer and its attendant microbiota are intact.

2 Materials

2.1 Culture of C. jejuni

1. Universal tubes, pipette gun, and pipettes.
2. Mueller-Hinton agar (MH agar), Mueller-Hinton broth (MH broth), and deionized H_2O (dH_2O).
3. Autoclave tape, airtight containers, petri dishes, sterile inoculation loops, and sterile swabs.
4. CampyGen™ microaerobic atmosphere generation system.

5. 25 cm^2 culture flasks, shaking incubator.
6. Skirrows antibiotic supplement (vancomycin 10 mg/L, polymyxin B 2–5 IU/mL, and trimethoprim 5 mg/L) (optional) [13].

2.2 Total Association and Invasion Assays

1. Tissue culture media (Table 1).
2. Tissue culture dishes (6, 12, or 24 wells).
3. 0.4 or 3 μm Transwell filters.
4. 0.4% trypan blue solution.
5. Hemocytometer.
6. MH agar and MH broth.
7. Petri dishes.
8. CampyGen™ gas packs for generation of microaerophilic conditions.
9. 400 μg/mL gentamicin sulfate.
10. Sterile dH_2O.
11. Sterile Dulbecco "A" Phosphate buffered saline pH 7.3 (137 mM NaCl, 3 mM KCl, 8 mM Na_2HPO_4, 1.5 mM KH_2PO_4).
12. 0.1% (vol/vol) Triton X-100 (dilute 10 μL Triton X-100 in 10 mL sterile PBS).
13. 12 mm Transwell® with 0.4 μm or 3.0 μm pore polycarbonate membrane insert.
14. Trypsin-EDTA.
15. EVOMX ohm meter and Endohm-12 probe to measure transepithelial resistance (World Precision Instruments).
16. 70% ethanol.

2.3 Differential Staining of Adherent and Intracellular *C. jejuni*

1. Tissue culture media (Table 1).
2. Tissue culture dishes.
3. Glass cover slips or Transwell filters (0.4 or 3.0 μm).
4. Glass slides.
5. Pencil for labeling glass slides.
6. MH agar and broth.
7. Log-phase culture of *C. jejuni*.
8. Carboxytetramethylrhodamine (TAMRA, Sigma) (*see* **Note 1**).
9. PBS Dulbecco "A" pH 7.3 (137 mM NaCl, 3 mM KCl, 8 mM Na_2HPO_4, 1.5 mM KH_2PO_4).
10. 4% (vol/vol) formalin solution (*see* **Note 2**).
11. Washing solution (PBST): PBS containing 0.05% (vol/vol) Tween 20.

Table 1
Cell lines used to study the interaction of *C. jejuni* with intestinal cells

Cells	Culture medium	Seeding conditions	Ref.
HCT-8 Human ileocecal adenocarcinoma cell line	RPMI 1640 supplemented with 10% FBS	Seeded onto 24-well PVDF plates at a density of 5×10^4 cells per well for infection assays. Cells grown overnight at 37 °C, with 5% CO_2 in a humidified atmosphere	[2]
INT 407 Human embryonic intestinal epithelial cells	MEM supplemented with 10% FBS	24-well tissue culture tray seeded with 1.5×10^5 cells per well and incubated for 18 h	[14]
CaCo2 Human colonic carcinoma cells	Eagle's minimal essential medium (EMEM) supplemented with 10% FBS	Transwell filter units containing 0.33-cm^2 porous membranes (3.0 μm pore size) seeded with 1.5×10^5 Caco-2 cells placed in a well of a 24-well tissue culture tray containing 1 mL of EMEM supplemented with 10% FBS. The media in the apical and basolateral cell chambers are changed every 3 days. The polarized epithelial cell monolayers are used after 10–14 days of incubation	[15]
T84 Human colonic epithelial cell line	A 1:1 mixture of Dulbecco's Modified Eagle Medium (DMEM) and Ham's F-12 medium containing 2.5 mM L-glutamine, 15 mM HEPES, and 0.5 mM sodium pyruvate supplemented with 5% (vol/vol) FBS	2×10^5 were seeded onto Transwell filters (0.33 cm^2 with 3.0 μm pore size). T84 cells used when transepithelial resistance reached 1400/cm^2 or higher	[16]
HT29 Human colon adenocarcinoma cell line	HT29 cells were grown in McCoy's 5A modified medium supplemented with 10% FBS	Seeded onto 24-well plates or onto Transwell inserts (0.4 μm pore size, 12 mm diameter) at a density of 1×10^5 cells/well for 48 h on 24-well plates prior to use and up to 21 days on Transwell inserts	[17]
HT29-MTX-E12 methotrexate (MTX)-adapted HT29 cells form tight junctions and adherent mucus layer	Dulbecco's Modified Eagle's Medium (DMEM) supplemented with 10% fetal bovine serum (FBS), 2 mM L-glutamine, and 1% nonessential amino acids	Seeded onto Transwell inserts (0.4-μm pore size, 12-mm diameter) at a density of 1×10^5 cells/well. Cells are grown up to 21 days on Transwell inserts	[18]

12. PBS containing 1% (wt/vol) bovine serum albumin (*see* **Note 3**).
13. Blocking solution: PBS containing 1% (wt/vol) BSA and 20% (vol/vol) goat serum (*see* **Note 3**).
14. Rabbit raised anti-*C. jejuni* antibody (*see* **Note 4**).
15. DAPI (4′,6-diamidino-2-phenylindole) (*see* **Note 5**).
16. Goat anti-rabbit secondary antibody conjugated to Alexa Fluor 488 (Life Technologies) (*see* **Note 6**).
17. Fluorescent mounting medium.

2.4 Preservation of Mucus Layer in the Liver for Microscopy

1. Sterile PBS, Dulbecco "A," pH 7.3 (137 mM NaCl, 3 mM KCl, 8 mM Na2HPO4, 1.5 mM KH2PO4).
2. Single edge razor blade.
3. Disposable scalpel.
4. Chicken liver (*see* **Note 7**).
5. Parafilm® M.
6. Liquid N_2.
7. 2-Methylbutane (isopentane).
8. Intermediate Cryomold® (Tissue-Tek).
9. Optimal cutting temperature (OCT) medium.

2.5 pIVOC Mounting Components

1. 55 mm qualitative filter paper.
2. Whatman cellulose nitrate membrane 0.3 μm filters.
3. Ibidi μ-Dish 35 mm, high (Cat. No. 81156).
4. Dissecting microscope.
5. Straight probe (F•S•T 10140-01, 0.25 mm tip diameter).
6. Tissue adhesive with cannula (Henkel Indermil xfine).
7. Surgical Scalpel Blade No.15 sterile (Swann-Morton).
8. Microdissecting forceps, slightly curved.

2.6 Biopsy Culture Media

1. DMEM/F12 at 4 °C for transport from the operating theater to the laboratory.
2. DMEM/F12 complemented with 5% FBS.
3. RPMI complemented with 3% FBS.

2.7 Materials for Immunofluorescent Staining of pIVOC Cultures

1. PBS with 0.2% Triton X for washing (PBST).
2. PBST supplemented with blocking agent (BSA, serum, etc.). The exact percentage of blocking agent is antibody specific and must be determined empirically.
3. Primary and secondary antibodies.
4. DNA staining reagent (DAPI, Hoechst, DRAQ5, etc. depending on the wavelength needed).

5. Formaldehyde solution 36–38% in H_2O.
6. Mowiol 4–88 solution for mounting (Sigma 81381).
7. Microscopy slides.
8. Microscopy cover slips 0.14 mm.
9. Optional: *Campylobacter* can be TAMRA stained prior to infection (*see* Subheading 3.9, **steps 2–7** for details) (*see* **Note 8**).

3 Methods

3.1 Culture of C. jejuni

C. jejuni strains are routinely grown on MH agar.

1. Using a sterile loop, take a small amount of bacteria from frozen stocks and streak onto a fresh agar plate under aseptic conditions (*see* **Note 9**).
2. Incubate the agar plate at either 37 or 42 °C for 48 h in a sealed airtight container under microaerobic conditions. These conditions can be generated by placing specialized CampyGen™ microaerobic gas packs in the airtight container. Alternatively, the inoculated plates can be placed in a microaerobic chamber (*see* **Note 10**).
3. For liquid cultures, *C. jejuni* may be inoculated from agar into a sterile 50 mL or universal tube containing approximately 10 mL of MH broth using a sterile cotton swab and placed in a shaking incubator at 37 °C or 42 °C at 200 rpm under microaerobic conditions (*see* **Notes 11** and **12**).
4. For growth in biphasic media, pipette approximately 7 mL of molten agar into a sterile 25 cm^2 tissue culture flask with a vented cap and prop it up at a 30–45° angle and allow to cool in a laminar flow hood. Once solidified, overlay with 7 mL of broth and inoculate from agar, as for liquid cultures above, and place standing in an incubator under microaerobic conditions.
5. If contamination is a problem, the medium (agar or liquid) can be supplemented with antibiotics as follows, vancomycin 10 mg/L, polymyxin B 2–5 IU/mL, and trimethoprim 5 mg/L (Skirrows supplement).

3.2 Routine Culture of Epithelial Cell Lines (e.g., HCT-8 Cells)

1. Grow cells in appropriate medium (Table 1) supplemented with 5–10% FBS. Cultures are incubated at 37 °C with 5% CO_2 in 25 cm^2 or 75 cm^2 plastic cell culture flasks.
2. When cells reach 90–100% confluence (*see* **Note 13**), remove cells for passaging by replacing culture medium with 0.25% trypsin-EDTA and incubate for 15–20 min at 37 °C.
3. After the cells are released from the plastic, pellet them by centrifugation for 5 min at 240×*g*, resuspend in tissue culture medium containing FBS, and passage by diluting them 1:10–1:20.

3.3 Cell Counting

1. Trypsinize cells from tissue culture flasks, and after resuspending the cells in tissue culture medium, count the number of viable cells you have using trypan blue staining and a hemocytometer.
2. Dilute the cell sample in trypan blue by preparing a 1:1 dilution of the cell suspension using a 0.4% trypan blue solution. Nonviable cells will be blue; viable cells will be unstained (*see* **Note 14**).
3. Carefully fill the hemocytometer chamber using a micropipette (*see* **Note 15**).
4. Focus on the grid lines of the hemocytometer using the 10× objective of the microscope. Focus on one set of 16 corner squares.
5. Count the number of cells in this area of 16 squares. When counting, only count live cells that are not stained blue (*see* **Note 16**). Move the hemocytometer to another set of 16 corner squares and carry on counting until all 4 sets of 16 corner squares are counted. The number of cells in one set of 16 corner squares is equivalent to the number of cells $\times 10^4$/mL. Therefore, to obtain the count:

   (a) Calculate the total number of cells from 4 sets of 16 corner squares.

   (b) Divide the count by 4 to get the average number of cells per corner square.

   (c) Then multiply by 2 to adjust for the 1:2 dilution in trypan blue.

   (d) Multiply the cell count by 10^4 to get the number of cells per mL.

3.4 Culture and Infection of Epithelial Cells Grown as Monolayers in Multiwell Dishes

1. After counting of cells, adjust the cell suspension to the required cell density (Table 1).
2. Seed cells onto wells of a 6-well, 12-well, or 24-well plate. To each well, add 2 mL of tissue culture medium for 6-well plate, 1 mL for 12-well plate, and 0.5 mL for 24-well plate. Incubate cells at 37 °C and 5% CO_2 until cells are 70–80% confluent. Prior to infection, ensure cells are in antibiotic-free medium (*see* **Note 17**).
3. Harvest bacteria from broth/biphasic medium, wash once in tissue culture media, and resuspend in tissue culture medium. Dilute bacterial culture suspensions using tissue culture medium so that the optical density of the suspension is 0.2 at 600 nm ($OD_{600} = 0.2$).
4. Dilute bacteria in tissue culture medium to achieve the target multiplicity of infection (MOI) (*see* **Note 18**). Typical MOIs for adhesion and invasion assays range from 100 to 1000.

5. Gently add tissue culture medium containing the bacterial inoculum to the cells.
6. Place cell dishes in an airtight container and incubate at 37 °C under microaerophilic conditions generated using a CampyGen™ gas pack.

3.5 Culture and Infection of Polarized Epithelial Cells on Transwell® Filter Supports

For infection with bacteria on the apical surface, the cells should be grown on either 0.4 or 3 μm filters. For infections with bacteria on the basolateral surface, cells must be grown on 3 μm filters.

1. To grow polarized monolayers, add cells at the required cell density (Table 1) in 0.5 mL of medium to the apical side of the Transwell filter placed in a well of a 12-well dish (*see* **Note 19**).
2. Add 1.5 mL of medium to the basolateral chamber and culture cells at 37 °C with 5% CO_2 until the cells have polarized as determined by measuring the transepithelial resistance using an EVOMAX meter coupled to an Endohm-12 chamber (World Precision Instruments) (*see* Subheading 3.6).
3. Harvest bacteria from broth/biphasic medium, wash once in tissue culture media, and resuspend in tissue culture medium. Dilute bacterial culture suspension using tissue culture medium so that the $OD_{600} = 0.2$.
4. Dilute bacteria in tissue culture medium to achieve the target MOI (*see* **Note 18**).
5. Remove tissue culture medium from the apical surface of the cells and replace with 150 μL of bacterial suspension. For infection of the basolateral surface, add the bacterial inoculum to the lower chamber.
6. Place cell dishes in an airtight container, and incubate at 37 °C under microaerophilic conditions generated using a CampyGen™ gas pack (*see* **Note 20**).

3.6 Measurement of Transepithelial Electrical Resistance

1. Sterilize the chamber of the Endohm-12 probe with 70% ethanol and then wash with sterile tissue culture media.
2. Add 1.5 mL of tissue culture media to the Endohm-12 chamber, transfer a sterile Transwell® filter to the chamber using a sterile forceps, and add 0.5 mL of media to the apical surface of the filter.
3. Replace the cap on the Endohm-12 and measure the electrical resistance of the blank filter.
4. Place a cell culture in the Endohm-12 chamber and add 0.5 mL media to the apical surface, replace the cap on the chamber, and measure the resistance of the culture (filter + cells).
5. Transfer the cell culture into its culture dish and place in incubator.

6. To calculate the TER of the monolayer, use the following formula:

(Resistance of culture – Resistance of blank filter) × surface area of filter. Results are presented as Ω/cm2

3.7 Gentamicin Protection Assay

1. After incubation of cells with bacteria for the desired time point, wash cells with tissue culture medium to remove nonadherent bacteria (*see* **Note 21**).
2. Replace tissue culture medium with medium containing 400 μg/mL gentamicin sulfate and incubate under microaerophilic conditions for 2–3 h to kill extracellular bacteria.
3. Wash cells with sterile PBS and lyse with 300 μL of 0.1% (vol/vol) Triton X-100 in PBS for 15 min at 37 °C.
4. Make serial tenfold dilutions of this lysate and spread 100 μL of the dilution on MH agar plates using a sterile inoculation spreader (*see* **Note 22**).
5. Incubate plates for 48–72 h at 37 °C under microaerobic conditions, after which bacterial colony-forming units will have formed, and the number of colony-forming units per mL can be enumerated.

3.8 Total Association Assay

1. After incubation of cells with bacteria for the desired time point (Subheading 3.7, **step 1** above), wash cells with sterile PBS and lyse with 300 μL of 0.1% (vol/vol) Triton X-100 in PBS for 15 min at 37 °C.
2. Make serial tenfold dilutions of this lysate and spread 100 μL of dilution on MH agar plates using a sterile inoculation spreader (*see* **Note 22**).
3. Incubate plates for 48–72 h at 37 °C under microaerophilic conditions, after which bacterial colony-forming units will have formed and the number of colony-forming units per mL can be enumerated.

3.9 Differential Staining of Adherent and Intracellular *C. jejuni* Bacteria

1. Seed cells on Transwell filters or on glass cover slips and grow until 70–80% confluent (*see* **Note 23**).
2. Culture *C. jejuni* on MH agar plates and in MH broth or biphasic medium as described above. Subculture bacteria on day of infection in order to obtain log-phase culture.
3. To label *C. jejuni*, harvest bacteria from broth/biphasic medium and dilute bacterial culture to $OD_{600} = 0.2$.
4. Take 1 mL of diluted bacterial culture and pellet by centrifugation at 6500 × *g* for five minutes.
5. Decant supernatant and resuspend the pellet in 1 mL sterile PBS.

6. Add 1 μL of 10 mg/mL TAMRA stock solution to the bacterial suspension in PBS. Mix thoroughly by pipetting and incubate the bacteria for 30 min at 37 °C in the dark (wrap sample in tube in tinfoil).
7. Spin down the bacteria by centrifugation at 2500 × *g* for 5 min, remove the supernatant, and resuspend the bacterial pellet in PBS. Repeat this step twice more and then finally resuspend the bacterial pellet in 1 mL tissue culture media.
8. Infect cells with TAMRA-labeled bacteria at desired MOI for selected time point at 37 °C under microaerophilic conditions generated using CampyGen™ gas packs.
9. Following infection, wash cells in sterile PBS to remove non-adherent bacteria and fix cells by incubating infected cells in 500 μL of 4% formalin for 1 h at room temperature.
10. Block cells by incubating in blocking solution for 1 h at room temperature. Add enough blocking solution to fully cover the monolayer.
11. To detect adherent bacteria, probe cells with a rabbit anti-*C. jejuni* antibody diluted in blocking solution for 1 h at room temperature (*see* **Note 24**)
12. Remove antibody solution and wash cells three times in PBST (*see* **Note 25**).
13. Incubate the cells with the secondary antibody solution for 1 h at room temperature in the dark (*see* **Note 26**).
14. Wash cells three times in PBST.
15. Mount the cover slips (*see* **Note 27**) or Transwell filters (*see* **Note 28**) on slides in fluorescent mounting medium.
16. Visualize the cells using a fluorescent microscope. Adherent bacteria will appear yellow as they have been stained with both TAMRA (red) and the anti-*C. jejuni* antibody (green), while intracellular bacteria appear red only.

3.10 Preservation of Mucus Layer on HT29 Cell Derivatives

1. Remove medium and gently wash the monolayer with PBS.
2. Cut wafer-thin strips of chicken liver with a razor blade (*see* **Note 7**).
3. Carefully remove the Transwell filter from its support using scalpel and sandwich between two strips of the liver.
4. Carefully wrap the liver/membrane sandwich in Parafilm® M and snap freeze in isopentane that has been precooled in liquid N_2.
5. Remove Parafilm® M and using a razor blade trim the edges of the frozen liver/membrane sandwich.
6. Place the liver/membrane sandwich in Intermediate Cryomold®, which contains some OCT, and cover with OCT. Take care to ensure that there are no bubbles.

7. Snap freeze in precooled isopentane.
8. Store samples at −20 °C until use.

3.11 Polarized In Vitro Organ Culture (pIVOC) with Intestinal Biopsies

1. *C. jejuni* should be grown in advance (*see* Subheading 3.1). Use fresh *C. jejuni* cultures harvested from agar plates to inoculate 50 mL of RPMI supplemented with 3% FBS. Growing conditions and size of the inoculum need to be adjusted to make sure to have *C. jejuni* in RPMI 3% FBS at 0.4 OD at the time of the intended pIVOC infection (*see* **Note 29**).
2. Standard mucosal biopsy forceps-derived biopsy specimens taken at the time of endoscopy should be put into transport medium and placed on ice. Biopsies should be transported directly to the laboratory. pIVOC should be prepared immediately following biopsy harvest (*see* **Note 30**).
3. Add 2 mL of DMEM/F12 with 5% FBS to one well of a sterile 6-well plate and keep it in microaerophilic conditions (*see* **Notes 31** and **32**).
4. Make a single orifice in the middle of the μ-Dish with the straight probe.
5. Place a filter paper on the dissecting microscope and add 1 mL of cold (4 °C) DMEM/F12.
6. Place one biopsy in the middle of the filter paper (*see* **Note 33**).

All the following steps are performed under the dissecting microscope:

7. Orient the mucosal side of the biopsy up by gently spreading it on the filter paper.
8. Place a Whatman membrane on the filter paper and place the oriented biopsy mucosal side upon the membrane (*see* **Note 34**).
9. Position the μ-Dish with its orifice above the biopsy, and with the No. 15 scalpel, adjust the opening to the size of the biopsy.
10. Irrigate the biopsy below periodically as you adjust the opening in the μ-Dish (*see* **Note 35**).
11. Once the opening is adequate, add some tissue adhesive on the μ-Dish about 1 mm from the opening in a circular pattern (*see* **Note 36**).
12. Briefly take the media surrounding the biopsy with a filter paper.
13. Immediately place the μ-Dish on top of the spread and oriented biopsy (*see* **Note 37**).
14. With the curved forceps, ensure proper sealing of the tissue between the μ-Dish and the Whatman membrane but gently pressing to the sides of the biopsy.

15. Add some cold media (4 °C) on the top of the μ-Dish as soon as the tissue adhesive has started to set.
16. Place the mounted biopsy on the 6-well plate profiled with DMEM/F12 and replace the top cold media with 37 °C preconditioned RPMI 3% FBS (2–3 mL), and return the 6-well plate to the microaerophilic incubator.
17. Repeat **steps 8 to 19** with each biopsy.
18. After all biopsies are mounted, replace the top media with 1 mL of fresh RPMI 3% FBS warmed to 37°C.
19. Prepare the *C. jejuni* for infection by adjusting the OD_{600} to 0.4 in RPMI 3% FBS.
20. Add 1 mL of these bacteria on top of the μ-Dish (1:1 dilution) for a final volume of 2 mL of *C. jejuni* at OD_{600} 0.2.
21. Place the 6-well plate on a centrifuge and spin it for 5 min at $200 \times g$.
22. Place the pIVOC back in microaerophilic conditions for the intended infection time (this will depend on the pathogenicity of the organism used—typically use a 3 h incubation for *C. jejuni*).
23. Fix the samples by pipetting 36% formaldehyde (~0.2 mL) directly into the top of the μ-Dish for a final concentration of 3–4% formaldehyde in microaerophilic conditions for 20 min.
24. Bring the samples to room temperature.
25. Take the μ-Dishes and cut out a circle in the bottom of the μ-Dish around the biopsy with a scalpel.
26. Wash five times with PBST making sure the pIVOC is completely submerged (*see* **Note 38**).
27. Permeabilize the tissue with PBS 0.2% Triton X for 20 min.
28. Wash five times with PBS then one time with PBST (*see* **Note 38**).
29. Block for 1 h at room temperature.
30. Apply primary antibody (incubation here is antibody dependent and needs to be determined empirically for each antibody and sample).
31. Wash five times with PBST.
32. Block for 1 h at room temperature with PBST supplemented with blocking agent.
33. Incubate with secondary antibody and DNA stain for 1 h at room temperature following the manufacturer recommendations as a starting point (*see* **Note 39**).
34. Wash five times with PBST ensuring enough PBST is used to completely cover biopsy for each wash (*see* **Note 38**).

35. If the Whatman filter is still attached, carefully remove it. At this point it is possible to unmount the biopsy from the remaining parts of the µ-Dish.
36. Place the biopsy mucosal side upon a microscopy slide.
37. Place a drop of Mowiol on top of the biopsy and a cover slip on top.
38. As soon as the Mowiol starts to set (approximately 2–3 min) but before the cover slip can compress the biopsy (this can be assessed by following the setting of Mowiol under the dissecting microscope), place the slide with the biopsy and cover slip upside down to allow the Mowiol to set without compressing the biopsy.
39. Image on an inverted microscope.

4 Notes

1. Prepare TAMRA by dissolving 10 mg powder in 1 mL methanol to make a stock concentration of 10 mg/mL. Store at −20 °C.
2. Dilute formaldehyde solution down to a 4% concentration in PBS. For example, if you have a stock solution of formaldehyde at a concentration of 37.5%, add 2.13 mL of this stock to 47.87 mL PBS to make 4% formalin.
3. Weigh out 100 mg of BSA. Add to 10 mL PBS and mix thoroughly until fully resuspended. Use this to make blocking solution, i.e., take 1.6 mL of PBS and 1% BSA solution and add 400 µL goat serum.
4. Different antibodies work at different concentrations which can range from 1/50 to 1/1000 for immunofluorescence. If working with a new antibody, perform optimization experiments first to determine the optimal concentration to use.
5. A nuclear stain is required to visualize the eukaryotic cells, and the most widely used stain is DAPI.

(a) To prepare the solution, weigh out 10 mg of DAPI powder and resuspend in 1 mL dH_2O to make a stock concentration of 10 mg/mL. Aliquot this out in 50 µL volumes and store at −20 °C.

(b) DAPI is used at a working concentration of 2 µg/mL. To prepare this, take 1 µL of 10 mg/mL stock and add to 4.999 mL of PBS and 1% BSA solution (1:5000 dilution).

6. To detect the primary antibody, cells then need to be probed with a secondary antibody conjugated to a fluorophore which fluoresces at a specific wavelength. The choice of secondary antibody is determined by what animal the primary antibody was raised in. For example, if the primary antibody is raised in

a rabbit, an anti-rabbit secondary antibody is required. Cells can be incubated with DAPI nuclear stain and the secondary antibody simultaneously. To prepare the secondary antibody at a 1:500 dilution, add 1 μL of goat anti-rabbit IgG secondary antibody conjugated to Alexa Fluor to 488–499 μL of PBS containing 1% BSA and 2 μg/mL DAPI.

7. Store the chicken liver at −20 °C and use while still frozen, as it is easier to cut wafer-thin slices when it is frozen.
8. If bacteria are TAMRA stained before infection, then there is no need to use primary and secondary antibodies to detect them after infection.
9. All bacterial culture procedures are carried out under aseptic conditions in a laminar flow hood or on the bench using a Bunsen burner.
10. Atmospheric generation gas packs take approximately 30 min to become fully activated. Microaerobic chambers are modified atmosphere workstations that allow for temperature, humidity, and atmospheric gas manipulation.
11. We find that broth-based cultures grown under agitation in universal tubes result in optimum growth of microaerophilic bacteria. This may be due to increased gaseous exchange taking place during shaking. To ensure good growth when inoculating from agar to liquid media, an initial OD_{600} of 0.05–0.2 is recommended.
12. When using universal or 50 mL tubes for liquid cultures, it is also advisable to loosen the cap and tape it in place to allow the shaking culture access to the surrounding microaerobic atmosphere. For infection and invasion assays, bacteria should be grown on MH agar for 48 h prior to inoculation into liquid media overnight and subcultured to obtain log-phase bacteria the following day.
13. Examine cells using an inverted microscope to determine confluency. Some cells grow rapidly and could be confluent within 24 h; others grow more slowly.
14. Make sure the cell suspension to be counted is well mixed. Mix 100 μL of cells with 100 μL of 0.4% trypan blue solution and mix gently. Allow cells and stain to mix for 1–2 min prior to counting to allow the cells to take up the stain.
15. Take care not to overfill the chamber.
16. You might want to use a hand tally counter for convenience. When counting, always count only live cells that look healthy (unstained by trypan blue).
17. Wash cells thoroughly with tissue culture medium and replace medium with antibiotic-free medium 24 h before infection.

18. To determine the MOI, the number of eukaryotic and bacterial cells must be determined. To calculate the number of eukaryotic cells present at time of infection, cells are cultured for required amount of time and then trypsinized and counted to determine the number of cells. To calculate the number of bacterial cells, the culture is diluted to the specific optical density, and then serial dilutions of the culture are plated on MH agar plates which are incubated at 37 °C under microaerophilic conditions for 48–72 h to determine the number of bacterial colony-forming units (CFUs) present. For example, if there are 1×10^6 epithelial cells present, an MOI of 500 requires 5×10^8 *C. jejuni* cells. An OD_{600} of 0.2 is approximately 1×10^9 *C. jejuni* cells/mL; therefore, cells are infected with 500 μL of the culture.
19. Use sterile forceps to handle the Transwell filter support when placing in the culture dish.
20. Anaeropak rectangular jars made by the Mitsubishi Gas Chemical Co. Inc. are ideal for this purpose.
21. Wash cells gently and use medium prewarmed to 37 °C.
22. Make sure that bacterial suspensions are well mixed prior to plating to prevent the bacteria clumping. Sometimes the bacteria (depending on the strain) will swarm on the agar plate. If this occurs, increase the percentage of agar in the plate.
23. When using cover slips, the slips will often need to be sterilized first. Pick up a cover slip with tweezers, dip in absolute ethanol, and then allow the ethanol to evaporate off in a laminar flow hood. When the cover slip is fully dry, place one cover slip in each well of the tissue culture dish.
24. If you don't have a lot of antibody, 50 μL of antibody solution will be enough to cover cells on a cover slip. Therefore, if you make up 200 μL of primary antibody solution, this will be enough to cover four cover slips. Gently pipette the antibody solution onto the cover slip, ensuring it covers all the cells, and then incubate the cells in a humidified environment to ensure that the antibody solution does not evaporate off.
25. If the cells are seeded on cover slips in a tissue culture dish, wash cells by pouring PBS down the side of the well and moving it gently over the cells. Some cells can easily detach from the cover slip under abrasive circumstances, such as if PBS is pipetted directly at the cells.
26. Once cells have been probed with a secondary antibody conjugated to a fluorophore, it is important to remember to always keep the cells in the dark, as the secondary antibody is light sensitive and its signal will fade if cells are exposed to light.
27. The use of fluorescent mounting medium ensures that the cells do not dry out. To mount cover slips, take the cover slip out of

the tissue culture dish. A sharp needle is often useful in initially dislodging the cover slip, and tweezers may then be used to handle it. Place a small drop of mounting medium on a glass slide and then place the cover slip with the cells facing down in the medium. Ensure no air bubbles are present. Seal the cover slip with clear nail polish and allow to dry for 10–15 min.

28. For cells grown on Transwell filters, cut the filter from the Transwell support using a scalpel blade. Place the Transwell filter cell side up onto a labeled glass slide. Place a drop of fluorescent mounting medium onto a glass cover slip and place the cover slip on top of the Transwell filter on the glass slide. Press gently on the cover slip to spread the mounting medium all over the cover slip. Seal the edges of the cover slip with a clear nail polish solution. Allow the nail polish and mounting medium to set before examining the cells under a fluorescent microscope.
29. *C. jejuni* cultures should be optimal (exponential growth phase) before infection. If needed, confirm purity with gram staining. Make sure that *C. jejuni* does not go below 25 °C at any time; otherwise, infection will not take place.
30. The time between collection of the biopsies and mounting should be kept to a minimum. Do not wait until all the biopsies are collected to start.
31. Mounting of the individual biopsies is time-consuming. An experienced researcher should expect 15–20 min per mounting.
32. As infection will take place in microaerophilic conditions, make sure to have enough media preconditioned in microaerophilic conditions for the culture and infection.
33. Make sure that the biopsies are irrigated during the whole procedure by periodically adding cold media (4 °C) on top whenever they are not completely immersed in media.
34. Avoid using forceps with pointing ends as handling the biopsies with them might injure the tissue compromising the integrity of the barrier in pIVOC.
35. Biopsies should be handled in cold media (4 °C) until they are mounted and ready for culture. Only then should they be put back in warm media.
36. If the tissue adhesive is too close to the opening of the μ-Dish and the opening is not very big, tissue adhesive might migrate into the opening.
37. For the biopsy opening, you should try to maximize the opening in the μ-Dish, but there has to be enough tissue on all sides as to ensure there are no openings into the basal side of the tissue.
38. The volume used for washing depends on the specific apparatus present in each lab. It can be as small as 1 mL if done on a tray or 10 mL if done by immersion.

39. All antibody concentrations and incubations will need to be optimized empirically for the specific tissue and blocking agent used.

Acknowledgments

This work was supported by grants from The National Children's Research Center Dublin.

References

1. Liévin-Le Moal V, Servin AL (2013) Pathogenesis of human enterovirulent bacteria: lessons from cultured, fully differentiated human colon cancer cell lines. Microbiol Mol Biol Rev 77(3):380–439. doi:10.1128/MMBR.00064-12
2. Alemka A, Whelan S, Gough R et al (2010) Purified chicken intestinal mucin attenuates *Campylobacter jejuni* pathogenicity *in vitro*. J Med Microbiol 59(Pt 8):898–903. doi:10.1099/jmm.0.019315-0
3. Corcionivoschi N, Alvarez LA, Sharp TH et al (2012) Mucosal reactive oxygen species decrease virulence by disrupting *Campylobacter jejuni* phosphotyrosine signaling. Cell Host Microbe 12(1):47–59. doi:10.1016/j.chom.2012.05.018
4. Augeron C, Laboisse CL (1984) Emergence of permanently differentiated cell clones in a human colonic cancer cell line in culture after treatment with sodium butyrate. Cancer Res 44(9):3961–3969
5. Pinto M, Appay MD, Simonassmann P et al (1982) Enterocytic differentiation of cultured human-colon cancer-cells by replacement of glucose by galactose in the medium. Biol Cell 44(2):193–196
6. Lesuffleur T, Barbat A, Dussaulx E et al (1990) Growth adaptation to methotrexate of HT-29 human colon carcinoma cells is associated with their ability to differentiate into columnar absorptive and mucus-secreting cells. Cancer Res 50(19):6334–6343
7. Gouyer V, Wiede A, Buisine MP et al (2001) Specific secretion of gel-forming mucins and TFF peptides in HT-29 cells of mucin-secreting phenotype. Biochim Biophys Acta 1539(1-2):71–84
8. Behrens I, Stenberg P, Artursson P et al (2001) Transport of lipophilic drug molecules in a new mucus-secreting cell culture model based on HT29-MTX cells. Pharm Res 18(8):1138–1145
9. Dolan B, Naughton J, Tegtmeyer N et al (2012) The interaction of *Helicobacter pylori* with the adherent mucus gel layer secreted by polarized HT29-MTX-E12 cells. PLoS One 7(10), e47300. doi:10.1371/journal.pone.0047300
10. Rollins DM, Coolbaugh JC, Walker RI et al (1983) Biphasic culture system for rapid Campylobacter cultivation. Appl Environ Microbiol 45(1):284–289
11. Mooney A, Byrne C, Clyne M et al (2003) Invasion of human epithelial cells by *Campylobacter upsaliensis*. Cell Microbiol 5(11):835–847
12. Schuller S, Lucas M, Kaper JB et al (2009) The *ex vivo* response of human intestinal mucosa to enteropathogenic *Escherichia coli* infection. Cell Microbiol 11(3):521–530. doi:10.1111/j.1462-5822.2008.01275.x
13. Skirrow MB (1977) Campylobacter enteritis: a "new" disease. Br Med J 2(6078):9–11
14. Christensen JE, Pacheco SA, Konkel ME (2009) Identification of a *Campylobacter jejuni*-secreted protein required for maximal invasion of host cells. Mol Microbiol 73(4):650–662. doi:10.1111/j.1365-2958.2009.06797.x
15. Konkel ME, Mead DJ, Hayes SF et al (1992) Translocation of *Campylobacter jejuni* across human polarized epithelial cell monolayer cultures. J Infect Dis 166(2):308–315
16. Louwen R, Nieuwenhuis EE, van Marrewijk L et al (2012) *Campylobacter jejuni* translocation across intestinal epithelial cells is facilitated by ganglioside-like lipooligosaccharide structures. Infect Immun 80(9):3307–3318. doi:10.1128/IAI.06270-11
17. Lane JA, Marino K, Naughton J et al (2012) Anti-infective bovine colostrum oligosaccharides: *Campylobacter jejuni* as a case study. Int J Food Microbiol 157(2):182–188. doi:10.1016/j.ijfoodmicro.2012.04.027
18. Alemka A, Clyne M, Shanahan F et al (2010) Probiotic colonization of the adherent mucus layer of HT29MTXE12 cells attenuates *Campylobacter jejuni* virulence properties. Infect Immun 78(6):2812–2822. doi:10.1128/IAI.01249-09

Chapter 13

Characterization of Ligand–Receptor Interactions: Chemotaxis, Biofilm, Cell Culture Assays, and Animal Model Methodologies

Rebecca M. King and Victoria Korolik

Abstract

Chemotactic motility is an essential virulence factor for the pathogenesis of *Campylobacter* spp. infection. In Chapter 6, we described technologies that enable initial screening and identification of ligands able to interact with chemoreceptor sensory domains. These include amino acid and glycan arrays, NMR, and SPR that are utilized to identify potential ligands interacting with *Campylobacter jejuni*. Here we describe techniques that enable the characterization and evaluation of ligand–receptor binding in chemotaxis through the assessment of motility and directed chemotactic motility as well as the associated phenotypes—autoagglutination behavior, biofilm formation, ability to adhere and invade cultured mammalian cells, and colonization ability in avian hosts.

Key words Campylobacter, Motility, Chemotaxis, Agglutination, Biofilm, Adherence, Invasion, Animal colonization

1 Introduction

Campylobacter flagella motility is an important factor for the intestinal colonization of avian and mammalian hosts and subsequently for the invasion of intestinal epithelial cells [1, 2]. Furthermore, flagella-mediated motility of *Campylobacter* influences host colonization by promoting migration through viscous milieu such as gastrointestinal mucus [3]. Chemotaxis is a process whereby a chemotactic signaling system allows bacteria to follow favorable chemical gradients [4]. Chemoeffectors attracting bacterial cells to a point of high effector concentration are considered attractors, and chemotaxis in the direction of increasing attractor concentrations is defined as "positive" chemotaxis. In contrast, chemorepellent gradients cause bacteria to swim to points of lower chemoeffector concentrations, and this is considered as "negative" chemotaxis [2]. The role of chemotaxis in infection has been

James Butcher and Alain Stintzi (eds.), *Campylobacter jejuni: Methods and Protocols*, Methods in Molecular Biology, vol. 1512,
DOI 10.1007/978-1-4939-6536-6_13,

studied using mutants that are non-chemotactic, due to loss of core signal transduction proteins or mutants that lack individual chemoreceptors [5–10]. The direct association of the chemotaxis system with the flagella apparatus affects bacterial motility, which in turn, is an essential factor for the pathogenesis of *Campylobacter* strains. Nonmotile mutants demonstrate the most severe in vivo colonization defects [7–9]. *C. jejuni* chemoreceptor mutants display more subtle deficits and can be associated with reduced motility and, thus, reduced infectivity [2, 9, 11].

There are a number of different methods in the literature to measure chemotactic motility of wild-type and mutant *Campylobacter* strains as well as associated phenotypes [7, 12–14]. Both qualitative and quantitative chemotaxis assays are difficult to perform in a consistent and reproducible manner, and controls used by various research groups are not always appropriate for the assays performed [15]. Ideally a uniform, consistent and reliable set of methods and controls would benefit progress in this area of research. Thus, this chapter is designed to describe methodologies that enable the investigation of these aspects associated with chemotactic motility and consequent affects in vitro and in vivo.

2 Materials

All solutions are prepared using deionized water and analytical grade reagents followed by sterilization through autoclaving at 120 °C for 20 min. The materials are stored at room temperature unless otherwise stated. All solutions and agar are allowed to cool to 55 °C prior to addition of supplements.

2.1 Universal Supplements and Components

1. Columbia blood agar (CBA): 19.5 g Columbia blood agar base dissolved in 475 mL water. After sterilization, 25 mL (5 %) defibrinated horse blood is added to cooled agar supplemented with 200 μL polymyxin B (1 mg/mL; 1250 IU), 200 μL vancomycin (25 mg/mL; 5 mg), 50 μL trimethoprim (50 mg/mL; 2.5 mg), and additional appropriate selection if required. Store at 4 °C.
2. 2 % Columbia blood agar (CBA): 19.5 g Columbia blood agar base, 5 g bacteriological agar dissolved in 475 mL water. After sterilization, 25 mL (5 %) defibrinated horse blood is added to cooled agar supplemented with 200 μL polymyxin B (1 mg/mL; 1250 IU), 200 μL vancomycin (25 mg/mL; 5 mg), 50 μL trimethoprim (50 mg/mL; 2.5 mg), and additional appropriate selection if required. Store at 4 °C.
3. Brucella broth: 11 g heart infusion (HI) broth, 5 g tryptone, 1 g yeast extract dissolved in 500 mL water.

2.2 Motility Assay Components

1. 0.35% Mueller–Hinton agar (MHA): 19 g Mueller–Hinton agar base dissolved in 500 mL water. Store at 4 °C (*see* **Note 1**).

2.3 Nutrient Depleted Chemotaxis Assay Components

1. 0.5% agar without nutritional supplements: 1.25 g bacteriological agar dissolved in 250 mL water. Store at 4 °C.
2. 0.1% agar without nutritional supplements: 0.1 g bacteriological agar dissolved in 100 mL water. Store at 65 °C until use (in order to keep 0.1% agar molten prior to tempering at 37 °C; no more than 4 days).
3. 200 mM ligand: ligand dissolved in presterilized water. Store at 4 °C.
4. 1% saline: 5 g NaCl dissolved in 500 mL water.

2.3.1 Apparatuses Required for Chemotaxis Assays

5. Stable bench top or table at 37 °C that has been checked using a spirit level.
6. Microcentrifuge.
7. Shaker.
8. Sterile disposable plastic transfer pipette.
9. Spectrophotometer.
10. Wet paper towels.
11. Large plastic container and smaller plastic container filled with water.

2.4 Autoaggluti nation Assay Components

1. Phosphate buffered saline (PBS) pH 7.4: dissolve 8 g NaCl, 0.2 g KCl, 1.44 g Na_2HPO_4, 0.24 g KH_2PO_4 in 1 L water, pH adjusted to 7.4 with HCl.
2. Scanning electron microscope.

2.5 Biofilm Assay Components

1. Mueller–Hinton (MH) broth: 10.5 g Mueller–Hinton broth dissolved in 500 mL water.
2. 1% crystal violet: 5 mL crystal violet solution diluted to 500 mL with water and filtered through Whatman filter paper to remove debris.
3. Modified biofilm dissolving solution (MBDS): Dissolve SDS to final concentration of 10% with 80% ethanol in water (*see* **Note 2**).
4. Flat bottom 96-well plate.
5. Spectrophotometer.

2.6 In Vitro Adherence and Invasion Assay Components

1. 25 or 75 cm^2 vented tissue culture flasks.
2. Fetal bovine serum (FBS): Heat inactivated at 56 °C for 30 min.
3. Tissue culture media supplemented with 10% heat inactivated FBS (*see* **Note 3**).

4. Phosphate buffered saline (PBS) pH 7.4: Dissolve 8 g NaCl, 0.2 g KCl, 1.44 g Na_2HPO_4, 0.24 g KH_2PO_4 in 1 L water, pH adjusted to 7.4 with HCl.
5. 0.1 % Triton X-100: 10 μL Triton X-100 diluted in 9.99 mL of water.
6. Gentamicin: Dissolve 1 g of gentamicin in 10 mL sterilized water to prepare 100 mg/mL stock. Store at −20 °C.

2.7 In Vivo Colonization Model Components

1. 1-day-old chicks (8–10 chicks per group).
2. Water.
3. Sterilized water and food feeders.
4. Irradiated chicken feed.
5. Sterilized housing cages/isolators.
6. 42 °C room.
7. Sterilized dissection equipment/kit.

3 Methods

3.1 Motility Assay

1. Culture *C. jejuni* strains on CBA with appropriate selection under microaerophilic (5 % O_2, 10 % CO_2, 85 % N_2) conditions for 24 h at 42 °C.
2. Subculture *C. jejuni* strains in 10 mL Brucella broth with appropriate selection in 25 cm^2 tissue culture flasks under microaerophilic conditions at 42 °C for 18 h with shaking (50 rpm) (*see* **Note 4**).
3. Collect *C. jejuni* cells by centrifugation at 3000 × *g* for 5 min; discard the supernatant.
4. Wash the cell pellet by resuspending in 1 mL Brucella broth and collect the cells by centrifugation at 3000 × *g* for 5 min.
5. Resuspend cells in 1 mL Brucella broth and determine optical density using spectrophotometry. Adjust cell numbers to 1×10^9 cells/mL (OD_{600nm} 0.5) with Brucella broth.
6. 5 μL of the cell suspension (5×10^6 cells/mL) is used to stab the middle of the 0.35 % MHA plate (*see* **Note 5**).
7. Incubate plates under microaerophilic conditions for 18–24 h at 42 °C.
8. Following growth, measure the halo diameter (cm) using a ruler and compare to wild-type strains (*see* **Note 6**).
9. Critical experimental controls. For all motility assays, use a positive control: highly motile *C. jejuni* 81–176 or 11168-O and negative control and nonmotile *C. jejuni* 81116 or 11168-O *fla*A$^-$/*fla*B$^-$ mutant.

3.2 Nutrient Depleted Chemotaxis Assay

1. Culture *C. jejuni* strains on CBA with appropriate selection under microaerophilic (5 % O_2, 10 % CO_2, 85 % N_2) conditions for 24 h at 42 °C.
2. Subculture *C. jejuni* strains in 10 mL *Brucella* broth with appropriate selection in 25 cm^2 tissue culture flasks under microaerophilic conditions at 42 °C for 18 h with shaking (50 rpm) (*see* **Note 4**).
3. Aliquot 0.1 % agar into 50 mL tubes and temper at 37 °C overnight.
4. Prepare 0.5 % agar plates by pouring agar (30–35 mL) into petri dishes to measure approximately 0.5 cm in depth for the base (*see* **Note 7**).
5. Using a marker pen, indicate the center of the agar plate, as this is the designated area to place the bacterial suspension.
6. Remove one 6 mm diameter plug from the 0.5 % agar petri dish. To remove the 6 mm diameter plugs, a sterile disposable plastic transfer pipette is cut at the 0.5 mL mark and the transfer pipette inserted into the 0.5 % agar approximately 1 cm from the edge of the plate, squeezing the bulb prior to insertion. Using the suction created by the transfer pipette, the plug is removed by both releasing the bulb of the transfer pipette simultaneously with removing the transfer pipette from the 0.5 % agar plate (*see* **Note 8**).
7. Once the plug is removed and discarded, using a marker pen, circle the well where the plug was removed and label with the ligand name (*see* **Note 9**).
8. Fill each well with approximately 6–8 drops of 0.5 % agar containing 2 mM of selected ligand (Fig. 1) (*see* **Note 10**).
9. Place the prepared 0.5 % agar plates containing ligands at 37 °C on a stable bench top or table (*see* **Note 11**).
10. Overlay the 0.5 % agar plates containing ligand with 5 mL of tempered 0.1 % agar (from step 3.2.3) and leave plates at 37 °C for 1–2 h to allow diffusion of the ligand in order to create a chemical gradient (*see* **Note 12**).
11. To create a humid environment for the chemotaxis assay, place a small container of water and stacks of wet paper towel around the plates on the bench/table and cover all components with a large container (Fig. 2), being careful not to bump or move the bench/table or plates at any time as this will disrupt the chemotactic movement of the bacteria.
12. Collect the subcultured *C. jejuni* cells from **step 2** by centrifugation at 3000 × *g* for 5 min and discard the supernatant.
13. Wash the cell pellet by resuspending in 1 mL Brucella broth using gentle pipetting so not to shear the flagella; do not vortex. Centrifuge at 3000 × *g* for a further 5 min.

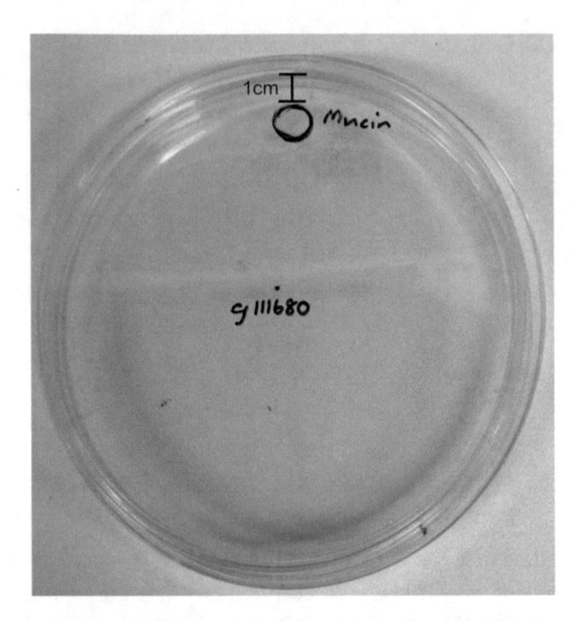

Fig. 1 Preparation of 0.5 % agar plates without nutritional supplements. The ligand is added to agar plates by replacing a section of the agar with an agar plug containing ligand (2 mM concentration). This allows for diffusion of the ligand to create a chemical gradient to determine "positive" or "negative" chemotaxis

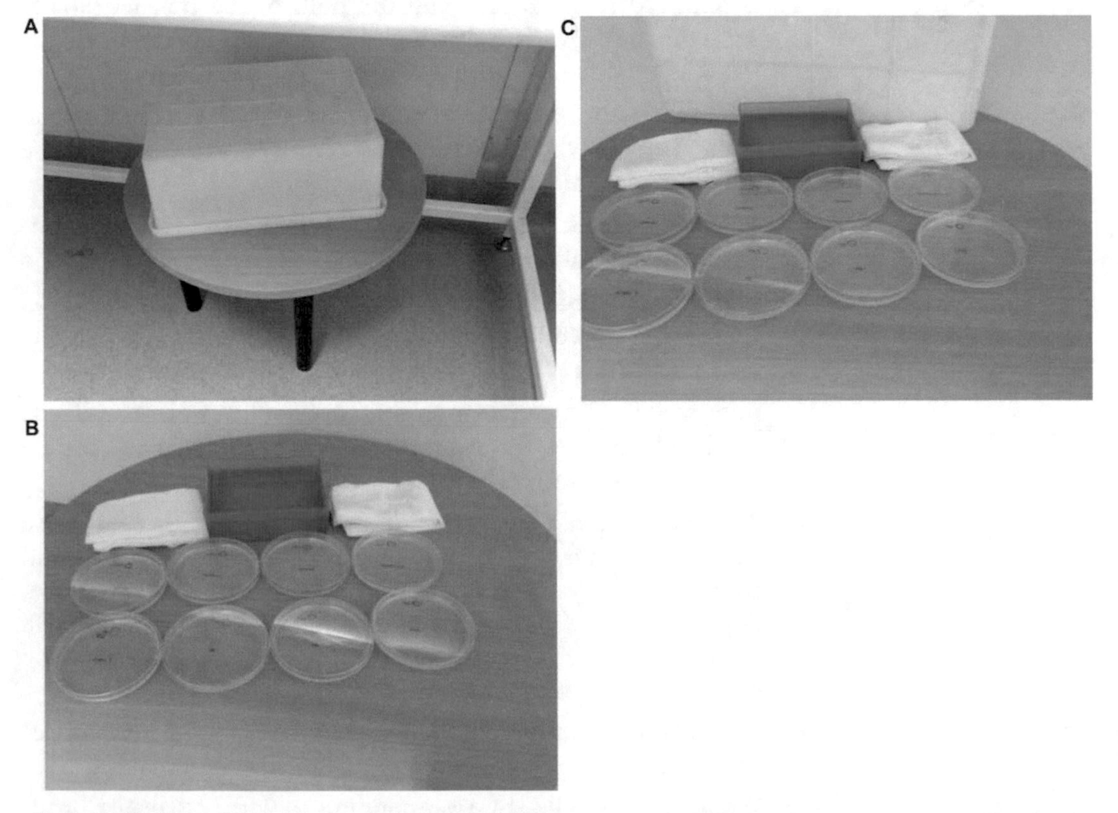

Fig. 2 Apparatus assembly for chemotaxis assays. To create a humidified atmosphere which is essential for viability of chemotaxis assays, containers of water and paper towels are placed evenly around the chemotaxis assay plates and covered

14. Determine optical density of the cells using spectrophotometry and adjust cell numbers to 1×10^9 cells/mL (OD_{600nm} 0.5) with 1% saline.
15. Gently place a 100 μL drop of bacterial suspension (1×10^8 cfu/mL) into the center of the 0.5% agar plate containing ligand that has been overlayed with 0.1% agar. Insert the pipette tip approximately 1–2 mm into the 0.1% agar layer to expel the bacterial suspension (*see* **Note 13**).
16. Incubate the inoculated plates without disturbing for 4 h to allow chemotactic migration.
17. After 4 h remove the bacteria associated with each ligand containing plug by removing a 5 mm area surrounding the plug and the plug itself. To remove the plug, a disposable plastic transfer pipette was cut at the 1.5 mL mark and UV sterilized for 20 min prior to use. The transfer pipette is inserted into the 0.5% agar squeezing the bulb prior to insertion. Using the suction created by the transfer pipette, the plug is removed by both releasing the bulb of the transfer pipette simultaneously with removing the transfer pipette from the 0.5% agar plate. Once the plug is removed, place into an Eppendorf tube containing 900 μL Brucella broth.
18. Additionally, a center plug where the bacterial suspension was placed is removed along with a plug from an area of the plate where no ligand is present in order to perform viability counts and ensure that the assay has not been biased by unintentional movement of the bacterial inoculum (*see* **Note 14**).
19. A hole is punctured in the lid of the Eppendorf tubes containing removed plugs and incubated under microaerophilic conditions for 1 h at 37 °C to allow bacteria to dissociate from the plug into the media.
20. Perform bacterial enumeration (1:10 serial dilutions) to determine cfu/mL. For the test ligand and positive control dilutions, culture 10^{-3} to 10^{-7} dilutions and for the negative control culture 10^{-1} dilution onto 2% CBA and incubate at 42 °C for 48 h for viable counts.
21. Critical controls. For all chemotaxis assays, a nonmotile, nonchemotactic control *C. jejuni* strain (81116 *fla*A$^-$/*fla*B$^-$ mutant) is used to determine no bias has influenced movement of bacteria during the assay. Additionally, mucin (10 mg/mL stock) and no ligand containing plugs are included in all assays as chemotactic controls. Mucin has been reported as a chemoattractant for *C. jejuni* [14], and favorable movement toward mucin should occur with all strains tested. Consequently, no chemotactic movement should occur to plugs that do not contain ligands.

3.3 Autoagglutination Assay

1. Culture *C. jejuni* strains on CBA with appropriate selection under microaerophilic conditions (5% O_2, 10% CO_2, 85% N_2) for 24 h at 42 °C.

2. Subculture *C. jejuni* strains on CBA with appropriate selection under microaerophilic conditions (5% O_2, 10% CO_2, 85% N_2) at 37 °C and 42 °C for 24 h and 18 h concurrently (*see* **Note 15**).
3. Harvest *C. jejuni* cells in duplicate with 1 mL PBS (pH 7.4) or Brucella broth (*see* **Note 16**).
4. Determine optical density of the cells using spectrophotometry and adjust cell numbers to 1×10^9 cells/mL ($OD_{600\ nm}$ 0.5) using PBS and/or Brucella broth.
5. Transfer 2 mL of the bacterial suspension into sterile 5 mL polystyrene 12×75mm round bottom tubes that have been UV sterilized prior to use and incubate at 25, 37, or 42 °C under microaerophilic conditions for 24 h.
6. Carefully remove 1 mL of the upper aqueous phase by aspiration and measure $OD_{600\ nm}$. Additionally, viable bacteria are enumerated by plate counts by performing 1:10 serial dilutions and culturing onto 2% CBA and incubating at 42 °C for 48 h to determine cfu/mL.
7. The lower 1 mL of solution containing the majority of the autoagglutinated cells is analyzed by scanning electron microscopy or photographic images [11, 16].

3.4 Biofilm Assay

1. Culture *C. jejuni* strains on CBA with appropriate selection under microaerophilic conditions (5% O_2, 10% CO_2, 85% N_2) for 24 h at 42 °C.
2. Subculture *C. jejuni* strains on CBA with appropriate selection under microaerophilic conditions (5% O_2, 10% CO_2, 85% N_2) for 18 h at 42 °C.
3. Harvest cells in 1 mL MH broth and determine optical density using spectrophotometry. Adjust cell numbers to 1×10^9 cfu/mL (OD_{600nm} 0.5).
4. Add 100 μL of bacterial suspension to a minimum of eight wells in a flat bottom 96-well plate.
5. Incubate under microaerophilic conditions at 42 °C for 48 h (*see* **Note 17**).
6. Discard media and rinse wells thoroughly with water three times to ensure removal of planktonic cells and media.
7. Add 125 μL filtered 1% crystal violet to stain formed biofilm. Allow to stain for 15 min [16].
8. Rinse wells thoroughly with water three times.
9. Add 150 μL MBDS to dissolve stained biofilm.
10. Analyze biofilm formation by measuring the 96-well plate at $OD_{600\ nm}$ using a microplate reader and media-only control.

3.5 *In Vitro* Adherence and Invasion Assay

1. Mammalian cells are grown in 25 cm^2 or 75 cm^2 flasks at 37 °C in 5% CO_2 humidified atmosphere and maintained in media supplemented with 10% FBS and 1% nonessential amino acids without the use of antibiotics.
2. Mammalian cells are seeded into 24-well plates at 10^5 cells/well and grown to >80% confluency at 37 °C in a 5% CO_2 humidified atmosphere. Prior to bacterial challenge, wash the cell monolayer once with PBS (pH 7.4).
3. Culture *C. jejuni* strains on CBA with appropriate selection under microaerophilic conditions (5% O_2, 10% CO_2, 85% N_2) at 42 °C for 24 h.
4. Subculture *C. jejuni* strains in 10 mL Brucella broth with appropriate selection in 25 cm^2 tissue culture flasks under microaerophilic conditions at 42 °C for 18 h with shaking (50 rpm) (*see* **Note 18**).
5. Collect *C. jejuni* strains by centrifugation at 3000 × *g* for 5 min, discard the supernatant, and resuspend in Brucella broth.
6. Determine optical density of the cells using spectrophotometry and adjust *C. jejuni* cell numbers to 1 x 10^7 cfu/mL in tissue culture media supplemented with 10% FBS. Perform viable counts to verify the bacterial suspension.
7. Infect confluent cells with 100 μL bacterial suspension containing 1×10^6 cfu of *C. jejuni* (mammalian to bacterial cell ratio = 1:10) supplemented with 400 μL tissue culture media supplemented with 10% FBS added into duplicate wells. Perform the coculture experiments in standard microaerophilic conditions (5% O_2, 10% CO_2, 85% N_2) for 1 h to allow bacterial adherence and internalization.
8. As an additional step, *C. jejuni* can be forced to interact with the epithelial cell monolayer using centrifugation for 5 min at 1000 × *g* (*see* **Note 19**).
9. Wash the cell monolayer three times with PBS (pH 7.4). To determine bacterial adherence, lyse the cell monolayer with 200 μL 0.1% Triton X-100, and total bacteria associated with the cells (intracellular and extracellular bacteria) are enumerated by viable counts on 2% CBA agar plates incubated under microaerophilic conditions at 42 °C for 48 h.
10. To measure bacterial invasion, wash the infected cells three times with PBS (pH 7.4) and add 1 mL of fresh tissue culture media containing 400 μg/mL gentamicin to the infected cells and incubate for 3 h under standard microaerophilic conditions (5% O_2, 10% CO_2, 85% N_2) to kill remaining viable extracellular bacteria.
11. Wash the cell monolayer three times with PBS (pH 7.4) and lyse the cells with 200 μL 0.1% Triton X-100; enumerate

intracellular bacteria by viable counts on 2% CBA agar plates incubated under microaerophilic conditions at 42 °C for 48 h.

12. Data representation: Adherent bacteria are expressed as the percentage of bacteria counted as compared to the infection dose. Invasive bacteria are expressed as the invasion index; the number of invaded bacteria expressed as a percentage of the number of adhered organisms. All assay results are calculated from the mean of at least three separate biological replicates. *E. coli* is used in all assays as a non-adherent, noninvasive control (*see* **Note 20**).

3.6 In vivo Colonization Assay

1. Culture *C. jejuni* strains on CBA with appropriate selection under microaerophilic conditions (5% O_2, 10% CO_2, 85% N_2) at 42 °C for 24 h.
2. Subculture *C. jejuni* strains on CBA with appropriate selection under microaerophilic conditions (5% O_2, 10% CO_2, 85% N_2) at 42 °C for 18 h.
3. Harvest cells with 1 mL Brucella broth and determine optical density (OD_{600nm}) using spectrophotometry and adjust cell numbers to 1×10^7 cfu/mL.
4. Inoculate one-day-old chicks orally with 100 μL bacterial suspension containing 1×10^6 cfu of *Campylobacter* spp.
5. Monitor colonization daily by sampling using cloacal swabs cultured onto selective CBA plates incubated under microaerophilic conditions for 24–48 h.
6. After 5 days, euthanize chickens by cervical dislocation and aseptically remove chicken caeca.
7. Weigh each caeca and resuspend contents in 1 mL Brucella broth.
8. Enumerate bacterial colonization by serial dilutions cultured onto 2% selective CBA incubated under microaerophilic conditions for 24–48 h.

4 Notes

1. Ensure 0.35% MHA plates are warmed to room temperature prior to use.
2. Use 65 °C incubation if unable to solubilize ~>85%.
3. The choice of cell line depends on the specific experimental aim investigated, and subsequently, an appropriate cell line is chosen for that purpose.
4. Alternatively, bacterial cells can be grown on solid agar plates or biphasically; no significant difference in subsequent motility has been observed between these methods of bacterial cell preparation for this assay.

5. Insert pipette tip slightly into the agar, approximately 1 mm before gently expelling bacterial suspension.
6. Perform all assays in triplicate in order to obtain accurate data to represent with statistical significance.
7. If the agar plate is too thin, the overlayed 0.1% agar tends to dry out reducing chemotactic motility and viability of bacteria.
8. Ensure the transfer pipette is cut as straight as possible to allow for efficient suction and removal of the plug.
9. Ensure the well where the plug was removed is labeled, as after the hole is refilled with 0.5% agar containing ligand to be tested; it is difficult to see where to remove the agar after the bacterial suspension has been added.
10. When replacing each well with 0.5% agar containing ligand, ensure that the agar fills the well so that it is flushed with the surface of the agar; if too little is added, the well becomes concave. 0.5% agar plates containing ligand can be stored for up to 12 h at 4 °C with no significant effect on chemotaxis assays.
11. Ensure that the bench top or table has been checked with a spirit level as any slight slope can influence the chemotactic movement of the bacterial suspension and bias the chemotactic result. Let the plates sit for 15 min or until they have reached 37 °C. If 0.5% agar containing ligand plates are still below 37 °C when the 0.1% agar is overlayed, the plate will dry out affecting viability of the bacterial suspension.
12. If the 37 °C area where the chemotaxis assay is to be performed does not retain high humidity, a humid atmosphere will need to be created; otherwise, the 0.1% agar will dry out and affect viability of the bacteria.
13. When inoculating plates with the bacterial suspension, ensure that a very gentle expulsion technique is used; otherwise, the suspension can bead and run over the surface of the 0.1% agar instead of being placed in the center of the plate as a neat circle and can subsequently influence results.
14. Collect the plugs in the following order: ligand-associated bacteria (ligand plug), center (inoculum control), and then random area (negative control).
15. To ensure there are no differences in autoagglutination due to growth temperature or culture time, *C. jejuni* strains are subcultured at 37 and 42 °C for both 24 and 18 h.
16. To ensure there is no difference in autoagglutination behavior due to bacterial viability, both PBS and Brucella broth are used for this assay.
17. Humidify if necessary by placing a container with water in the gas chamber.

18. An important disadvantage of using agar plate-grown *C. jejuni* for infection experiments is that the bacteria vary significantly in shape, flagellation, motility, and viability [17]. Thus, before infection, the subcultivation of *C. jejuni* in liquid culture is highly recommended as it has been shown that *C. jejuni* growth in liquid broth leads to a high percentage of highly motile *C. jejuni* cells with similar shape [18].
19. Centrifugation represents a suitable method to synchronize the infection process in order to overcome variations in the interaction of *C. jejuni* strains with cultured epithelial cells that are due to variable motility properties rather than altered adherence or invasion characteristics [18]. However, when assessing chemotaxis mutants, centrifugation forces the bacteria onto the cells and introduces bias; therefore, performing both a centrifugation and non-centrifugation assay will eliminate any bias results.
20. Data representation of adherent bacteria after a centrifugation step may alter the number of bacteria adhered to culture cells; however, the calculations remain unchanged with adherent bacteria expressed as the percentage of bacteria counted compared to the infection dose, and invasive bacteria expressed as a percentage of adhered organisms.

References

1. Young KT, Davis LM, Dirita VJ (2007) *Campylobacter jejuni*: molecular biology and pathogenesis. Nat Rev Microbiol 5(9):665–679. doi:10.1038/nrmicro1718
2. Zautner AE, Tareen AM, Gross U et al (2012) Chemotaxis in *Campylobacter jejuni*. Eur J Microbiol Immunol 2(1):24–31. doi:10.1556/EuJMI.2.2012.1.5
3. Lertsethtakarn P, Ottemann KM, Hendrixson DR (2011) Motility and chemotaxis in *Campylobacter* and *Helicobacter*. Annu Rev Microbiol 65:389–410. doi:10.1146/annurev-micro-090110-102908
4. Wadhams GH, Armitage JP (2004) Making sense of it all: bacterial chemotaxis. Nat Rev Mol Cell Biol 5(12):1024–1037. doi:10.1038/nrm1524
5. Hendrixson DR, DiRita VJ (2004) Identification of *Campylobacter jejuni* genes involved in commensal colonization of the chick gastrointestinal tract. Mol Microbiol 52(2):471–484. doi:10.1111/j.1365-2958.2004.03988.x
6. Yao R, Burr DH, Guerry P (1997) CheY-mediated modulation of *Campylobacter jejuni* virulence. Mol Microbiol 23(5):1021–1031
7. Takata T, Fujimoto S, Amako K (1992) Isolation of nonchemotactic mutants of *Campylobacter jejuni* and their colonization of the mouse intestinal tract. Infect Immun 60(9):3596–3600
8. Chang C, Miller JF (2006) *Campylobacter jejuni* colonization of mice with limited enteric flora. Infect Immun 74(9):5261–5271. doi:10.1128/IAI.01094-05
9. Hartley-Tassell LE, Shewell LK, Day CJ et al (2010) Identification and characterization of the aspartate chemosensory receptor of *Campylobacter jejuni*. Mol Microbiol 75(3):710–730. doi:10.1111/j.1365-2958.2009.07010.x
10. Ottemann KM, Miller JF (1997) Roles for motility in bacterial-host interactions. Mol Microbiol 24(6):1109–1117
11. Rahman H, King RM, Shewell LK et al (2014) Characterisation of a multi-ligand binding chemoreceptor CcmL (Tlp3) of *Campylobacter jejuni*. PLoS Pathog 10(1):e1003822. doi:10.1371/journal.ppat.1003822
12. Golden NJ, Acheson DW (2002) Identification of motility and autoagglutination *Campylobacter jejuni* mutants by random transposon mutagenesis. Infect Immun 70(4):1761–1771
13. Friis LM, Pin C, Pearson BM et al (2005) *In vitro* cell culture methods for investigating *Campylobacter* invasion mechanisms. J Microbiol Methods 61(2):145–160. doi:10.1016/j.mimet.2004.12.003

14. Hugdahl MB, Beery JT, Doyle MP (1988) Chemotactic behavior of *Campylobacter jejuni*. Infect Immun 56(6):1560–1566
15. Kanungpean D, Kakuda T, Takai S (2011) False positive responses of *Campylobacter jejuni* when using the chemical-in-plug chemotaxis assay. J Vet Med Sci 73(3):389–391
16. Tram G, Korolik V, Day CJ (2013) MBDS solvent: an improved method for assessment of biofilms. Adv Microbiol 3(02):200
17. Ng LK, Sherburne R, Taylor DE et al (1985) Morphological forms and viability of *Campylobacter* species studied by electron microscopy. J Bacteriol 164(1):338–343
18. Backert S, Hofreuter D (2013) Molecular methods to investigate adhesion, transmigration, invasion and intracellular survival of the foodborne pathogen *Campylobacter jejuni*. J Microbiol Methods 95(1):8–23. doi:10.1016/j.mimet.2013.06.031

Chapter 14

Using *Galleria mellonella* as an Infection Model for *Campylobacter jejuni* Pathogenesis

Momen Askoura and Alain Stintzi

Abstract

Nonmammalian model systems of infection have been employed recently to study bacterial virulence. For instance, *Galleria mellonella* (the greater wax moth) has been shown to be susceptible to infection by many bacterial pathogens including the enteric pathogen *Campylobacter jejuni*. In contrast to the traditional animal models for *C. jejuni* such as the chick colonization model and ferret diarrheal model, the *Galleria mellonella* infection model has the advantages of lower cost, ease of use and no animal breeding is required. However, injecting the larvae with bacteria requires care to avoid killing of larvae, which could lead to misleading results. Here, we describe the infection of *G. mellonella* larvae by *C. jejuni* and how to record/interpret results.

Key words *Galleria mellonella*, *Campylobacter jejuni*, Infection, Pathogenesis, Survival, Hemolymph, Melanization

1 Introduction

G. mellonella has been employed as a model to evaluate the infectivity of many pathogens, such as *Listeria* spp. [1] and *S. pyogenes* [2]. Similarly, previous studies reported that *G. mellonella* could provide a valuable model to elucidate *C. jejuni* virulence factors [3, 4]. Infection of *G. mellonella* by *C. jejuni* was first demonstrated by Champion et al. [3]. In that study, Champion et al. demonstrated that three strains of *C. jejuni*, 11168H, 81-176 (human diarrheal isolates), and G1 (human Guillain-Barré syndrome isolate), were capable of killing *G. mellonella* effectively [3]. Importantly, *G. mellonella* harbors phagocytic cells known as hemocytes [5, 6]. Hemocytes phagocytose and kill pathogens using antimicrobial peptides and reactive oxygen species generated during a respiratory burst [5, 7]. Therefore, *G. mellonella* hemocytes share many properties with mammalian phagocytes [5, 7]. We were able to successfully infect *G. mellonella* larvae with different *C. jejuni* doses [8] following the methodology described elsewhere [2, 3] with some

James Butcher and Alain Stintzi (eds.), *Campylobacter jejuni: Methods and Protocols*, Methods in Molecular Biology, vol. 1512,
DOI 10.1007/978-1-4939-6536-6_14,

modifications. The results are taken by comparing the survival of infected larvae with that of uninfected controls. In addition, changes in larva appearance such as black melanization as well as microscopical examination of infected larvae hemolymph can be recorded to supplement the viable/nonviable results.

2 Materials

All solutions were prepared using distilled water. Prepared reagents were stored at room temperature (unless indicated otherwise).

2.1 Mueller-Hinton (MH) Broth

Suspend 21 g of the medium in 1 L of distilled water. Mix well and dissolve by heating with frequent agitation. Dispense into appropriate containers and sterilize in autoclave at 121 °C for 15 min. Final pH 7.4 ± 0.2 at 25 °C. Store prepared media in tightly sealed container at room temperature until use.

2.2 Mueller-Hinton (MH) Agar

Suspend 38 g of the medium in 1 L of distilled water. Heat with frequent agitation to completely dissolve the medium. Autoclave at 121 °C for 15 min. Cool to 50 °C. Mix gently and dispense Mueller-Hinton agar into sterile Petri dishes on a level, horizontal surface to give uniform depth. Allow to cool to room temperature. Prepared dishes are stored at 4 °C until use.

2.3 1× Phosphate Buffered Saline (PBS) Buffer

Dissolve 8 g of NaCl, 0.2 g of KCl, 1.44 g of Na_2HPO_4, and 0.24 g of KH_2PO_4 in 800 mL distilled H_2O. Adjust pH of solution to 7.4 with 1 M HCl. Adjust volume to 1 L with additional distilled H_2O. Sterilize by autoclaving at 121 °C for 15 min. Store prepared solution at room temperature.

2.4 70% [vol/vol] Ethanol

Measure 70 mL of 99% ethyl alcohol in a measuring cylinder and complete to 100 mL by distilled water. Store in screw-capped container at room temperature.

2.6 Safranin Gram Stain

Add 10 mL of the stock solution (2.5 g safranin O dissolved in 100 mL of 95% ethyl alcohol) to 90 mL of distilled H_2O. Store at room temperature (25 °C).

2.5 Hamilton 10 μL 901RN Syringe (Hamilton® Microliter™) or Insulin Syringe 29 G

3 Methods

Carry out all procedures at room temperature unless otherwise specified.

3.1 Bacterial Stocks

1. Grow *C. jejuni* from frozen 50% glycerol stock stored at −80 °C.
2. Inoculate *C. jejuni* on Mueller-Hinton agar and incubate at 37 °C for 2–3 days under microaerophilic conditions (*see* **Note 1**).
3. Inoculate isolated colonies of *C. jejuni* into Mueller-Hinton broth in biphasic culture flask and incubate under microaerophilic conditions for 24 h at 37 °C (*see* **Note 2**).
4. Let *C. jejuni* grow until it reaches the mid-exponential phase (*see* **Note 3**).
5. Adjust bacterial concentration to required doses (*see* **Note 4**).
6. Determine the infectious doses of *C. jejuni* (CFU/mL) by serial dilution in PBS and colony counting on MH agar plates.

3.2 Preparation of G. mellonella Larvae

1. Obtain *G. mellonella* larvae from a trusted vendor.
2. Use only larvae which appear healthy and active in infection experiments.
3. Allow *G. mellonella* larvae to acclimate in the laboratory after delivery for at least 24 h.
4. Maintain *G. mellonella* larvae with wood chips in a plastic container at 10–15 °C and use within 5–10 days after receipt. Under such conditions, larvae remain alive and healthy and do not need food or water supplementation.

3.3 G. mellonella Larvae Infection

1. Place groups of ten larvae in sterile petri dishes. Selected larvae should be of equal size and weight, having a cream-colored cuticle with minimal speckling or discoloration (Fig. 1).

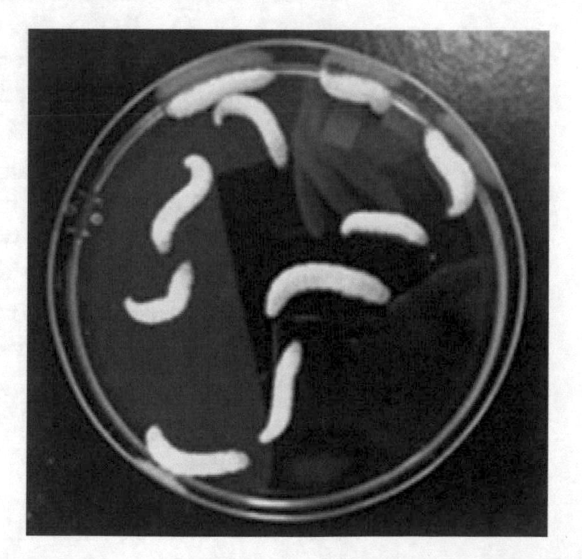

Fig. 1 Characteristics of *G. mellonella* larvae used in infection experiments. Larvae should be of equal size and weight, active, and having a cream-colored cuticle with minimal speckling or discoloration

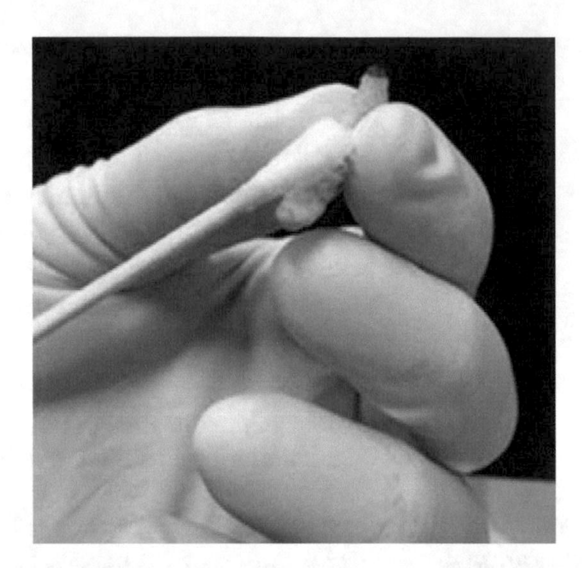

Fig. 2 Decontamination of *G. mellonella*. Larvae should be decontaminated prior to injection by soaking a sterile cotton swab with 70% ethanol and rubbing it over the site of injection

2. Decontaminate *G. mellonella* larvae by soaking a sterile cotton piece with 70% ethanol and rubbing it over the site of injection. Allow the alcohol to dry near a Bunsen burner (Fig. 2).
3. Carefully inject cohorts of ten larvae with 10 μL of tenfold serial dilutions of *C. jejuni* (10^7–10^5 CFU in PBS) into the hemocoel through the left hindmost proleg using a Hamilton 10 μL 901RN syringe (Hamilton® Microliter™) or an insulin syringe 29 G (BD™; Becton, Dickinson and Company) (Fig. 3). The infectious doses of both the test and control (CFU/mL) are determined by serial dilution and colony counting on MH agar plates.
4. Sterilize the needle after each injection; wash once in 1× PBS, twice in 70% ethanol, and finally once in 1× PBS (washing is done by filling the syringe with PBS or 70% ethanol and dispensing the contents into a waste container).
5. Inject another cohort of 10 larvae with 10 μL of sterile PBS, and leave another ten larvae with no injection as negative controls.
6. Maintain *C. jejuni*-infected, PBS-injected, and uninfected larvae in vented Petri dishes at room temperature under aerobic conditions.
7. Monitor mortality, survival, and appearance of *G. mellonella* larvae every 24 h for 6 days following inoculation (*see* **Note 5**).
8. Prepare smear slides for both *C. jejuni*-infected and uninfected larvae for microscopical examination.

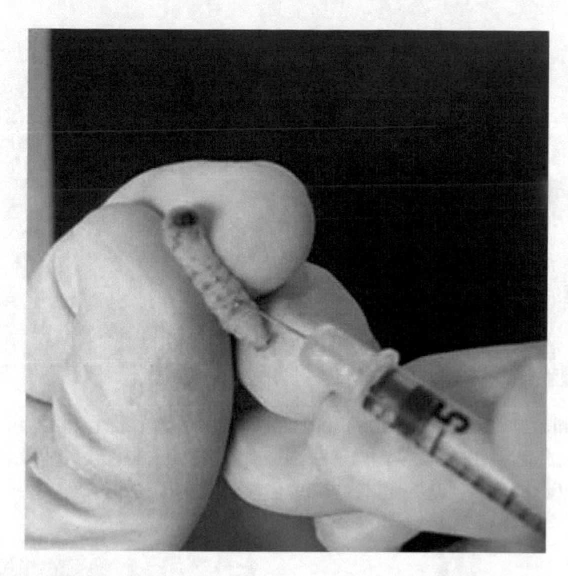

Fig. 3 Larvae infection by *C. jejuni*. Larvae are carefully injected with 10 μL of tenfold serial dilutions of *C. jejuni* into the hemocoel through the left hindmost proleg using a Hamilton 10 μL 901RN syringe or an insulin syringe 29 G

   (a) Aseptically remove the rear 2 mm of tested larva with a sterile surgical blade (Fig. 4a).

   (b) Collect the hemolymph into a sterile microcentrifuge tube by gently squeezing the larva body (Fig. 4b, c).

9. Subject the smear slides to dye staining.

   (a) Spread ~250 μL of larva hemolymph on glass slide and air-dry for 1–2 h (Fig. 5a).

   (b) Add 1–2 mL safranin dye, leave for 30 s, and then rinse slide with a gentle stream of water.

   (c) Allow slide to air-dry for 15 min and examine slides using a microscope at 40× magnification (to see black melanization) and at 100× magnification using oil immersion objective (to see *C. jejuni* cells).

10. Use an oil immersion objective to see *C. jejuni* cells.

   (a) Focus at low power on a region of a smeared and stained specimen.

   (b) Move the oil objective halfway to the next objective.

   (c) Place a drop of immersion oil directly over the area of the specimen to be examined.

   (d) Slowly rotate the 100× oil immersion objective to come in contact with the oil and clicks into place.

11. Photograph representative images for both control uninfected and *C. jejuni*-injected larvae.

12. Results are taken by monitoring mortality, survival, and appearance of *G. mellonella* larvae every 24 h for 6 days following

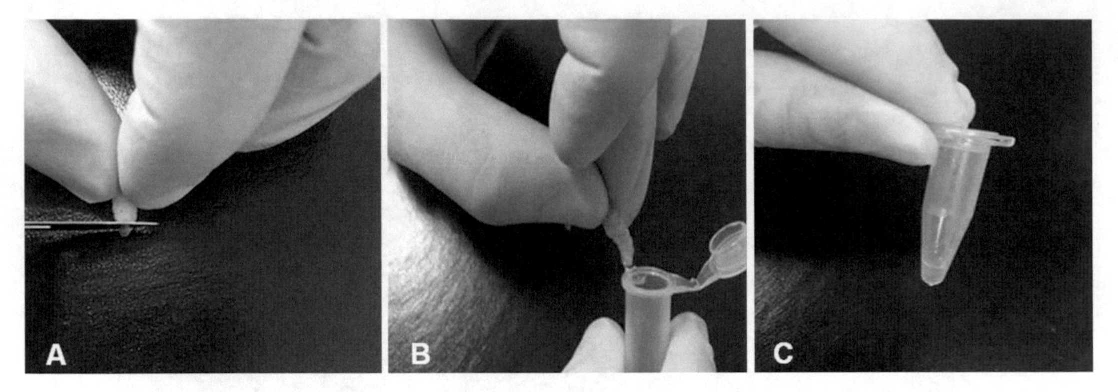

Fig. 4 Preparation of smear slides for tested larvae for microscopical examination. (**a**) The rear 2 mm of tested larva is aseptically removed with a sterile surgical blade. Larvae hemolymph is collected into a sterile microcentrifuge tube by gently squeezing the larva body (**b**) and (**c**)

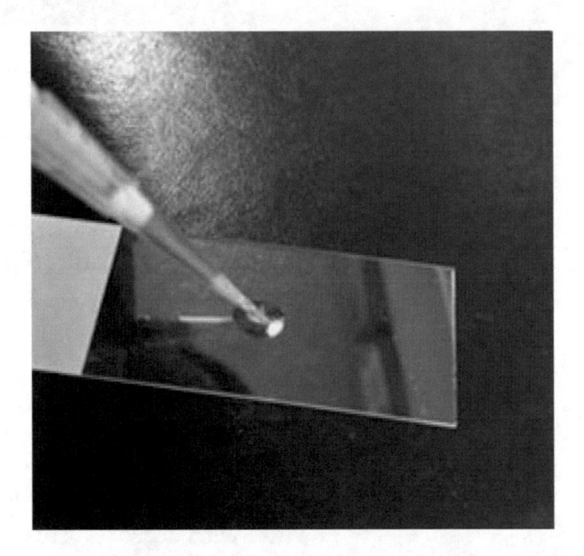

Fig. 5 Preparation of smear slides for tested larvae for microscopical examination. Spread a small volume of larva hemolymph on glass slide and let air-dry for 1–2 h before staining with Gram stain

inoculation. Plot survival curves of both the control and bacteria-infected larvae using the Kaplan-Meier method. Moreover, the 50% lethal dose (LD_{50}) of injected *C. jejuni* can also be calculated as previously described [2].

4 Notes

1. To achieve microaerophilic conditions, we use a MACS-VA500 microaerophilic workstation (Don Whitley, West Yorkshire, England) with gas mixture composed of 83% N_2, 4% H_2, 8% O_2, and 5% CO_2.
2. Biphasic culture flask contains 7 mL of Mueller-Hinton agar overlaid with 7 mL of Mueller-Hinton broth.

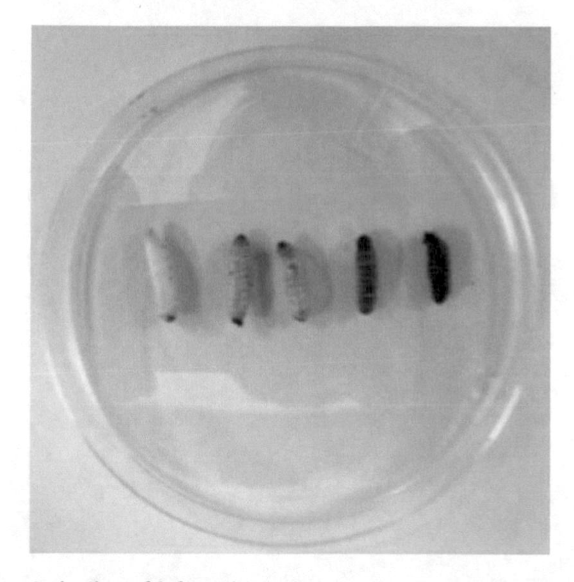

Fig. 6 Characterization of infected larvae. In contrast to healthy live larvae that have a cream-colored cuticle (*far left*), infected larvae exhibit brown to black discoloration (melanization) on their cuticle as the infection progresses (*increasing toward the right*)

3. *C. jejuni* growth is considered at the mid-exponential phase when the optical density ($O.D_{600}$) of bacterial culture equal to 0.8–1.
4. Centrifuge bacterial culture at $8000 \times g$ for 5 min at room temperature, wash bacterial pellets twice in PBS (pH 7.4), and resuspend in PBS to the desired bacterial concentration (CFU/mL).
5. Larvae are considered dead when they are not able to move even after pricking with a needle tip. Dead larvae also exhibit brown to black discoloration (melanization) on their cuticle, which is an important sign of larvae infection by pathogen (Fig. 6).

References

1. Mukherjee K, Altincicek B, Hain T et al (2010) *Galleria mellonella* as a model system for studying *Listeria pathogenesis*. Appl Environ Microbiol 76(1):310–317
2. Olsen RJ, Watkins ME, Cantu CC et al (2011) Virulence of serotype M3 Group A *Streptococcus* strains in wax worms (*Galleria mellonella* larvae). Virulence 2(2):111–119, 14338 [pii]
3. Champion OL, Karlyshev AV, Senior NJ et al (2010) Insect infection model for *Campylobacter jejuni* reveals that O-methyl phosphoramidate has insecticidal activity. J Infect Dis 201(5):776–782
4. Senior NJ, Bagnall MC, Champion OL et al (2011) *Galleria mellonella* as an infection model for *Campylobacter jejuni* virulence. J Med Microbiol 60(Pt 5):661–669
5. Lavine MD, Strand MR (2002) Insect hemocytes and their role in immunity. Insect Biochem Mol Biol 32(10):1295–1309
6. Kavanagh K, Reeves EP (2004) Exploiting the potential of insects for *in vivo* pathogenicity testing of microbial pathogens. FEMS Microbiol Rev 28(1):101–112
7. Bergin D, Reeves EP, Renwick J et al (2005) Superoxide production in *Galleria mellonella* hemocytes: identification of proteins homologous to the NADPH oxidase complex of human neutrophils. Infect Immun 73(7):4161–4170
8. Askoura M (2015) Acid adaptive mechanisms of *Campylobacter jejuni* in the gastrointestinal tract. Thesis, University of Ottawa, Ottawa, Canada

Chapter 15

Mouse Models for *Campylobacter jejuni* Colonization and Infection

Martin Stahl, Franziska A. Graef, and Bruce A. Vallance

Abstract

Relevant animal models for *Campylobacter jejuni* infection have been difficult to establish due to *C. jejuni*'s inability to cause disease in many common animal research models. Fortunately, recent work has proven successful in developing several new and relevant mouse models of *C. jejuni* infection, including the SIGIRR-deficient mouse strain that develops acute enterocolitis in response to *C. jejuni*. Here we describe how to properly infect mice with *C. jejuni*, as well as a number of accompanying histological techniques to aid in studying *C. jejuni* colonization and infection in mice.

Key words *Campylobacter jejuni*, Mouse models, Histology, Staining, Immunofluorescence, Pathology

1 Introduction

A prominent hurdle that has plagued researchers attempting to gain a better understanding of *Campylobacter jejuni* infection has been a lack of potential animal models that accurately reflect human campylobacteriosis [1]. Furthermore, many of the animal models that have been used for *Campylobacter* infection (such as piglets [2] and ferrets [3]) lack many of the necessary research tools used for more detailed studies of host-pathogen interactions. As the most commonly used animal model employed for research, mice have a wide variety of research tools available for studying their immune responses, including knockout strains in many key genes, along with antibodies and other products that are tailored to mouse biology and genetics. However, utilizing mice in *Campylobacter* research has proved problematic since most strains of mice are not overtly susceptible to *Campylobacter* infection and often their intestines cannot be efficiently colonized [4]. Recently, researchers have taken several steps to turn mice into a viable model for *Campylobacter* infection [4–6].

James Butcher and Alain Stintzi (eds.), *Campylobacter jejuni: Methods and Protocols*, Methods in Molecular Biology, vol. 1512, DOI 10.1007/978-1-4939-6536-6_15, © Springer Science+Business Media New York 2017

The first step is to open the mouse intestinal tract to *C. jejuni* colonization. The colonization resistance of the mouse GI tract to *C. jejuni* is thought to be the result of competition from the mouse intestinal microbiota [4, 6]. Supporting this concept are early studies finding that *C. jejuni* will readily colonize germ-free or limited-flora mice [7], but not mice with normal microbiota. More recently, mice pretreated with antibiotics or with a "humanized" microbiota were shown to be readily colonized following oral gavage with *C. jejuni* [4, 6]. Although there may be many complicated interactions between *C. jejuni* and commensal microbes, the most prominent mechanism at play is likely the availability of open nutrient niches. While *C. jejuni* may have difficulty in competing with resident microbes for nutrients, a substantial disruption in the microbiota likely opens niches within the intestinal ecosystem that can be taken over by *C. jejuni*.

A further complication, beyond that of colonization resistance, has been that even when *C. jejuni* efficiently colonize mice, these mice rarely show any overt or significant signs of disease. Evidence would suggest that the innate immune system of mice is highly tolerant to the presence of *C. jejuni*, even when it colonizes their intestines at high levels [6]. To overcome this obstacle, researchers have turned to genetically manipulated mice deficient in key genes regulating innate immune responses. An early model that proved effective was the IL-10-deficient mouse [5], which is known to be very susceptible to a variety of infections. While these mice can be infected by *C. jejuni* and develop enterocolitis as a result, a drawback is their extreme sensitivity to gut microbes can lead to spontaneous colitis in response to the mouse's own microbiota. As a result, this has forced researchers to keep these mice under germ-free conditions prior to *C. jejuni* infection [8]. IL-10 is also a key factor in the resolution of inflammation, so IL-10-deficient mice are usually unable to resolve these infections which ultimately prove terminal.

Recently, we have employed SIGIRR-deficient($^{-/-}$) mice to model *C. jejuni* infection [6, 9]. SIGIRR is an inhibitor of MyD88-dependent signaling that is highly expressed in the intestinal epithelium of both mice and humans. In its absence, pro-inflammatory signaling via MyD88-dependent receptors is substantially increased [10]. This has rendered *Sigirr*$^{-/-}$ mice more susceptible to *Citrobacter rodentium* and *Salmonella enterica* Typhimurium infections [9], as well as to *C. jejuni* [6]. The *C. jejuni* study also found that TLR4 was the main driver of inflammation in *Sigirr*$^{-/-}$ mice, whereas TLR2 had a protective role, with TLR2/SIGIRR-double-deficient mice being particularly susceptible to the development of severe disease [6].

In this chapter, we will outline how to use mice as a model for *C. jejuni* infection, as well as several downstream techniques for the analysis of mouse tissues, including histological techniques for the evaluation of colonization and inflammation. We will utilize *Sigirr*$^{-/-}$ mouse infection as a model, as they exhibit a striking enterocolitis; however, most techniques should be applicable to any mouse strain being employed.

2 Materials

2.1 Maintenance, Infection, and Euthanization of Mice

1. Supplies for the proper housing of mice based on the requirements and policies of your local animal facility.
2. Oral gavaging needle.
3. 1-mL syringes.
4. 70% ethanol.
5. PBS (137 mM NaCl, 2.7 mM KCl, 10 mM Na_2HPO_4, 1.8 mM KH_2PO_4, pH 8.0).
6. 50 mg/mL vancomycin in sterile PBS (or desired antibiotic for mouse pretreatment).
7. Mueller-Hinton broth (casein hydrolysate 17.5 g/L, beef extract 2.0 g/L, starch 1.5 g/L). Autoclave prior to use.
8. Proper equipment for the cultivation of *C. jejuni*.
9. Surgery tools (forceps, scissors).
10. Isoflurane.

2.2 Enumerating Colony-Forming Units

1. Karmali *Campylobacter* agar plates (e.g., Oxoid): Mix and autoclave 21.5-g Karmali *Campylobacter* agar base in 500 mL dH_2O. Dilute powdered Karmali supplement in 2 mL 50% ethanol (or follow manufacturer instructions) and add it to the autoclaved Karmali *Campylobacter* agar base once it has cooled to approximately 50 °C. Pour into petri plates and allow to cool and solidify entirely (*see* **Note 1**).
2. Tissue homogenizer.
3. Metal beads (appropriately sized for the tubes and tissue homogenizer being used).
4. PBS (137 mM NaCl, 2.7 mM KCl, 10 mM Na_2HPO_4, 1.8 mM KH_2PO_4, pH 8.0).

2.3 Histological Tissue Fixation

1. 10% formalin.
2. Carnoy's solution (60% formalin, 30% chloroform, 10% glacial acetic acid).
3. 4% paraformaldehyde.
4. 70, 80, 95, and 100% ethanol.
5. Paraffin.
6. Histological grade xylene.
7. Tissue cassettes.
8. Microscope slides (charged).
9. PBS-azide (PBS buffer + 0.05% sodium azide).
10. 20% sucrose solution.

11. Peel-away base mold (*see* **Note 2**).
12. Optimum cutting temperature (OCT) cryoprotective medium.
13. Isopentane.
14. Dry ice or liquid nitrogen.
15. Microtome and/or cryostat.
16. Oven capable of reaching 50 °C.
17. Water bath capable of reaching 80 °C.
18. Metal or glass slide holder.

2.4 H&E staining

1. Gill's hematoxylin solution (For 1 L: 6.0 g hematoxylin, 4.2 g aluminum sulfate, 1.4 g citric acid, 0.6 g sodium iodate, 269 mL ethylene glycol, 680 mL distilled water).
2. 1% eosin yellowish.
3. 1% lithium carbonate or Scott's tap water: (20 g sodium bicarbonate, 3.5 g magnesium sulfate).
4. Distilled water.
5. 1% acid alcohol: (99 mL 70% ethanol, 1 mL concentrated hydrochloric acid).
6. Commercial mounting medium.

2.5 Periodic Acid-Schiff (PAS) Staining

1. 0.5% periodic acid solution (0.5 g periodic acid in 100 mL dH_2O) (make fresh for each experiment).
2. Schiff's reagent (*see* **Note 3**).
3. Gill's hematoxylin solution (same as Subheading 2.4).

2.6 Immunofluorescence Staining

1. Antigen retrieval buffer (10 mM sodium citrate, 0.05% Tween 20, pH 6.0) (*see* **Note 4**).
2. Vegetable steamer.
3. Plastic transfer pipettes.
4. Metal or glass slide holder.
5. Grease/DAKO Pen for drawing hydrophobic border.
6. Tissues.
7. Blocking buffer (2% goat or donkey serum (*see* **Note 5**), 1% BSA, 0.1% Triton X-100, 0.05% Tween 20, 0.05% sodium azide, in PBS, pH 7.2, filter sterilized).
8. Antibody dilution buffer (0.1% Triton X-100, 0.1% BSA, 0.05% sodium azide, 0.04% EDTA, in PBS pH 7.2, filter sterilized).
9. Antibodies (*see* **Note 6**).
10. PBS (137 mM NaCl, 2.7 mM KCl, 10 mM Na_2HPO_4, 1.8 mM KH_2PO_4, pH 8.0).

11. ProLong Gold anti-fade mounting medium (containing DAPI) or equivalent.
12. Coverslips.
13. Clear nail polish.

3 Methods

3.1 Considering Experimental Design

1. *Consider the number of mice necessary for the experiment*: The number of mice used in an experiment can of course be highly variable. The need for statistical analysis requires higher numbers of mice per experiment. However, limitations due to cost, availability, and the imperative to not harm more mice than necessary require a delicate balance. The sample size required should be determined based on your particular hypothesis and experimental design.
2. *Consider the sex of the mice*: Although often overlooked during initial experimental designs, the sex of the mice can have a profound impact on results. Aside from the obvious effects of hormones and reproductive biology, numerous studies have found subtle as well as overt differences in immune responses and gut permeability based on the sex of the mouse. For example, male mice are more susceptible to DSS-induced colitis [11], and some studies have indicated they may be more susceptible to some enteric infections, including *C. jejuni* [12]. The mechanism behind this is largely unknown; however, it may be linked to intestinal permeability.
3. *Consider the age of the mice*: The age of mice can also significantly affect experimental results. The intestinal microbiota, cell turnover, and immune responses can all shift as the mice age. This may or may not impact overall results; however, it can contribute to variability between mice of different ages. A common age range to use for the purposes of *C. jejuni* infection experiments is 6–8 weeks; however, there is no set age requirement.
4. *Consider the source of the mice*: Although mice from different facilities may have essentially identical genetic backgrounds, their intestinal microbiota can vary significantly, not only between facilities but also between litters, and subtle differences can appear even between cages [13, 14]. Once again, depending on the experimental design, this may or may not be a significant factor; however, the intestinal microbiota plays a substantial role in the maturation of the immune system and may significantly affect colonization by enteric pathogens. Steps should be taken to minimize potential differences in the microbiota between experimental and control mice.
5. *Consider caging numbers and sanitation*: The numbers of mice in each cage can have a psychological impact on mice, but for

the purpose of studying enteric infections, the main impact of the number of mice in a cage, and the frequency at which the cage is changed/cleaned, is the degree to which the mice continue to reinoculate themselves and their cage-mates. Coprophagic behavior and general contamination lead to a steady stream of new bacteria being inoculated into the mouse, which can affect overall colonization numbers and the rate at which the bacteria are cleared.

6. *Consider mouse strains and controls*: In the protocols outlined here, we will describe infections using *Sigirr*$^{-/-}$ mice, as was published by Stahl et al. [6]. However, previous studies have used mice deficient in IL-10 [5], MyD88 [15], TLR4 [6], TLR2 [6], Muc1 [16], and other knockout strains not listed here, plus a variety of different wild-type mouse strains. Each strain may impact experimental conditions and the necessary experimental protocols for studying them.

3.2 Inoculating Mice with C. jejuni

1. Take a sterile 1-mL syringe and attach an oral gavage needle to the end of the syringe (*see* **Note 7**). Rinse the needle and syringe in sterile PBS, and then load the needle with the necessary amount of 50 mg/mL vancomycin (100 μL per mouse) (*see* **Note 8**).

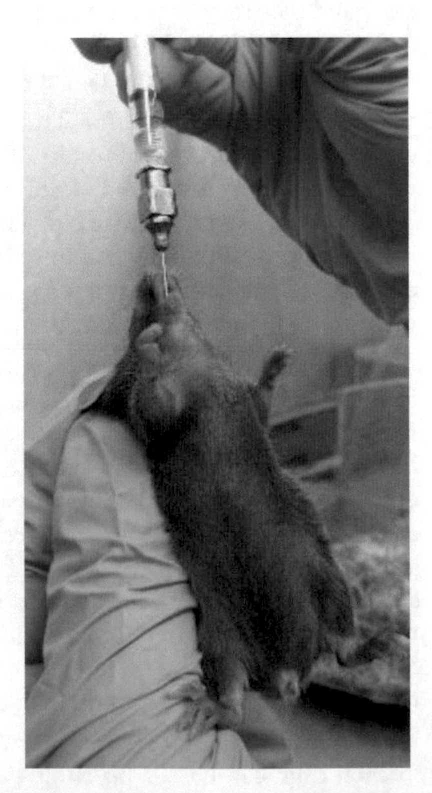

Fig. 1 Scruffed C57BL/6 mouse being gavaged with 100 μL of 50 mg/mL vancomycin in sterile PBS

2. Scruff the mouse tightly so that its head is upright and secure (*see* **Note 9**) (Fig. 1). The mouse's forelimbs should be unable to reach its mouth, its head should not be able to move side to side, and its tail and lower half should be supported and secure. Gently insert the feeding needle into the mouse's throat. Gently and quickly inject 100 μL of the diluted vancomycin and then remove the oral gavage needle from the mouse. Release the mouse into a fresh cage.
3. Roughly four hours following antibiotic treatment (*see* **Note 10**), prepare the inoculating culture of *C. jejuni* (*see* **Note 11**). Typically, this culture should contain approximately 10^8 colony-forming units (CFUs)/mL to allow for an inoculum of ~10^7/100-μL gavage; however, this number can be adjusted as needed based on the desired experimental conditions.
4. In a similar fashion to the initial antibiotic treatment (**steps 1** and **2**), take a 1-mL syringe, attach an oral gavage needle, and then wash both the needle and syringe in sterile PBS. Load the *C. jejuni* inoculum into the syringe, scruff the mouse, and orally gavage the inoculum into the mouse.

3.3 Monitoring the Mice for Signs of Infection

Typically, mice do not display significant outward signs of infection in response to *C. jejuni*. Even in the case of *Sigirr*$^{-/-}$ mice, outward symptoms are usually mild and non-life threatening. This is likely due to infection being largely limited to the intestine, with little systemic involvement or damage. To evaluate infection, monitor several key symptoms.

1. *Weight loss*: This is the most easily monitored category; however, it is not necessarily very informative. A simple oral gavage with antibiotics will trigger weight loss in the days following treatment, and the initial stress of gavaging an inoculum of *C. jejuni* will also trigger weight loss in the days following inoculation. Neither response necessarily indicates that the infection has proven successful. The mice typically regain weight after an initial drop, and there may or may not be any correlation with the severity of intestinal inflammation. Only in the most severe cases will the mouse's weight remain low or continue to drop. In these circumstances, a humane end point should be set based on guidelines set by your local facility and animal care committee guidelines.
2. *Behavioral signs*: A few behavioral signs are typically seen in mice undergoing *C. jejuni* infection. A noticeable decrease in activity might be evident, and in more severe infections, some hunching and piloerection might be observable. Severe signs of distress that may require intervention should be very rare.
3. *Stool consistency*: Although not always a reliable indicator of the extent of inflammatory damage in the intestine, if at any point during infection, the mouse develops watery or bloody

diarrhea, this is a symptom of significant inflammation within the intestine. The development of diarrhea to this extent however is relatively rare. More commonly, the stool may become soft, or will be accompanied by a significant amount of mucus, which are also strong signs of active inflammation within the cecum or colon. Conversely, mild-to moderate inflammation, especially if it is limited to the cecum, may not significantly affect stool consistency in the mice.

3.4 Necropsy

1. The euthanization of mice should be done according to guidelines recommended by your local animal facility and/or animal care committee. A typical example would be to anesthetize mice with isoflurane, followed by cervical dislocation.
2. Once the mice have been euthanized, moisten their abdomens with 70% ethanol, and then use a sharp pair of surgical scissors to make a small cut in the lower, central abdomen (Fig. 2a). From there, cut through the peritoneum to either side of the abdomen, to expose the whole abdominal cavity (Fig. 2b).
3. With a pair of tweezers, gently grip the mouse colon as close to the rectum as possible, cut it with a pair of scissors, and then gently pull on the colon. The colon and cecum, up to the small intestine should easily separate from the mesentery. To dislodge the small intestine without losing the mesenteric lymph

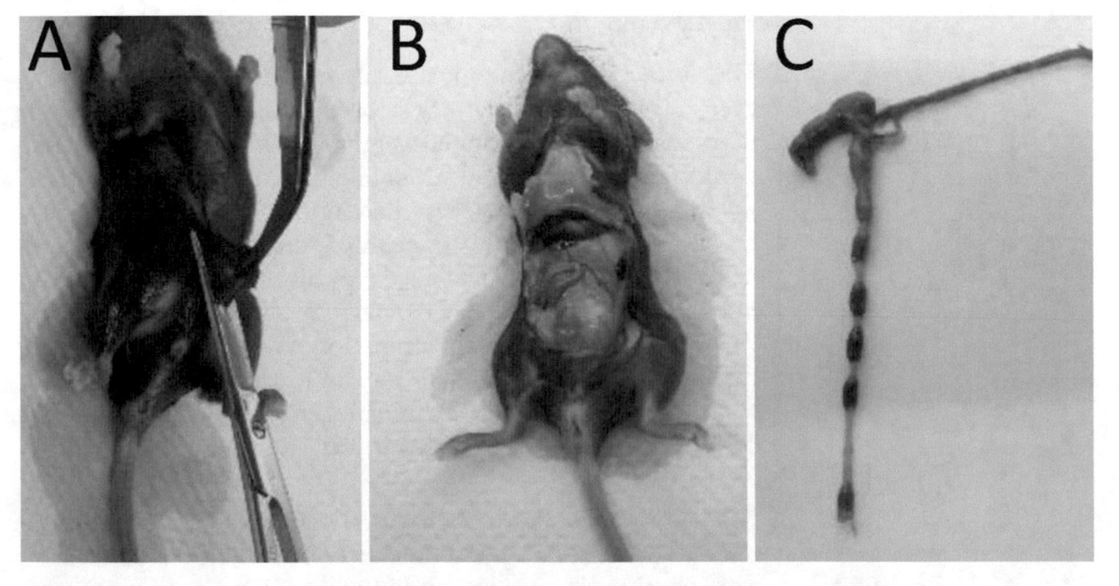

Fig. 2 Mouse necropsy. (**a**) The first cut at the base of the abdomen should cut through the peritoneum. (**b**) From the first incision, cutting around the edges of the abdomen will expose the abdominal cavity. (**c**) By gripping and cutting near the end of the colon, then giving it a firm tug, the intestinal tract can be unraveled and separated from the mesentery. Visible are the colon, cecum, ileum, and mesenteric lymph nodes

nodes, locate the lymph node nearest the intersection of the ileum and cecum (they are often enlarged in response to infection). Cut the mesentery, and then pull on the ileum to dislodge it and cut the ileum several centimeters proximal from the cecum. This should allow for the colon, cecum, ileum, and mesenteric lymph nodes to be removed from the abdominal cavity as a single unit (Fig. 2c).

3.5 Assessing Pathogen Burden (Counting Colony-Forming Units, CFU)

To analyze the severity and extent of *C. jejuni* infection, pathogen translocation to other organ systems can be evaluated. The tissues of interest (spleen, liver, mesenteric lymph nodes) should be homogenized, serially diluted, and plated onto selective agar plates. Counting how many single bacterial colonies form from each organ homogenate allows the calculation of the number of *C. jejuni* per gram of tissue.

1. Prepare Karmali-selective agar plates.
2. Weigh a sterile, 2-mL microcentrifuge tube containing 1 mL of PBS and a sterile metal bead for each organ you want to collect.
3. Remove target mouse organs under sterile conditions as described under the necropsy procedure above. Place them in each pre-weighed tube and weigh again to calculate the total weight of the tissue (current weight, initial empty weight = tissue weight).
4. Place tubes in tissue homogenizer and shake for 5 min (or a sufficient speed/duration to completely homogenize tissue).
5. Serially dilute the homogenate and plate in triplicate on selective plates under sterile conditions (*see* **Note 12**).
6. Incubate at 37 °C for 24–48 h under microaerobic conditions or until colonies are easily visible (*see* **Note 13**).
7. Count the colony forming units (CFUs) in the first dilution where defined single colonies appear.
8. Determine the number of bacteria per gram of tissue, factoring in the dilution and the total tissue weight.

3.6 Histological Analysis

Assessing changes in tissue morphology and cellular integrity on a microscopic level is invaluable to evaluate host susceptibility and to judge the severity of inflammation and other immune responses triggered by *C. jejuni* in the intestine. Additionally, staining techniques can allow for the determination of *C. jejuni* localization within the intestine. Numerous methods for tissue fixation and staining exist. Choosing the appropriate fixation method depends on the subsequent staining protocol and the cellular target of interest. Additionally, different regions of the intestine provide distinct information on the course of infection and exhibit dissimilar colonization patterns. For all methods, the sampling should be performed as quickly as possible, and unless otherwise indicated, tissues should be placed on ice or kept on 4 °C immediately upon collection, until they are ready for further processing.

3.6.1 Formalin Fixation

Formaldehyde-based fixations are the most widely used tissue fixation method. For most tissues a 10% formalin solution is a sufficient fixative. However, as formalin may contain up to 10% methanol, which can interfere with cellular targets, fixation with a 4% paraformaldehyde solution may be preferred in some cases (*see* Subheading 3.6.2 for details). The volume ratio for fixative to tissue should be no less than 15–20:1.

1. Collect small (less than 0.5 cm) cross sections of the intestine as quickly as possible following euthanization of the mouse (*see* **Note 14**).
2. Completely immerse the tissue in formalin-filled tubes for at least one day or up to several weeks and keep it at room temperature or 4 °C. Carefully place the tissue into a tissue cassette and soak it in 70% ethanol for 24 h (*see* **Note 15**).
3. Dehydrate the tissue prior to paraffin embedding using the following steps:
   (a) 2 × 1 min in 70% ethanol
   (b) 2 × 1 min in 80% ethanol
   (c) 2 × 1 min in 95% ethanol
   (d) 1 × 30 s in 100% ethanol
   (e) 2 × 40 s in 100% ethanol
   (f) 3 × 1 min in xylene
   (g) 1 × 30 s in paraffin wax heated to 56–58 °C
   (h) 3 × 1 min in heated paraffin wax and finally embed tissues into paraffin blocks
4. Trim paraffin blocks as necessary, mount onto a microtome, and cut at 3–10 μm thickness (6 μm usually gives the best results). This will produce a ribbon of sequentially cut tissue sections.
5. Place paraffin ribbon in water bath at 40 °C.
6. Mount sections onto a fresh glass microscope slide.
7. Allow sections to air-dry for 30 min and then bake sections in the oven at 50 °C overnight.
8. Proceed with deparaffinization (Subheading 3.7) and desired staining protocol (Subheadings 3.8–3.10).

3.6.2 4% Paraformaldehyde Fixed Frozen

1. Collect small (less than 0.5 cm) cross sections of the intestine as quickly as possible following euthanization of the mouse (*see* **Note 14**).

2. Immerse tissue in 4% paraformaldehyde-filled tubes for a minimum of 1 h on ice or at 4 °C. The volume ratio for fixative to tissue should be no less than 15–20:1.
3. Wash the tissue three times for 2 min in PBS-azide (tissue can be stored at −20 °C at this point).
4. Submerge the sample in 20% sucrose until it sinks to the bottom of the vial (4 h to overnight).
5. Fill cryoprotective embedding medium (OCT) in a base mold.
6. Transfer tissues into OCT (*see* **Note 16**).
7. Fill container with isopentane and place it on dry ice.
8. Use forceps to hold base mold edge and place it into isopentane solution until tissue solidifies (30 s to 1 min) (*see* **Note 17**).
9. Store frozen tissue block at −80 °C until sectioning.
10. For sectioning, attach frozen tissue block to the cryostat chuck.
11. Allow tissue to equilibrate to the cryostat temperature (−20 °C) before cutting sections.
12. Cut sections at 5–10 μm and place onto slide.
13. Dry at room temperature until sections are firmly adhered to the slide.
14. Store in −80 °C freezer until use.

3.6.3 Carnoy's Fixation

This fixation process preserves the mucus layer of the intestine and can be used to stain the carbohydrate-rich mucosubstances (glycosaminoglycans) with periodic acid-Schiff staining.

1. Collect small portions of the intestine, preferably still containing the luminal content so as to preserve the mucus layer (*see* **Note 14**).
2. Immerse the tissue in Carnoy's solution and keep it on ice for 2 h.
3. Rinse the tissue in 100% ethanol. Tissue can be stored in 100% ethanol at 4 °C until ready for further processing.
4. Place the tissue in a tissue cassette and embed in paraffin as described above (Subheading 3.6.1 **steps 3–8**)

3.7 Deparaffinization and Rehydration

For all paraffin-embedded tissue sections, the same basic deparaffinization and rehydration protocol has to be followed:

1. Deparaffinization: Place slides in a sealed slide holder. Place slide holder in an 80 °C water bath for 20 min (*see* **Note 18**).
2. Rehydration: Place slides in metal or glass slide holder. Rehydrate by placing slide holder in the following solutions:

   (a) 4 × 3 min in xylene

   (b) 2 × 3 min in 100% ethanol

   (c) 1 × 3 min in 95% ethanol

(d) 1 × 3 min in 70% ethanol

(e) 1 × 5 min in distilled water

3.8 H&E Staining

Hematoxylin stains basophilic structures purple-blue, while eosin stains acidophilic substances red-pink. This results in the following staining pattern: nuclei, blue-black; cytoplasm, different shades of pink; muscle fibers, deep pinkish red; fibrin, dark pink; red blood cells, orange/cherry-red. H&E stain is thus a suitable method to visualize overall tissue structure and integrity.

1. Deparaffinize and rehydrate sections as described above (Subheading 3.7).
2. Place sections in Gill's hematoxylin solution for 5 min, and then wash in tap water for 1 min.
3. Stain section blue (i.e., set hematoxylin stain) in either lithium carbonate or Scott's tap water for 1 min, and then wash in tap water for 1 min.
4. Place sections in 1% acid alcohol for 30 s, and then wash in tap water for 1 min.
5. Place sections in 1% eosin yellowish for 5 min, and then wash in tap water for 1 min.
6. Dehydrate using the following steps:
   (a) 2 × 10 min in 95% ethanol
   (b) 2 × 3 min in 100% ethanol
   (c) 3 × 5 min in xylene
7. Mount sections with mounting medium.

3.9 Periodic Acid-Schiff Staining

Periodic acid-Schiff stains carbohydrate-rich structures red resulting in the following staining pattern: glycogen, mucin, fungi = red/purple; nuclei/background = blue

1. Deparaffinize and rehydrate sections as described above (Subheading 3.7).
2. Oxidize in 0.5% periodic acid solution for 5 min and rinse with distilled water.
3. Place sections in Schiff reagent for 20 min (sections will become light pink).
4. Wash in tap water for 5 min (sections will become dark pink).
5. Counterstain in Gill's hematoxylin solution for 90 s and wash in tap water for 5 min.
6. Dehydrate using the following steps:
   (a) 2 × 10 min in 95% ethanol
   (b) 2 × 3 min in 100% ethanol
   (c) 3 × 5 min in xylene
7. Mount section using mounting medium.

3.10 Immunofluorescent Staining

In order to define molecular mechanisms in the context of disease, it is helpful to localize the expression of specific proteins within the diseased tissue. One approach to this goal is through immunofluorescent staining. To accomplish this, the different fixation methods described above require modified staining protocols.

Formalin and Carnoy's Fixed:

1. For all paraffin-embedded slides, follow deparaffinization and rehydration steps outlined in Subheading 3.7.
2. Antigen retrieval (optional): Place the slides in a slide holder containing antigen retrieval buffer preheated to roughly 90 °C. Place in steamer for 30 min to keep samples at 90–100 °C but to prevent boiling. Allow the slides to cool down to room temperature and wash for 2 min in distilled water.
3. Draw hydrophobic border around tissues with a Dako™ or grease pen.
4. Using a plastic transfer pipette, drop a small amount of blocking buffer onto the tissue. (It should be contained within the hydrophobic border.) Incubate for 1 h at room temperature.
5. Pour off the blocking buffer and carefully, without touching the tissue, wipe off the remaining buffer with a lint-free tissue.
6. Incubate with the primary antibody at 4 °C overnight or for 1–2 h at room temperature (*see* **Note 19**).
7. Wash with PBS 3 × 5 min (following the same procedures as described under **steps 4** and **5**.)
8. Add secondary antibody (diluted 1:1000 in antibody dilution buffer) for 1 h at room temperature.
9. Wash with PBS 2 × 5 min and distilled water 1 × 5 min.
10. Completely dry off the slides and remove any remaining grease stains from the Dako Pen.
11. Mount sections with a small drop of mounting media containing DAPI.
12. Place coverslip on top, carefully avoiding bubbles. Press the coverslip with a pipette tip to remove bubbles should any form around the tissues.
13. Seal sides with clear nail polish.

PFA-Fixed Frozen Sections:

1. Allow the slides to thaw for 30 s at room temperature.
2. Wash with PBS to rinse off the remaining OCT.
3. Proceed with **step 2** (or **3**) in the protocol outlined above for formalin- and Carnoy's-fixed samples.

3.10.1 Pathology Scoring

To compare the degree of inflammation between mice, a standard pathology scoring protocol needs to be developed. The protocol described below is the scoring method published by Stahl et al. [6]; however, there is no single ideal standard for evaluating tissue inflammation. Additional criteria can be added, or the relative weighting of criteria can be altered based on what is expected and what is considered important for the purpose of the experiment. Pathology scoring is typically used with H&E-stained tissues and should be conducted with a minimum of two or three researchers familiar with the scoring system, but blinded to the experimental conditions of each slide. After evaluating the tissue for each scoring criteria, add them together for a final score. Statistics can be done to compare total scores, or to evaluate potential differences within each criteria. An example of each criterium is provided in Fig. 3.

1. In the submucosa, inflammation is often accompanied by submucosal edema, i.e., a localized collection of fluid (0 = no edema, 1 = mild edema, 2 = moderate edema, 3 = severe edema).
2. Crypt hyperplasia (an increase in crypt depth due to rapid epithelial cell proliferation). Scores represent a percent increase over the average crypt depth of uninfected tissue (0 = no change, 1 = 1–50 %, 2 = 51–100 %, 3 = >100 %).
3. The number of goblet cells per crypt has been known to decrease as inflammation becomes more severe. This may be due to mucus hypersecretion, goblet cell loss, reduced goblet cell differentiation, or some combination of the three. Depletion should be assessed relative to an uninfected control (0 = no change, 1 = mild depletion, 2 = severe depletion, 3 = absence of goblet cells).

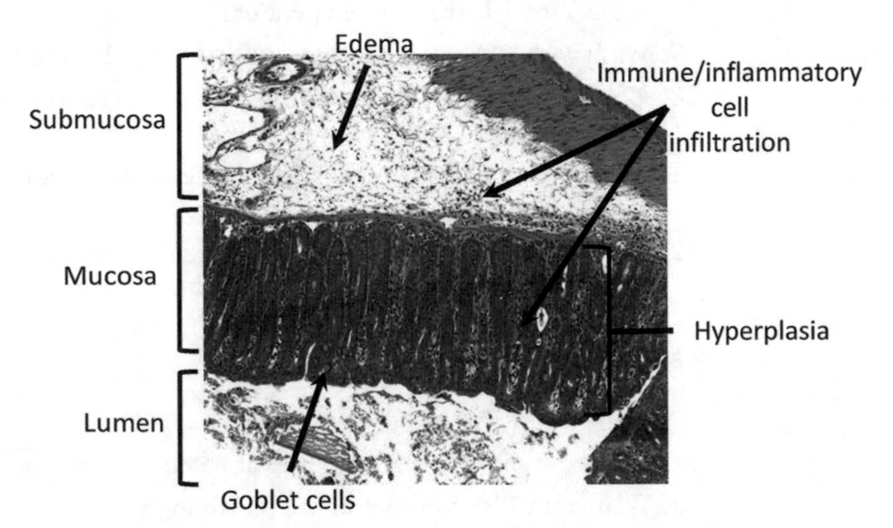

Fig. 3 H&E stained cecal section of a *Sigirr*$^{-/-}$ mouse colonized with *C. jejuni* 81–176. Indicated are submucosal edema, mucosal and submucosal immune cell infiltrates, crypt hyperplasia, and the location of the remaining goblet cells

4. One of the most important factors in assessing the severity of inflammation is the degree of damage to the intestinal epithelium (0 = no pathological changes detectable, 1 = epithelial desquamation [few cells sloughed, epithelial surface rippled], 2 = erosion of the epithelial surface [epithelial surface is rippled, damaged], 3 = epithelial surface severely disrupted/damaged, large amounts of cell sloughing. 4 = ulceration). Where ulceration has occurred, an additional 1 point is added for each 25% fraction of the tissue that is ulcerated. For example, a very small ulcer would have a score of 4 (4 + 0), and an ulcer comprising half the tissue section would have a score of 6 (4 + 2). A maximum score of 8 (4 + 4) will represent a near complete loss of the epithelium.
5. Count the number of polymorphonuclear cells (PMNs) and mononuclear cells present within the submucosa. They should be counted per 400× magnification field and scored as follows: 0 = <5 cells, 1 = 5–20 cells, 2 = 21–60 cells, 3 = 61–100 cells, 4 = >100 cells.
6. Inflammatory cells will also invade the mucosa. As with the submucosa, PMNs and mononuclear cell infiltration (per 400× magnification field) should be evaluated and scored: 0 = <5 cells, 1 = <20 cells, 2 = 20 to 50 cells, 3 = > 50 cells.

4 Notes

1. Plating tissue homogenates on a selective plate for *Campylobacter* is necessary to eliminate the growth of contaminating bacteria from the intestinal microbiota. There are a number of potential options, but a common and effective selective choice is to use *Campylobacter* agar plates, with Karmali-selective supplements.
2. An easy and cost-effective way is to prepare a mold by wrapping tinfoil around a tube (choosing the tube size depending on the size/number of your samples) and filling the tinfoil mold with the cryoprotective medium.
3. Ensure that the Schiff reagent has not deteriorated by pouring 10 mL 37% formalin into a reagent tube. Add a few drops of the Schiff reagent to be tested. A working Schiff reagent will turn a reddish-purple color almost immediately. A delayed reaction showing a deep blue-purple stain indicates a deteriorated Schiff reagent.
4. Other methods of antigen retrieval can also be done. The heated sodium citrate buffer method listed here is often, but not always effective, depending on the circumstances.
5. The animal source of the serum for the blocking buffer should match the animal source of the secondary antibody.

6. The animal source of the primary and secondary antibodies should be selected carefully to avoid cross-reactions. When well planned, as many as three or four protein targets can be stained for simultaneously.
7. It is usually recommended to store the gavaging needle in 70% ethanol to prevent bacterial contamination. Rinsing in sterile PBS thoroughly will remove the ethanol prior to inoculating the mice.
8. Vancomycin was used since it has the ability to kill most Gram-positive bacteria, and will have little impact on *C. jejuni* itself. Other types of antibiotics could be substituted under the following conditions: It should not be toxic to the mice, it should have a broad enough impact to open niches within the GI tract for *C. jejuni* colonization, and if it is toxic for *C. jejuni* as well, enough time should pass between antibiotic treatment and inoculation for the antibiotic to fully dissipate.
9. There are a number of different techniques for scruffing mice. The main necessity is to keep the mouse relatively still and its head immobile. If not, the mouse may be injured by the gavaging needle as it struggles during the injection.
10. The 4-h period between antibiotic pretreatment and inoculation is somewhat arbitrary. It is meant to give the antibiotic enough time to take effect, and not enough time for any other Gram-negative bacteria that may be resistant to the vancomycin to begin to overgrow the intestine. More or less time may be desirable, especially if a different antibiotic is being used.
11. There are also numerous ways of growing a culture of *C. jejuni* for inoculation depending on the materials available. It is recommended to initially grow the bacteria on a motility plate (0.4% agar in MH broth) to confirm motility and then grow a culture overnight in either a liquid or biphasic culture. It is also optimal to have the inoculating culture in a logarithmic stage of growth (~10^7–10^8 CFUs) and to confirm that the *C. jejuni* are in fact motile if it is supposed to be a motile strain.
12. It is helpful to serially dilute the tissue homogenates in a 96-well plate and subsequently plate them with an adjustable-spacer multichannel pipette.
13. Growth of *C. jejuni* should be done under microaerobic conditions if possible; however, *C. jejuni* will also grow on Karmali-selective agar plates at higher oxygen conditions for the purpose of CFU counting, although the bacteria will be less viable after 48 h under high oxygen conditions.
14. Do not try to push out the stool. The sample can be fixed together with the luminal contents. The subsequent histological analysis will be easier as the in situ location of the bacteria in the lumen/mucus layer will be preserved.

15. Multiple tissue portions can be placed in one cassette. This makes comparing groups, e.g., control versus treated groups, easier as they will be placed together on one microscope slide.
16. Carefully arrange tissues so they do not touch each other, are all in one plane, and are located toward the bottom, so the tissue is easily exposed when sections are cut.
17. If the block is left in isopentane too long, it may crack.
18. This step is intended to melt the paraffin before adding the slide to the first xylene wash. It can be omitted if the incubation time in the xylene is increased to compensate.
19. Dilute the antibody in antibody dilution buffer. Typical dilutions range from 1:50 to 1:500; however, this will need to be optimized for each specific antibody; 1:100 is a common default dilution if the optimum dilution is unknown. Usually 100 μL of antibody solution per tissue sample/marked area will suffice.

References

1. Newell DG (2001) Animal models of *Campylobacter jejuni* colonization and disease and the lessons to be learned from similar *Helicobacter pylori* models. Symp Ser 30:57S–67S
2. Babakhani FK, Bradley GA, Joens LA (1993) Newborn piglet model for campylobacteriosis. Infect Immun 61(8):3466–3475
3. Fox JG, Ackerman JI, Taylór N et al (1987) *Campylobacter jejuni* infection in the ferret: an animal model of human campylobacteriosis. Am J Vet Res 48(1):85–90
4. Bereswill S, Fischer A, Plickert R et al (2011) Novel murine infection models provide deep insights into the “menage a trois” of *Campylobacter jejuni*, microbiota and host innate immunity. PLoS One 6(6):e20953. doi:10.1371/journal.pone.0020953
5. Mansfield LS, Bell JA, Wilson DL et al (2007) C57BL/6 and congenic interleukin-10-deficient mice can serve as models of *Campylobacter jejuni* colonization and enteritis. Infect Immun 75(3):1099–1115. doi:10.1128/IAI.00833-06
6. Stahl M, Ries J, Vermeulen J et al (2014) A novel mouse model of *Campylobacter jejuni* gastroenteritis reveals key pro-inflammatory and tissue protective roles for Toll-like receptor signaling during infection. PLoS Pathog 10(7):e1004264. doi:10.1371/journal.ppat.1004264
7. Chang C, Miller JF (2006) *Campylobacter jejuni* colonization of mice with limited enteric flora. Infect Immun 74(9):5261–5271. doi:10.1128/IAI.01094-05
8. Lippert E, Karrasch T, Sun X et al (2009) Gnotobiotic IL-10; NF-kappaB mice develop rapid and severe colitis following *Campylobacter jejuni* infection. PLoS One 4(10):e7413. doi:10.1371/journal.pone.0007413
9. Sham HP, Yu EY, Gulen MF et al (2013) SIGIRR, a negative regulator of TLR/IL-1R signalling promotes Microbiota dependent resistance to colonization by enteric bacterial pathogens. PLoS Pathog 9(8):e1003539. doi:10.1371/journal.ppat.1003539
10. Khan MA, Steiner TS, Sham HP et al (2010) The single IgG IL-1-related receptor controls TLR responses in differentiated human intestinal epithelial cells. J Immunol 184(5):2305–2313. doi:10.4049/jimmunol.0900021
11. Mahler M, Bristol IJ, Leiter EH et al (1998) Differential susceptibility of inbred mouse strains to dextran sulfate sodium-induced colitis. Am J Physiol 274(3 Pt 1):G544–G551
12. Strachan NJ, Watson RO, Novik V et al (2008) Sexual dimorphism in campylobacteriosis. Epidemiol Infect 136(11):1492–1495. doi:10.1017/S0950268807009934
13. Rogers GB, Kozlowska J, Keeble J et al (2014) Functional divergence in gastrointestinal microbiota in physically-separated genetically identical mice. Sci Rep 4:5437. doi:10.1038/srep05437
14. Thoene-Reineke C, Fischer A, Friese C et al (2014) Composition of intestinal microbiota in immune-deficient mice kept in three different housing conditions. PLoS One 9(11):e113406. doi:10.1371/journal.pone.0113406

15. Watson RO, Novik V, Hofreuter D et al (2007) A MyD88-deficient mouse model reveals a role for Nramp1 in *Campylobacter jejuni* infection. Infect Immun 75(4):1994–2003. doi:10.1128/IAI.01216-06

16. McAuley JL, Linden SK, Png CW et al (2007) MUC1 cell surface mucin is a critical element of the mucosal barrier to infection. J Clin Invest 117(8):2313–2324. doi:10.1172/JCI26705

Chapter 16

Metabolomic Analysis of *Campylobacter jejuni* by Direct-Injection Electrospray Ionization Mass Spectrometry

Robert M. Howlett, Matthew P. Davey, and David J. Kelly

Abstract

Direct-injection mass spectrometry (DIMS) is a means of rapidly obtaining metabolomic phenotype data in both prokaryotes and eukaryotes. Given our generally poor understanding of *Campylobacter* metabolism, the high-throughput and relatively simple sample preparation of DIMS has made this an attractive technique for metabolism-related studies and hypothesis generation, especially when attempting to analyze metabolic mutants with no clear phenotype. Here we describe a metabolomic fingerprinting approach with sampling and extraction methodologies optimized for direct-injection electrospray ionization mass spectrometry (ESI-MS), which we have used as a means of comparing wild-type and isogenic mutant strains of *C. jejuni* with various metabolic blocks.

Key words Electrospray ionization mass spectrometry, Metabolome, Metabolomics, Direct injection, DIMS, Metabolite extraction, Fingerprinting

1 Introduction

Metabolomics is an approach to understand cellular metabolic pathways, either at the level of individual pathways or a specific set of metabolites (targeted approach) or through an overall assessment of a cell's global metabolite content (nontargeted approach). While methods aimed at identifying specific metabolites and analyzing their cellular concentrations are not new, it has only recently been possible to analyze and process large numbers of metabolites simultaneously using high resolution nuclear magnetic resonance (NMR) or mass spectrometry (MS)-based techniques. This has enabled methods to be developed for the preliminary analysis of "silent mutations" in genes involved in central and noncentral metabolic pathways, by comparing the global metabolite profiles of wild-type versus mutated cells [1].

James Butcher and Alain Stintzi (eds.), *Campylobacter jejuni: Methods and Protocols*, Methods in Molecular Biology, vol. 1512, DOI 10.1007/978-1-4939-6536-6_16,

Dunn et al. [2] noted the many techniques are utilized for sampling the metabolome, and at present no single technology dominates; each has its own biased metabolite detection and strengths for certain applications. Electrospray ionization time-of-flight mass spectrometry (ESI-TOF-MS) is a technique that has been used for high-throughput fingerprinting in both plants and bacteria [1, 2]. Although ESI-TOF-MS-based metabolomics can suffer from a lack of mass accuracy, its high sensitivity and high throughput make it an ideal candidate for establishing a strategy for the initial analysis of a large number of cellular metabolites. Similar methodologies were originally utilized for high-throughput clinical screening; however, there have now been several microbial applications of this technique. Initially used in bacterial strain discrimination [3, 4], Kaderbhai et al. [5] showed that direct-injection ESI-TOF-MS techniques were capable of discriminating between tryptophan metabolism mutants in *E. coli*. Our recent work has shown that putative metabolite allocation to ESI-TOF-MS outputs can enable visualization of metabolic blocks in *C. jejuni* amino-acid metabolism mutants [1]. ESI-TOF-MS techniques are advantageous over other techniques such as NMR when attempting a nontargeted analysis, due to their far higher sensitivity; however, the absence of a chromatographic step makes unambiguous metabolite identification impossible. Li et al. [6] have since built on the DIMS fingerprinting techniques in [1] by separating and identifying the key metabolites of antibiotic-resistant mutants of *C. jejuni* using UHPLC-MS. For this reason it is important that DIMS techniques are used as a "first step" for hypothesis formation toward the function of putative metabolic genes before rigorous metabolite identification or phenotypic analysis is performed. The potential for analyzing altered metabolic fluxes as organisms experience different conditions, such as changes to temperature or growth phase, also exists [7].

Quenching of metabolism and metabolite extraction are areas of intense debate and concern in metabolomics, as many metabolite pools turn over on a time scale of the same order (or shorter) as the time taken to quench metabolic reactions. When using the adenylate energy charge parameter (ECP) as an indicator of the maintenance of adenine nucleotide levels during quenching, some techniques have been shown to be too slow, as well as resulting in leakage of many cellular metabolites [8]. ESI-TOF-MS direct-injection methods enable a quenching methodology to be employed that results in no metabolite loss, through the freezing of whole cell pellets in liquid nitrogen (−196 °C). However, in the case of bacterial cells like *C. jejuni*, this does require a short centrifugation step [1]. Although some metabolic changes will inevitably occur during such processing, for the particular approach we describe here, where wild-type and mutant strains are subjected to the same quenching and extraction conditions, valid comparisons are possible. The efficacy of metabolite extraction techniques has been shown to be organism specific [8], but many comparative studies have shown cold methanol-based methodologies to be the most efficient for extracting the broadest range of metabolites [8–10]. For this reason a simple cold methanol-based extraction protocol was adopted in our work, enabling high reproducibility and extraction of a broad range of metabolites [1, 11]. Below, we provide details of the methods we have used to obtain metabolite profiles from *C. jejuni* strains using DIMS, based on our recent publication [1].

2 Materials

2.1 Growth Media

1. Columbia agar containing 5% (v/v) lysed horse blood and 10 μg ml^{-1} amphotericin B and vancomycin (CA-AV) for subculturing of strains.
2. For liquid growth of isogenic mutants, use brain-heart infusion with 5% (v/v) fetal calf serum (BHI-FCS). No selective antibiotics are used so as not to affect metabolism (*see* **Note 1**).

2.2 Metabolite Extraction

1. 2 mL polypropylene tubes or other solvent-resistant types.
2. Liquid nitrogen.
3. HPLC grade methanol/chloroform 1:1 (v/v) (*see* **Note 2**).
4. Stainless steel ball bearings (5 mm diameter) (*see* **Note 3**).
5. Ultra-high purity (UHP) water.

2.3 Equipment

1. An MACS cabinet (Don Whitley Scientific, Shipley, UK) or other micro-aerobic incubator, to maintain micro-aerobic growth conditions [we routinely use 10% (v/v) O_2, 5% (v/v) CO_2, and 85% (v/v) N_2].
2. A temperature-controlled microcentrifuge.
3. A cell disruptor/homogenizer.
4. A liquid chromatography (LC) electrospray ionization (EI) time-of-flight (TOF) spectrometer (such as an LCT spectrometer, Waters Ltd, Manchester, UK) with an automated Waters 2695 Separations Module combining a high-performance liquid chromatography (HPLC) pump and autosampler (Waters, Hemel Hempstead, UK) with a LockSpray™ interface.

3 Methods

Throughout all stages, it is important to record as much metadata as possible regarding differences in sample preparation (i.e., different days of extraction, different media batches) as this will be used in the final multifactorial analysis to look for the influence of methodological variables. The effects of this can be minimized by ensuring as much randomizing of processing where possible.

3.1 Preparation and Sampling of Wild-Type Parent and Isogenic Mutant Strain Cultures

Assume all growth conditions are at 37 °C and under micro-aerobic conditions unless otherwise stated. Good aseptic technique must be maintained while working with cultures.

1. Maintain a culture streaked to single colonies on CA-AV plus selective antibiotics. Use a colony to inoculate a 5 mL BHI-FCS starter culture in 25 mL conical flasks and grow for 16 h (until stationary phase), shaking at 180 rpm. Ensure a minimum of 5 replicates for each strain to be analyzed.
2. Spin down starter cultures (6000 × *g*, room temperature ~22 °C) and resuspend in 1 mL BHI-FCS. Use to inoculate 25 mL cultures of BHI-FCS in 250 mL conical flasks to a starting OD_{600} of 0.1.
3. Grow cultures under micro-aerobic conditions with adequate shaking until they are at the required growth phase for sampling (e.g., mid log phase or stationary). This can be checked by taking the OD_{600} of the culture and comparing to a known growth curve for the strain being sampled (*see* **Note 4**).
4. Using the OD_{600}, calculate the volume required to take a standardized cell density for sampling. For example, sample the equivalent of 1 mL at OD_{600} 1.00 for all samples (in this instance a sample measuring OD_{600} 1.12 would require 893 μL for sampling).
5. Take the calculated volume and quickly place in a 2 mL polypropylene microfuge tube. Spin down the sample in a bench microfuge (14,000 × *g*, 1 min, room temperature) and quickly aspirate off all of the liquid media before flash freezing the pellet in liquid nitrogen. During this stage it is important to be set up so processing time is minimized as much as possible (*see* **Note 5**).
6. Pellets can and must be stored at −80 °C prior to metabolite extraction.

3.2 Metabolite Extraction

Unless otherwise stated, samples, tubes, and solvents are kept on ice at all times.

1. Remove samples from the −80 °C freezer and place the samples in ice bucket, open lids, and quickly add one prechilled 5 mm stainless steel ball bearing (*see* **Note 3**).
2. Add 1 mL of ice-cold HPLC Grade methanol/chloroform 1:1 (v/v) (stored at −20 °C overnight prior to extraction) to each tube.
3. Briefly vortex each tube to homogenize the pellet.

4. Place tubes at −80 °C for 1 h.
5. Remove the samples from the −80 °C freezer and place into a cell disruptor at 4 °C for 1 min.
6. Return samples to −80 °C for 1 h.
7. Remove the samples from −80 °C and briefly vortex.
8. Transfer the sample, minus the steel ball bearing, to a new pre-cooled 2 mL polypropylene tube.
9. Add 400 μL of ice-cold UHP water to the sample and briefly vortex before centrifugation (4000 × *g*, 1 min, 4 °C).
10. Transfer the upper aqueous phase to a new 1.5 mL polypropylene tube.
11. Repeat **step 9** before removing the upper aqueous phase and combining with the previous extraction.
12. Store both the lower nonpolar (chloroform) phase and upper polar (methanol/water aqueous) phase at −80 °C for the subsequent analysis below. The analysis method below can be used on either phases.

3.3 Mass Spectrometry-Based Quantification of Metabolites

We previously performed ESI-TOF-MS on an LCT spectrometer operating with 3.6 GHz time to digital conversion under the control of a MassLynx data system (version 4) running on Windows NT on an IBM compatible PC. The mass spectrometer was operated at a resolution of 4000 (FWHM) at mass 200 *m/z* in positive and negative ion modes at a capillary voltage of 2800 V (positive) and 2500 V (negative), extraction cone at 3 V and sample cone at 20 V with a rangefinder lens voltage of 75 V chosen for detection of masses from 50 to 800 Da. Source temperature was 110 °C and desolvation temperature was 120 °C. Flow rates were 100 L/h for nebulization and 400 L/h for desolvation (*see* **Note 6** and Refs. [1, 11, 12] for further details).

1. Take the solvent phase from the −80 °C and thaw on ice.
2. Samples may be loaded by one of two methods: Either using a syringe pump at a flow rate of 20 μL/min or with an automated Waters 2695 Separations Module combining a HPLC pump and an autosampler with an inject volume of 100 μL at a flow rate of 50 μL/min (*see* **Note 7**).
3. A LockSpray™ interface should be used to give an external standard and allow automated correction of mass measurements (5 ng/μL sulfadimethoxine gives a lock mass of 309.0653 or 311.0814 for negative and positive ionization modes, respectively). Samples can be run in positive or negative ionization mode as described above (*see* **Note 8**). Run three technical replicates for each sample. Spectra should be collected in centroid mode at a rate of one spectrum s^{-1} (0.95 s scan time, 0.05 s

interscan delay) with 180 spectra summed over a 3-min period before being exported from the MassLynx data system as text file peak lists (accurate mass to 4 decimal places vs. ion count) without background subtraction or smoothing.

4. Import text files into Microsoft Excel where the accurate masses of three replicate analyses of each sample can be compared.
5. The accepted range proposed to give the maximum number of peaks with minimum false positives was proposed by Overy et al. [11] to be defined best by plotting the acceptable mass variance as a linear function of the m/z values. The same equations were used in our work: positive mode, $y < 0.00003x + 0.0033$; for negative mode $y < 0.00003x + 0.044$ where y is the standard deviation of the three masses and x is the mean of the three masses. Microsoft Excel spreadsheets should be arranged to automate this process, and those masses found to have a y value within acceptable limits can then have their mean accurate mass and mean response (as % total ion count in order to minimize sample-sample variation and normalize data sets) exported to a separate table (*see* **Note 9**).

3.4 Data Analysis

At this stage each sample should have a list of mean accurate mass and mean response as % total ion count. This can undergo various forms of analysis, and the bioinformatics tools available are in constant development. Here is a brief description of two forms of analysis performed by Howlett et al. [1].

3.4.1 Principal Component Analysis (PCA)

PCA is a multivariate data analysis technique that can be used to simplify the complexity of the dataset, enabling visualization of patterns between samples. In the case of analyzing isogenic mutants, it is used to visualize if the metabolome is altered such that it can be differentiated from the parent or wild-type strain. It is important to also include metadata in any analysis to ensure strains are not separating due to processing variables, such as the day of extraction or the media batch used. Detected masses are used as observations across the strains, treated as variables. Various commercial software packages are available to perform PCA (SPSS, IBM Corp® or SIMCA-P, Umetrics®), but an increasing number of excellent free online packages can be used such as MetaboAnalyst http://www.metaboanalyst.ca/ [13]. These software packages will accept data produced in a similar format and enable the complexity of the data to be broken down into a series of orthogonal variables termed principal components. As well as analyzing strain separation and the effects of methodological variation, these techniques can also be used to detect and remove outliers from future analysis. Scatter plots produced through PCA are also useful as a means of detecting specific mass/molecular weight bins that may be the cause of the separating samples. A large literature is available that gives an in-depth description of PCA, for example [14, 15].

3.5 Assigning of Putative Metabolites

Mean accurate masses can be assigned putative metabolite identities. This has previously been performed through the comparison of monoisotopic masses likely to be present in extracts (such as $[M+H]^+$, $[M+Na]^+$, and $[M-H]^-$ for positive and negative ionization modes, respectively) against a list of over 1900 metabolites from an in-house database compiled from KEGG (http://www.genome.jp/kegg/), BioCyc (http://biocyc.org/), and HumanCyc (http://humancyc.org/) to an accuracy of 0.2 Da [1] (*see* **Note 10**). This was performed automatically using an in-house produced macro in the Microsoft Excel software package (Prof Mike Burrell, The University of Sheffield, UK); however, other online resources such as METLIN (https://metlin.scripps.edu/) and XCMS (https://xcmsonline.scripps.edu/) can also perform similar functions. More information on this process can be found in Davey 2011 [16]. Following the allocation of putative metabolites, relevant statistical analysis can be performed to highlight changes of interest. In order to reduce the number of false-positive results, Benjamini-Hochberg- or Bonferroni-corrected significance levels should be used to isolate the most significant metabolite differences between strains and account for possible type 1 errors [17]. Tools such as the BioCyc (www.biocyc.org) omics viewer can then be used to visualize fold changes in the metabolites detected across the metabolic pathways of *C. jejuni*, where information on genetic and proteomic information for the pathways are also presented [16].

4 Notes

1. Rich growth medium is used when global analysis of metabolism is performed. In these instances it is hoped that as many metabolic pathways are active as possible. Also where isogenic mutants may suffer from growth deficiencies, it is important to optimize growth rate and phase of harvesting to be as close to the isogenic parent as possible. In different circumstances, such as in the analysis of specific growth conditions or when analyzing specific metabolic pathways, a defined growth medium would be preferable.
2. Mix methanol/chloroform 1:1 (v/v) within a fume hood and cool in a −20 °C spark-free laboratory freezer prior to use. WARNING: methanol and chloroform are highly hazardous chemicals; please read all MSDS and wear appropriate personal protective equipment when handling these chemicals throughout the procedure.
3. Ball bearings must be washed in methanol/chloroform 1:1, autoclaved, and oven dried. Before use they should be placed in a −20 °C freezer overnight.

4. Where strains may have differing growth rates/dynamics, it is important to ensure that samples are taken at matched points in their growth cycle. To do this you should perform a growth curve prior to analysis, or take the OD_{600} throughout to monitor growth.
5. Our method employs a 1 min centrifugation before quenching in liquid nitrogen completely halts metabolism, enabling retention of the maximum number of metabolites. However, it is important to note that changes in metabolite levels during centrifugation will undoubtedly occur. For this reason practice is required to ensure rapid manipulation for maximum reproducibility between samples and minimum time spent processing.
6. Other direct-injection mass spectrometers are also suitable for this technique, though MS conditions would have to be slightly altered accordingly. Furthermore, similar analyses have been performed using nuclear magnetic resonance methods [18] and MALDI-TOF [19]; however, sample processing would have to be altered.
7. Samples should be analyzed in a randomized order to minimize effects of day-to-day machine variation. Other autosamplers would also be suitable for this step.
8. Previous analyses have shown that both positive and negative ionization modes are capable of distinguishing between bacterial strains [8]. However, the presence of multiple positive ionization species compared to the relatively simple negative ionization mode data sets makes negative ionization mode preferable for putative metabolite allocation [1].
9. This methodology negates the need for a noise threshold and enables true low-intensity metabolite peaks to be kept.
10. Lists of predicted *C. jejuni*-specific metabolites can be compiled from genome sequence information and exist within, e.g., the KEGG database; however, we have found these to be too restrictive, leaving many mass bins with no metabolite allocated [1]. The use of metabolites lists from multiple databases results in some mass bins receiving multiple metabolite identities. In these instances knowledge of the likely metabolome is required to decipher the most likely identity. Independent confirmation of metabolite identities is required to confirm any hypothesis made based on observed changes in putatively identified metabolite pools.

Acknowledgments

This work was supported by a Biotechnology and Biological Sciences Research Council (BBSRC) Doctoral Training Award, to R.M.H. M.P.D. acknowledges the support of a Wellcome Trust

Value in People award—reference 083772/Z/07/Z. Work in D.J.K.'s laboratory is supported by a grant from the BBSRC. We thank Prof. Paul Quick and Prof. Mike Burrell for help with data processing and Heather Walker for technical help.

References

1. Howlett R, Davey M, Quick WP et al (2014) Metabolomic analysis of the food-borne pathogen *Campylobacter jejuni*: application of direct injection mass spectrometry for mutant characterisation. Metabolomics 10:887–896
2. Dunn WB, Overy S, Quick WP (2005) Evaluation of automated electrospray-TOF mass spectrometry for metabolic fingerprinting of the plant metabolome. Metabolomics 1:137–148
3. Goodacre R, Timmins EM, Burton R et al (1998) Rapid identification of urinary tract infection bacteria using hyperspectral whole-organism fingerprinting and artificial neural networks. Microbiology 144:1157–1170
4. Vaidyanathan S, Rowland JJ, Kell DB et al (2001) Discrimination of aerobic endospore-forming bacteria via electrospray-ionization mass spectrometry of whole cell suspensions. Anal Chem 73:4134–4144
5. Kaderbhai NN, Broadhurst DI, Ellis DI et al (2003) Functional genomics via metabolic footprinting: monitoring metabolite secretion by *Escherichia coli* tryptophan metabolism mutants using FT-IR and direct injection electrospray mass spectrometry. Comp Funct Genomics 4:376–391
6. Li H, Xia X, Li X et al (2015) Untargeted metabolomic profiling of amphenicol-resistant *Campylobacter jejuni* by ultra-high-performance liquid chromatography–mass spectrometry. J Proteome Res 14:1060–1068
7. Howlett RM (2013) Analysis of *Campylobacter jejuni* amino acid metabolism and solute transport systems. PhD Dissertation. The University of Sheffield
8. Faijes M, Mars AE, Smid EJ (2007) Comparison of quenching and extraction methodologies for metabolome analysis of *Lactobacillus plantarum*. Microb Cell Fact 6:27
9. Park C, Yun S, Lee SY et al (2012) Metabolic profiling of *Klebsiella oxytoca*: evaluation of methods for extraction of intracellular metabolites using UPLC/Q-TOF-MS. Appl Biochem Biotechnol 167:425–438
10. Maharjan RP, Ferenci T (2003) Global metabolite analysis: the influence of extraction methodology on metabolome profiles of *Escherichia coli*. Anal Biochem 313:145–154
11. Overy SA, Walker HJ, Malone S et al (2005) Application of metabolite profiling to the identification of traits in a population of tomato introgression lines. J Exp Bot 56:287–296
12. Walker H (2011) Metabolic profiling of plant tissues by electrospray mass spectrometry. In: de Bruijn FJ (ed) Handbook of molecular microbial ecology I—Metagenomics and complementary approaches. Wiley, Hoboken. ISBN 9780470644799
13. Xia J, Wishart D (2011) Web based inference of biological patterns, functions and pathways from metabolomics data using MetaboAnalyst. Nat Protoc 6:743–760
14. Jolliffe I (2002) Principal component analysis. Wiley Stats, New York
15. Eriksson L, Byrne T, Johansson E, et al. (2013) Multi- and megavariate data analysis basic principles and applications. MKS Umetrics AB. ISBN-10: 9197373028
16. Davey MP (2011) Metabolite identification, pathways, and omic integration using online databases and tools. In: de Bruijn FJ (ed) Handbook of molecular microbial ecology: Metagenomics and complementary approaches. Wiley, Hoboken. ISBN 9780470644799
17. Broadhurst DI, Kell DB (2006) Statistical strategies for avoiding false discoveries in metabolomics and related experiments. Metabolomics 2:171–196
18. Raamsdonk LM, Teusink B, Broadhurst D et al (2001) A functional genomics strategy that uses metabolome data to reveal the phenotype of silent mutations. Nat Biotechnol 19:45–50
19. Bessede E, Solecki O, Sifre E et al (2011) Identification of *Campylobacter* species and related organisms by matrix assisted laser desorption ionization–time of flight (MALDI-TOF) mass spectrometry. Clin Microbiol 17:1735–1739

Chapter 17

Methods for Genome-Wide Methylome Profiling of *Campylobacter jejuni*

Kathy T. Mou, Tyson A. Clark, Usha K. Muppirala, Andrew J. Severin, and Paul J. Plummer

Abstract

Methylation has a profound role in the regulation of numerous biological processes in bacteria including virulence. The study of methylation in bacteria has greatly advanced thanks to next-generation sequencing technologies. These technologies have expedited the process of uncovering unique features of many bacterial methylomes such as characterizing previously uncharacterized methyltransferases, cataloging genome-wide DNA methylations in bacteria, identifying the frequency of methylation at particular genomic loci, and revealing regulatory roles of methylation in the biology of various bacterial species. For instance, methylation has been cited as a potential source for the pathogenicity differences observed in *C. jejuni* strains with syntenic genomes as seen in recent publications. Here, we describe the methodology for the use of Pacific Biosciences' single molecule real-time (SMRT) sequencing for detecting methylation patterns in *C. jejuni* and bioinformatics tools to profile its methylome.

Key words Methylation, Motif, Methylome, Restriction modification, *Campylobacter jejuni*, SMRT sequencing, Pacific Biosciences, PacBio, REBASE, NGS

1 Introduction

Prior to next-generation sequencing (NGS), technologies for studying DNA modification on a genome-wide scale were limited to bulk methods that were costly and time consuming and lacked the necessary strand-specificity and nucleotide-level resolution to identify base modifications [1]. However, the advent of NGS technologies such as Pacific Biosciences' single molecule real-time (SMRT) DNA sequencing greatly accelerated and expanded our understanding of bacterial epigenomics. These high-throughput methods allow for simultaneous acquisition of not only genomic but also epigenomic information at the base level giving way for quicker characterization of bacterial methylomes [2].

The methods in this chapter utilize SMRT sequencing technology to study the methylome of *Campylobacter jejuni*. In essence,

James Butcher and Alain Stintzi (eds.), *Campylobacter jejuni: Methods and Protocols*, Methods in Molecular Biology, vol. 1512, DOI 10.1007/978-1-4939-6536-6_17, © Springer Science+Business Media New York 2017

SMRT sequencing observes the catalytic activity of individual DNA polymerase molecules in real time, while the enzyme incorporates fluorescently labeled nucleotides to the complementary template strand [3]. The catalytic activity of the SMRT sequencing polymerase is also sensitive to primary and secondary structures of the DNA [4]. When the polymerase encounters chemical modifications such as a methylated base, its kinetics are changed. The synthesis rate of translocation and binding of the fluorescently labeled nucleotide to the template strand changes in a unique manner associated with the specific type of modification [5]. This is immediately detected by the sequencer, and subsequent analysis would label the "anomaly" for a specific base or chemical modification in the DNA template. Subsequent analysis can also detect sequence motifs in the genome, which are short sequences recognized and modified by methyltransferases [6]. The quick and direct mapping of genome-wide methylation patterns that SMRT sequencing provides has enabled the profiling and study of methylomes in multiple bacterial species [7].

In the field of *Campylobacter* research, several publications have already characterized the methylome profile of four *C. jejuni* strains including the most commonly studied laboratory strains 11168 and 81–176 [6, 8], the abortifacient clone IA3902 that has become the major cause of *Campylobacter*-associated sheep abortions in the United States [8], and the F38011 strain which was isolated from a human bloody diarrhea case [9]. Bioinformatics tools described in this chapter have allowed even deeper analysis of *C. jejuni* methylomes, including: identifying what genes have which specific sequence motif(s), areas of the genome that have higher or lower than average frequency of methylations (hyper- or hypomethylation, respectively), as well as the specific genes found in these regions [8]. Such analyses have never before been performed and will provide future researchers ample information for characterizing their respective *C. jejuni* strain's methylome. Current information on all known and predicted genes and enzymes involved in DNA restriction and modification on the whole genome level in *C. jejuni* and all other sequenced bacterial organisms are available on the web-based Restriction Enzyme dataBASE (REBASE) [10].

This chapter will describe protocols for determining the methylome profile of *C. jejuni*. In short, DNA is extracted from *C. jejuni* cultures and assessed for quality. Following Pacific Biosciences guidelines, the DNA sequencing library is prepared for SMRT sequencing and sequenced on SMRT Cells using the Pacific Biosciences RS sequencer. Post-sequencing analysis will readily call what base(s) have been methylated and identify the sequence motifs associated with the methylated base. Further bioinformatics analyses using our protocols will also identify where regions of hyper- and hypomethylation occur in the genome, the genes located in these regions in addition to what sequence motifs are found in which genes, and how to create graphics showing various aspects of the methylome profile.

2 Materials

2.1 Bacterial Culture Preparation Components

1. *Campylobacter jejuni* stock culture.
2. Difco Mueller Hinton (MH) broth and agar (Becton Dickinson, Franklin Lakes, NJ).
3. 100 × 15 mm petri dishes (Fisher Scientific, Pittsburgh, PA).
4. L-shaped cell spreaders (Fisher Scientific, Pittsburgh, PA).
5. Disposable inoculating loops (Fisher Scientific, Pittsburgh, PA).
6. 16 × 125 mm disposable lime glass culture tubes (VWR, Radnor, PA).
7. Mitsubishi AnaeroPack 2.5 L Rectangular Jar (Thermo Scientific, Wilmington, DE).
8. 1.5 mL microcentrifuge tubes.
9. Custom-made microaerophilic gas (5 % O_2, 10 % CO_2, 85 % N_2) from Airgas purchased through Iowa State University Chemistry Stores.
10. Incubator set to 42 °C.

2.2 Preparation and Quality Assessment of DNA for SMRT Sequencing

1. DNA extraction kit such as Wizard Genomic DNA Purification Kit (Promega, Madison, WI).
2. Nuclease-free water.
3. Spectrophotometry equipment such as NanoDrop ND-1000 spectrophotometer (Thermo Scientific, Wilmington, DE).
4. Fluorimetry equipment such as the Qubit 1.0 fluorometer and Qubit dsDNA BR Assay Kit (Life Technologies, Grand Island, NY).
5. Gel electrophoresis supplies (suggested):
   (a) Molecular biology agarose (Bio-Rad, Hercules, CA).
   (b) 1 Kb DNA Ladder (Promega, Madison, WI).
   (c) Mini-Sub cell GT cell gel electrophoresis system (Bio-Rad, Hercules, CA).
   (d) 1X Tris–Acetate–EDTA electrophoresis buffer: 100 mL 10× Tris–Acetate–EDTA colorless liquid solution (Fisher Scientific, Pittsburgh, PA), 900 mL water.
   (e) Blue/Orange 6× loading dye (Promega, Madison, WI).

2.3 Sequencing Library Preparation and SMRT Sequencing

1. Template Preparation and Sequencing Guide (Pacific Biosciences, Menlo Park, CA).
   (a) DNA shearing device such as gTUBEs (Covaris, Inc., Woburn, MA).
   (b) BluePippin (Sage Science, Beverly, MA).

(c) SMRTbell Template Prep Kits, DNA Sequencing Reagents, and SMRT Cells (Pacific Biosciences, Menlo Park, CA).

(d) AMPure® XP Beads (Beckman Coulter, Brea, CA).

2. SMRT Analysis suite v2.0 [11] with the HGAP algorithm for assembling genome into single contigs and identifying sequence motifs (Pacific Biosciences, Menlo Park, CA).

2.4 Bioinformatic Analysis of Methylation Motifs and Whole-Genome Methylation Motif Distribution Plots

1. Laptop (PC or Macintosh) or a Desktop machine. No need for high-performance computing machines.
2. Statistical software package R Version 3.0 or higher. No special packages are required. A standard installation is sufficient.
3. BEDTools suite v2.18.2.
4. Perl programming language Perl 5. No special packages are required.

3 Methods

3.1 Bacterial Culture in Preparation for Genomic DNA Extraction

1. Streak *C. jejuni* stock with an inoculating loop onto MH agar plate and incubate at 42 °C in a gas jar containing microaerophilic gas for 48 h.
2. Select a colony with an inoculating loop and passage into 5 mL MH broth. Incubate at 42 °C in gas jar with microaerophilic gas for 16–24 h.
3. Passage 100 μL of broth culture per MH agar plate. Spread using L-shaped spreader and incubate in same gas and temperature conditions as previous steps for no more than 24 h (or less time if you need to collect DNA of cells growing in log phase versus stationary phase at 24 h) (*see* **Note 1**).
4. Collect lawn cultures by adding 1 mL MH broth to the lawn plates, use L-shaped spreader to gently scrape the lawn cultures off the agar plate, and then carefully tilt the plate to an angle to pipette out 100 μL of lawn culture and dispense in a sterile microcentrifuge tube. Proceed to the next section immediately.

3.2 Genomic DNA Extraction and Assaying Quality of DNA

1. Follow manufacturer's protocol of preferred kit for isolating DNA. We used the Wizard Genomic DNA purification kit (Promega, Madison, WI) to extract DNA in nuclease-free water from the 100 μL lawn culture. A few modifications to the Wizard purification kit protocol are described in **Note 2**.
2. To measure DNA concentration, use a fluorometer such as the Qubit fluorometer.
3. To assess DNA quality, use a spectrophotometer like the NanoDrop® spectrophotometer (*see* **Note 3**). *See* **Note 4** for more information on specific DNA quality guidelines for PacBio sequencing (taken from Pacific Biosciences Template Preparation and Sequencing Guide).

4. Use gel electrophoresis (1 % agarose gel using 1X Tris–Acetate–EDTA and agarose) to make sure the size of the DNA is greater than 10 kb. Load at least 0.5 μg DNA onto gel.
5. Prepare 10 μg genomic DNA per strain for library preparation and sequencing.

3.3 Sequencing Library Preparation and De Novo SMRT Sequencing

1. Follow the Template Preparation and Sequencing Guide by Pacific Biosciences. Basic steps include:
   (a) Fragment genomic DNA to approximately 15 Kb using gTUBEs following manufacturer's protocol.
   (b) Concentrate DNA using AMPure® XP magnetic beads per manufacturer's protocol.
   (c) Perform DNA damage repair on the fragmented DNA and end repair to prepare the DNA for ligation using Pacific Biosciences' DNA Template Prep Kit.
   (d) Purify DNA again using AMPure® XP magnetic beads following manufacturer's protocol.
   (e) Ligate hairpin adapters to each end of the DNA to form SMRTbell templates.
   (f) Add exonuclease to remove failed ligation products and then purify with AMPure® XP magnetic beads to remove enzymes.
   (g) Size-select SMRTbell templates with sizes greater than 5–10 Kb using Blue Pippin. Follow manufacturer instructions on how to use Blue Pippin equipment to conduct size selection. Size selection will increase the number of long continuous reads and allow for easier de novo assembly.
   (h) Anneal sequencing primers to SMRTbell templates.
   (i) Bind sequencing polymerase to SMRTbell templates.
   (j) Prepare your reagent plates and sample plate. Then load reagent, mixing, and sample plates onto the PacBio RS sequencer.
   (k) Load SMRT cells and tips onto the sequencer. Use 1 SMRT cell per sample. If the coverage is not sufficient to fully close the genome, run additional SMRT cells as needed (*see* **Note 5**).
2. Start the sequencing run following PacBio's standard protocol. For long inserts (greater than 3 Kb), 3–4 h collection times are recommended.
3. Assemble genomes into single contigs using HGAP algorithm as part of the SMRT analysis suite v2.0. The assembled sequence will be used as a reference during the base modification analysis.

4. Detection of modified bases can also be carried out using the PacBio SMRT Analysis Suite. Analyze your SMRT Sequencing data using the "RS_Modification_and_Motif_Analysis.1" protocol. Details are available at http://www.pacb.com/pdf/TN_Detecting_DNA_Base_Modifications.pdf. Analysis will generate several reports specific to base modification detection (*see* **Note 6**).
5. Use SeqWare computer resource (http://seqware.github.io/) to identify RM system genes from the complete genome sequence of *C. jejuni* [6]. SeqWare is an open-source suite of software tools that, combined with internal databases, provides comprehensive analysis of sequence data. To identify RM system genes, scan sequence data locally for annotated RM systems from REBASE (http://rebase.neb.com/rebase/rebase.html) [10] as well as homologues from BLAST searches, and check for genomic contexts as indicators of potential new RM system components. SeqWare will localize motifs and domains, assign probable recognition specificities, classify accepted hits, and mark Pfam relationships (compare protein domains, families, etc.). All candidates are then manually inspected before being assigned to an RM system and entered into REBASE.
6. To visualize the methylation patterns identified in *C. jejuni*, one commonly used tool is the Circos software package, a command-line application which displays genomic data in a circular layout [12] (*see* **Note 7**). Figure 1 contains example

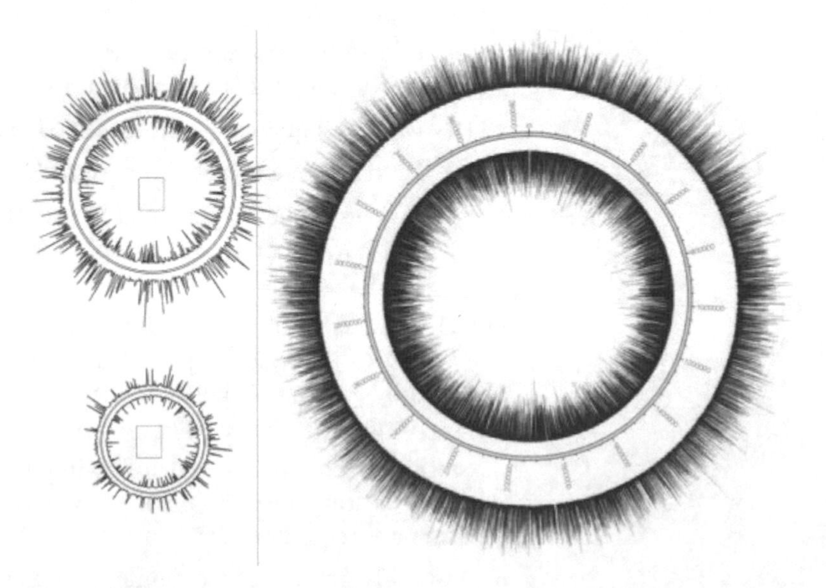

Fig. 1 Example Circos plots of a chromosome of a strain of *Sphingomonas* chromosome (*large circle, right side*) and two plasmids (*two smaller circles, left side*). The kinetic variation events on positive and negative strands of the genome are plotted as red peaks on the outer and inner loops of the plot, respectively. The gray loop in between the positive and negative strands shows the scale of the genome in base pairs

Circos plots showing distinct kinetic variation profiles (variation in the rate for DNA polymerase to incorporate bases onto the template strand during synthesis) indicative of DNA methylation in the chromosome and plasmids.

7. The scatter plot in Fig. 2 is another way of showing kinetic variation profiles identified in the genome (*see* **Note 8**). This plot was created through SMRT Portal, an open-source browser-based application from Pacific Biosciences that is part of the SMRT Analysis software run on a local server. An example of using SMRT Portal to analyze SMRT sequencing data and methylation analysis is shown in Fig. 3 (*see* **Note 9**).

3.4 Bioinformatic Analysis of Methylation Motifs and Whole-Genome Methylation Motif Distribution Plots

1. Convert sequence motif score text files into general feature format (GFF). This is done by simply rearranging the columns as shown below.

```
cat Methylation_Motif_Scores.txt | grep -v
"start" | awk '{print $5"\t" "." "\t" "nucleo-
tide_motif" "\t" $1 "\t" $1+length($4)-1 "\t"
"." "\t" $2 "\t" "." "\t" "Name="$4";ID="$4"_"$1
";base="$6";score="$7";ipdRatio="$8";cover
age="$9""}' > Methylation_Motifs.gff
```

2. Split the GFF file into multiple files, one for each motif. Select only the lines containing the motif and write into a separate file for each motif. For example,

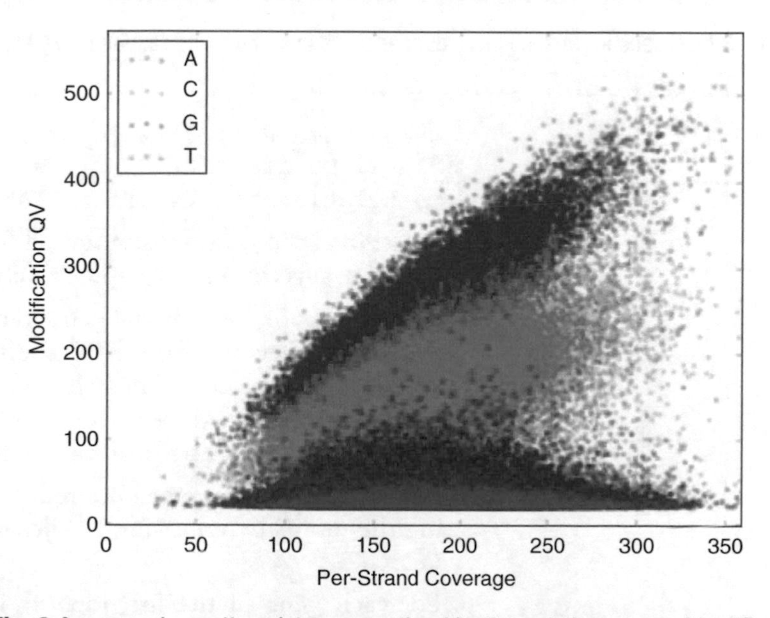

Fig. 2 An example scatter plot representing kinetic variation events identified across the genome of *Desulfobacca acetoxidans* strain DSM11109. The *colored dots* in the plot represent bases, which are defined in the *upper left panel*. The *x*-axis shows the amount of sequence coverage for all genomic positions. The *y*-axis shows the kinetic scores for each base

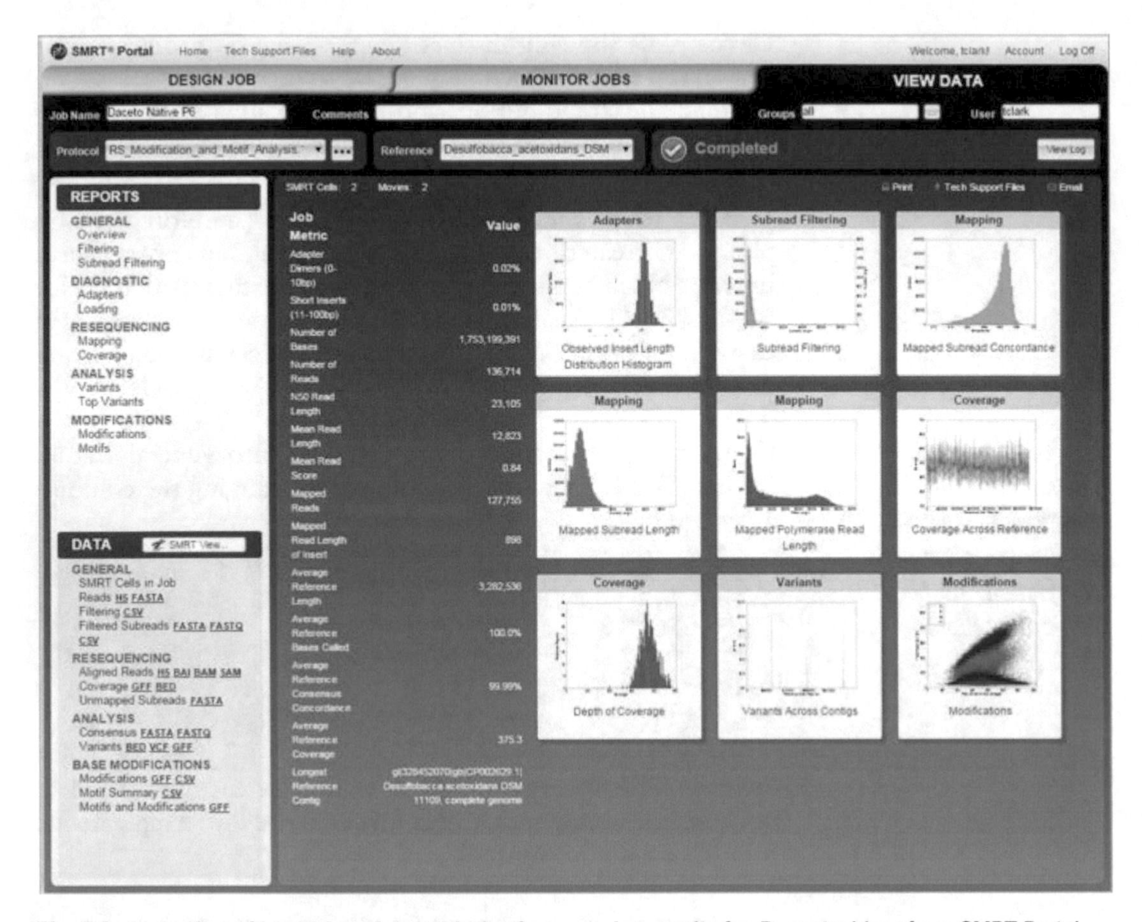

Fig. 3 A screenshot of secondary data analysis of sequencing results for *D. acetoxidans* from SMRT Portal

```
grep 'Name=RAATTY' Methylation_Motifs.gff >
RAATTY_Motifs.gff
Methylation_Motifs.gff > TAAYNNNNNTGC_Motifs.gff
```

3. Download the genome annotation file (genes.gff) for your *C. jejuni* strains from ftp://ftp.ncbi.nih.gov/genomes/.
4. For each motif, identify the genes that contain that motif. Use "intersect" function from BEDTools suite as shown below to extract the genes containing the motif.

```
intersectBed -wa -s -a genes.gff -b RAATTY_
Motifs.gff > Genes_containing_RAATTY.gff
```

5. For each motif, generate a nonredundant list of genes containing the motifs by removing duplicates from the file generated in **step 5**.
6. For each gene in the list, identify its functional category if available. Keep track of the number of genes that are missing functional annotation.
7. Create a summary of the percent number of genes and functional gene groups that have the associated methylation motif(s) in Microsoft Excel.

8. Confirm motif identifications of *C. jejuni* strain by accessing REBASE: http://rebase.neb.com/rebase/rebase.html [10].
   (a) Select "REBASE Genomes" and find your respective *C. jejuni* strain in the index.
   (b) Choose either "circular" or "list" for the RM schematics of the *C. jejuni* strain.
   (c) The "summary" link at the top of the page will give you a comprehensive list of all RM systems, motifs, genes, and names of the enzymes in the strain.
   (d) Follow naming schemes from REBASE for predicted or confirmed RM genes/enzymes found in your *C. jejuni* strain to keep nomenclature consistent with the database. Contact REBASE staff for any questions you may have.
9. Using hist() in R, calculate the appropriate bin size and plot the distribution of methylated motifs across the *C. jejuni* genome (e.g., *see* Fig. 4 and **Note 10**). For example, here is how to plot the distribution for a bin size of 1000.

```
met <- read.table("Methylation_Motif_Scores.
txt", header=T)
bins=seq(0,1700000,1000)
tiff('IA3902_Allmotifs_Bothstrands.tiff',
width=7.09, height=3.35, units="in",res=1000)
hist(met[,1],breaks = bins, xlab = "Position",
main = "IA3902 Both Strands")
dev.off()
tiff('IA3902_Allmotifs_Bothstrands.tiff',
width=7.09, height=3.35, units="in",res=1000)
hist(met[,1],breaks = bins, xlab = "Position",
main = "IA3902 Both Strands")
dev.off()
```

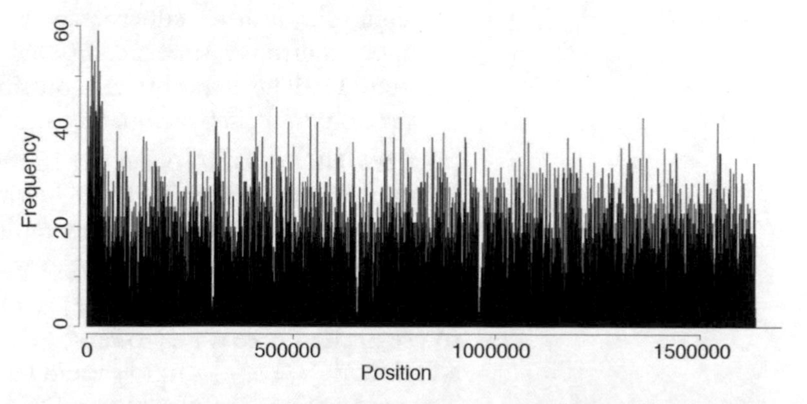

Fig. 4 Example of a genome-wide methylation distribution plot. The methylation distribution plots of all motifs for this *C. jejuni* IA3902 were combined with a bin size of 1000 across the genome. Each bin is represented as a *black bar* on the graph. The *x*-axis begins at the *ori*C region near position 0 and extends through the entire 1.6 megabases genome of the strain. The *y*-axis shows the frequency of methylation motifs per bin

10. Identify hypomethylated (or hypermethylated) regions of the genome that have statistically significant lower (or higher) number of methylated motifs.

```
met <- read.table("Methylation_Motif_Scores.
txt", header=T)
bins=seq(0,1700000,1000)
aa<-hist(met[,1],breaks = bins,plot=F)
mean(aa$counts)
sd(aa$counts)
which(aa$counts<5) ## Hypomethylation
aa$counts[which(aa$counts<5)] ##counts
aa$breaks[which(aa$counts<5)] ##bins
which(aa$counts>45) ## 45 is obtained by Mean +
(3 * SD) ## Hypermethylation
aa$counts[which(aa$counts>45)] ## counts
aa$breaks[which(aa$counts>45)] ## bins
```

11. Generate a comprehensive list of all motifs and associated hypo- or hypermethylated genes in each *C. jejuni* strain including the gene name, locus tag, gene description, and gene start and end positions.

4 Notes

1. When making lawn plates, make at least two plates per strain to be used for DNA isolation in case one of the plates does not grow well.
2. When following the Promega protocol, some modifications to the protocol include:
   (a) **Step 9**: incubate at 37 °C for 1 h.
   (b) **Steps 12** and **13**: remove supernatant after the first centrifuge spin, transfer supernatant to a new microcentrifuge tube, centrifuge supernatant again, and then proceed to **step 13**. This prevents any possible carryover of protein precipitant from previous steps.
   (c) **Step 18**: dry DNA pellet for 10 min, remove residual ethanol, dry again for another 1.5 min, add 100 μL nuclease-free water, and store DNA samples at 4 °C overnight to allow DNA pellet to dissolve in water.
3. When using a spectrophotometer like the NanoDrop, make sure to use nuclease-free water to serve as a reference or "blank." In addition, the $OD_{260/280}$ ratio should be in the range of 1.8–2.0. The ratio of Qubit to NanoDrop DNA concentrations can also sometimes help identify RNA contamination. The NanoDrop also provides DNA concentration but is not specific to double-stranded DNA. Therefore, we prefer to base our decisions for how much DNA to use for sequencing according to DNA concentration values determined by the fluorometer (Qubit) over the spectrophotometer (such as the NanoDrop).

4. According to Pacific Biosciences template preparation guidelines, ensure that your DNA sample is double stranded, has undergone a minimum number of freeze-thaw cycles, has not been exposed to high temperatures (>65 °C for 1 h can cause a detectable decrease in sequence quality), has not been exposed to extreme pH (<6 or >9), and has an $OD_{260/280}$ ratio in the range of 1.8–2.0.
5. We ran four SMRT cells for each strain. However, the number of SMRT cells ran for each strain was unnecessary as a pilot test showed that we could get the same assembly for each strain using just 1 SMRT cell of data.
6. The "modifications.csv" file contains information for every position in the reference sequence (both strands). The average polymerase speed (interpulse duration or IPD) for each position is compared to an in silico reference built from unmodified DNA. A modification score is given to each position in the reference which indicates the likelihood that that position contains a base modification. The analysis algorithm has been trained to identify the type of modification for the most common base modifications in bacterial samples (m6A, m4C, m5C). Once the modified positions are detected, the software automatically searches for common sequence motifs that harbor base modifications. This is particularly useful in bacterial and archaeal samples where the modifications are imparted by restriction-modification systems. A summary of the identified motifs is also generated ("motif_summary.csv"). General features format (GFF) files are also generated that identify the position of all modified bases (above an adjustable cutoff) in the reference sequence.
7. For more information on how to create this image or other variations using Circos, see this reference [12] as well as the Circos website: http://circos.ca/.
8. Although not shown in this example plot, a cutoff value for the kinetic score (*y*-axis) would normally be shown as a dashed line across the plot that is used as a threshold to qualify a nucleotide as either methylated or non-methylated. However, this example shows a higher proportion of adenine (red dots) and cytosine (green dots) residues with high kinetic scores compared to the other bases in *Desulfobacca acetoxidans* strain DSM11109. This suggests that the major types of base modifications in this strain are N6-methyladenine (m6A) and N4-methylcytosine (m4C).
9. Pacific Biosciences also hosts a wide breadth of how-to resources on GitHub for analyzing methylation in prokaryotes. This link (https://github.com/PacificBiosciences/motif-finding) features step-by-step guides and scripts for creating plots like Fig. 2 along with how to identify methylation motifs and assess the quality of such calls.

10. For determining statistically significant increase/decrease in methylation per bin, we used the mean ± 3 times standard deviation

Acknowledgments

This work was supported by the Agriculture and Food Research Initiative Fellowships Grant Program of the USDA National Institute of Food and Agriculture grant 2013-67011-21155, National Institute for Allergy and Infectious Diseases grant K08AI07052303, and Iowa State University start-up funds.

References

1. Korlach J, Turner SW (2012) Going beyond five bases in DNA sequencing. Curr Opin Struct Biol 22(3):251–261. doi:10.1016/j.sbi.2012.04.002, http://dx.doi.org/
2. Davis BM, Chao MC, Waldor MK (2013) Entering the era of bacterial epigenomics with single molecule real time DNA sequencing. Curr Opin Microbiol 16(2):192–198. doi:10.1016/j.mib.2013.01.011, http://dx.doi.org/
3. Flusberg BA, Webster DR, Lee JH et al (2010) Direct detection of DNA methylation during single-molecule, real-time sequencing. Nat Methods 7(6):461–465
4. Eid J, Fehr A, Gray J et al (2009) Real-time DNA sequencing from single polymerase molecules. Science 323(5910):133–138. doi:10.1126/science.1162986
5. McCarthy A (2010) Third generation DNA sequencing: Pacific Biosciences' single molecule real time technology. Chem Biol 17(7):675–676. doi:10.1016/j.chembiol.2010.07.004, http://dx.doi.org/
6. Murray IA, Clark TA, Morgan RD et al (2012) The methylomes of six bacteria. Nucleic Acids Res 40(22):11450–11462. doi:10.1093/nar/gks891
7. Sánchez-Romero MA, Cota I, Casadesús J (2015) DNA methylation in bacteria: from the methyl group to the methylome. Curr Opin Microbiol 25:9–16. doi:10.1016/j.mib.2015.03.004, http://dx.doi.org/
8. Mou KT, Muppirala U, Severin A et al (2015) A comparative analysis of methylome profiles of *Campylobacter jejuni* sheep abortion isolate and gastroenteric strains using PacBio data. Front Microbiol 5:1–15. doi:10.3389/fmicb.2014.00782
9. O'Loughlin JL, Eucker TP, Chavez JD et al (2015) Analysis of the *Campylobacter jejuni* Genome by SMRT DNA Sequencing Identifies Restriction-Modification Motifs. PloS One 10(2):e0118533. doi:10.1371/journal.pone.0118533
10. Roberts RJ, Vincze T, Posfai J et al (2010) REBASE – a database for DNA restriction and modification: enzymes, genes and genomes. Nucleic Acids Res 38(Database issue):21
11. Chin CS, Alexander DH, Marks P et al (2013) Nonhybrid, finished microbial genome assemblies from long-read SMRT sequencing data. Nat Methods 10(6):563–569
12. Krzywinski M, Schein J, Birol İ et al (2009) Circos: an information aesthetic for comparative genomics. Genome Res 19(9):1639–1645. doi:10.1101/gr.092759.109

Chapter 18

Characterizing Glycoproteins by Mass Spectrometry in *Campylobacter jejuni*

Nichollas E. Scott

Abstract

The glycosylation systems of *Campylobacter jejuni* (*C. jejuni*) are considered archetypal examples of both *N*- and *O*-linked glycosylations in the field of bacterial glycosylation. The discovery and characterization of these systems both have revealed important biological insight into *C. jejuni* and have led to the refinement and enhancement of methodologies to characterize bacterial glycosylation. In general, mass spectrometry-based characterization has become the preferred methodology for the study of *C. jejuni* glycosylation because of its speed, sensitivity, and ability to enable both qualitative and quantitative assessments of glycosylation events. In these experiments the generation of insightful data requires the careful selection of experimental approaches and mass spectrometry (MS) instrumentation. As such, it is essential to have a deep understanding of the technologies and approaches used for characterization of glycosylation events. Here we describe protocols for the initial characterization of *C. jejuni* glycoproteins using protein-/peptide-centric approaches and discuss considerations that can enhance the generation of insightful data.

Key words Glycoprotein, *Campylobacter jejuni*, Mass spectrometry

1 Introduction

1.1 Protein Glycosylation in Bacteria

Initially assumed to only exist within eukaryotic systems, protein glycosylation is now recognized to occur in all domains of life [1–5]. Within eubacteria, glycosylation is increasingly being recognized as a common posttranslational modification, especially within pathogens [6, 7]. Although the function of bacterial glycosylation is still largely unknown, glycosylation appears to be required for wild-type levels of fitness and host colonization within species harboring these systems [8–11]. This association with pathogenesis and commensalism has driven increasing interest in the characterization of bacterial glycosylation, with multiple unique glycosylation system identified to date [5, 7, 12]. Unlike the highly conserved *N*- and *O*-linked glycosylation pathways of eukaryotic system [13, 14], studies of bacterial glycosylation have noted the presence of unique eubacteria-specific glycans. These include glycans composed of bacillosamine [15], pseudaminic, and legionaminic acid [16] which

James Butcher and Alain Stintzi (eds.), *Campylobacter jejuni: Methods and Protocols*, Methods in Molecular Biology, vol. 1512, DOI 10.1007/978-1-4939-6536-6_18,

when combined generate diverse glycan structures which differ even between closely related strains [17] or species [18]. This diversity in glycan composition and structures results in bacterial glycosylation being less predictable in nature than most eukaryotic glycosylation systems. This lack of predictability represents a significant challenge not only in the analysis of glycosylated proteins and peptides but also in the selection of approaches to detect and identify glycosylated proteins. In the study of eukaryotic, glycosylation methods that exploit the predictable chemical nature of glycans dominate the field, allowing the analysis of glycosylation sites [19], glycan composition [20], or both within the same experiment [21–23]. It is important to note, however, that such approaches are generally incompatible with prokaryotic glycosylation due to the inherent diversity of bacterial glycans.

1.1.1 Protein Glycosylation Within C. jejuni

The two glycosylation systems possessed by *C. jejuni* are classified as *N*- and *O*-linked glycosylations based on the functional group of the amino acid used to attach the glycan. Both systems attach different glycans, with the *N*-linked system leading to the attachment of a heptameric structure composed of Gal*N*Ac-α1,4-Gal*N*Ac-α1,4-[Glcβ1,3]-Gal*N*Ac-α1,4-Gal*N*Ac-α1,4-Gal*N*Ac-α1,3-Bac-β1, where Bac is bacillosamine [2,4-diacetamido-2,4,6 trideoxyglucopyranose] [15], while the *O*-linked system leads to the attachment of either pseudaminic acid derivatives (such as Pse5Ac7Ac) [24, 25] or legionaminic acid derivatives (such as Leg5Am7Ac) [26, 27] to its substrate. Both systems are required for wild-type levels of virulence, with *O*-linked glycosylation being essential for flagellar filament assembly [28], autoagglutination [29], and adherence and invasion [30]. Similarly, *N*-linked glycosylation has been shown to be necessary for host colonization with an adherence and invasion of epithelial cells [10] and reduced colonization in the chicken gastrointestinal tract [31].

O-linked glycosylation was the first glycosylation event identified within *C. jejuni*, and this process appears to solely target the flagellin [5, 32, 33]. The biosynthetic enzymes required for *O*-glycan synthesis are variable between strains [30, 34] and are typically located in the flanking region of the *flaA* and *flaB* structural genes [35]. The synthesis of Pse5Ac7Ac requires genes Cj1293, Cj1294, Cj1311, Cj1312, Cj1313, and Cj1317 (all gene numbers listed follow the *C. jejuni* NCTC 11168 naming scheme) [28, 36–38], and an additional seven enzymes are needed for Leg5Am7Ac [39] with further modification of both *O*-linked glycans possible by additional enzymes [11, 30]. To date, the glycotransferase responsible for the addition of the *O*-linked glycans to the protein is still unknown. Glycosylation of the flagellin does not appear to require a sequence-specific recognition motif, but it is essential that the site be accessible to the protein surface, with serine and threonine both being occupied [25, 29, 40]. It is suggested that at least 19 glycosylation sites are occupied in flagellin

[25, 29, 40]. The high proportion of glycosylation events coupled to diverse glycans generated by the pseudaminic or legionaminic acid pathways results in thousands of possible proteoforms that may all contribute to *C. jejuni* virulent in subtly different ways.

N-linked glycosylations in contrast target at least 70 proteins [41–43] but was not identified until 1999 [44]. The protein glycosylation locus (*pgl*; Cj1120c to Cj1130c) [44] is responsible for the synthesis of the *N*-linked glycan and is conserved within *C. jejuni* and the related species, such as *C. coli* [5, 45]. Within this loci Cj1126 (also known as PglB) is the sole enzyme required for glycan transfer to protein substrates [46]. The *N*-linked glycosylation process has similar substrate requirements to that of *N*-linked glycosylation in eukaryotic and requires the sequence motif N-X-S/T (where asparagine [N] is the site of attachment and "X" refers to any amino acid except proline). However, this motif alone is not sufficient for glycosylation in *C. jejuni* [47] with a further D/E-X sequence typically observed upstream of glycosylation events [48]. Recently multiple studies have observed that glycosylation can occur within sites that only partially contain the bacterial *N*-linked sequon D/E-X-N-X-S/T. To date noncanonical glycosylation has been observed in *C. lari* [49], *Desulfovibrio desulfuricans* [50], *C. jejuni* [43], and *C. fetus fetus* [18] with the extended sequon appearing to enhance PglB binding and kinetics of glycosylation [51, 52] but not strictly being required for glycosylation.

1.1.2 Approaches to Characterize C. jejuni Glycoproteins

Although it is still a challenge, the availability of knowledge concerning glycan composition makes glycosylation events on *C. jejuni* proteins significantly easier to identify and characterize than other bacterial systems [15, 25, 53]. This information enables the selection of reagents to probe and enrich glycosylation events on both the protein and peptide level. This knowledge enables the use of chemical, enzymatic, and affinity-based approaches for detection and enrichment. Chemical means, such as the presence of vicinal diol in both the *O*- and *N*-linked glycans, enables the generation of aldehyde functionalities using Periodic acid-Schiff (PAS) chemistry to allow the selective labeling of sugars with probes for the detection of glycoproteins in a nonspecific manner. This approach was the first used by Doig et al. to confirm the heavily glycosylated status of the flagellin [33] where the presence of multiple glycans on the flagellin enabled the differentiation of glycosylation. Enzymatic means for detection and enrichment, such as the use of galactose oxidase for labeling *N*-linked glycoproteins and peptides, have now also been shown [54]. This *N*-linked-specific approach is highly versatile as it enables a range of chemical or affinity handles to be attached to *N*-linked glycoproteins and peptides [54]. The final strategy that can be used for detection and enrichment of glycopeptides is an affinity-based approach, which will either exploit the hydrophilic properties of glycan linked to protein substrates such as in the case of normal-phase [55] or hydrophilic interaction liquid chromatography [42] or utilize biological affinity reagents such as glycan-specific antibodies [18, 33] or lectins [15, 56]. Irrespective of the approach for glycoprotein and glycopeptide enrichment, confirmation of the glycosylated status is essential for assigning the glycosylation status of a protein.

1.1.3 Mass Spectrometry-Based Characterization C. jejuni Glycoproteins/Peptides

Currently one of the most versatile approaches for the identification and confirmation of the glycosylated status of a protein or peptide is the use of mass spectrometry (MS). The use of MS technologies provides a number of advantages over other analytical approaches for the confirmation of glycosylation including enabling the confirmation of the chemical linkage of glycan to the peptide in which the specific residue modified is able to be determined via a range of fragmentation approaches [40, 42, 43, 53], the utilization of ~1/1000 of the material needed for approaches such as NMR, the ability to quantify changes in glycosylation sites and glycoproteome [43, 57], and the ability to detect small changes to glycan structures even within a complex samples [41, 55]. In this chapter we will describe the use of liquid chromatography-tandem mass spectrometry (LC-MS/MS) for the characterization of *C. jejuni* glycoproteins. The key focus of this section will be to outline robust approaches for the identification of glycopeptides from isolated single proteins, as well as introducing global approaches for the selective enrichment of glycopeptides from complex lysates (Fig. 1).

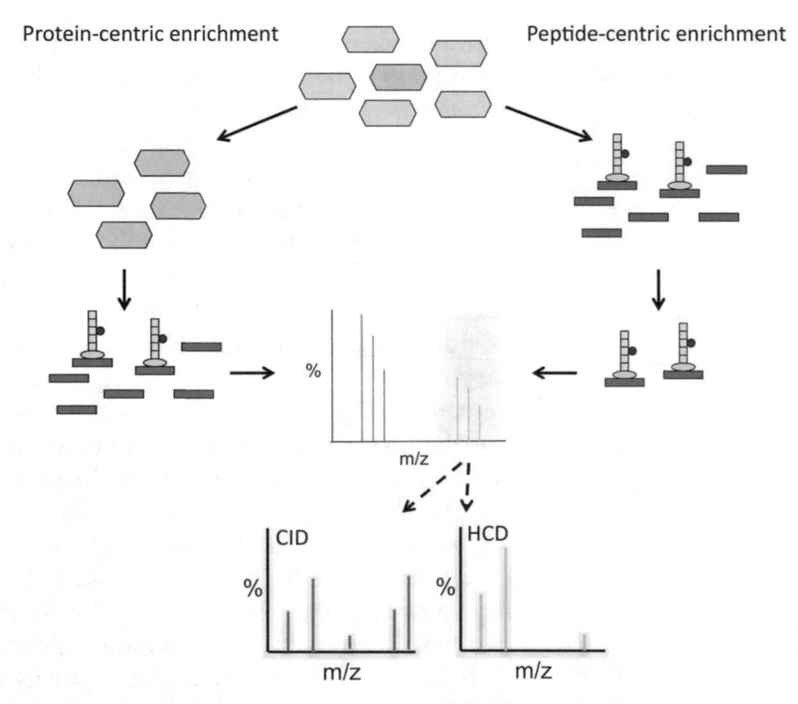

Fig. 1 Workflow for the analysis of *C. jejuni* glycosylation using protein- and peptide-centric approaches. In protein-centric approaches the protein(s) are enriched from a complex lysate prior to digestion. In peptide-centric approaches, the enrichment of glycopeptide occurs after digestion. After digestion, both workflows converge to utilize the same techniques and MS approaches to preform analysis

In addition we will address recent developments in quantitative proteomics that can be implemented in *C. jejuni* to enhance the study of *N*- and *O*-linked glycosylations.

2 Materials

2.1 Generation of Bacterial Cells

1. *C. jejuni* (example strain NCTC11168, 81–176 or clinical strain of interest).

   Growth medium prepared as liquid broth or 1.5% agar plates (*see* **Notes 1** and **2**). Growth medium can be obtained from commercial sources or prepared as follows:

   - Skirrow media: 2 g/L heart muscle infusion, 13 g/L pancreatic digest of casein, 5 g/L yeast extract, 5 g/L sodium chloride, 0.01 g/L vancomycin, 0.005 g/L trimethoprim, 2500 units/L polymyxin B, 7% defibrinated horse blood (*see* **Note 3**).
   - Mueller-Hinton media: 2 g/L beef infusion, 1.5 g/L starch, 17.5 g/L casein hydrolysate.
   - Brain-heart infusion media: 17.5 g brain-heart infusion, 10 g/L gelatin hydrolysate, 2 g/L dextrose, 5 g/L sodium chloride, 2.5 g/L disodium phosphate.
   - Dissolve media in water and autoclave.
2. Phosphate-buffered saline (PBS, 137 mM sodium chloride, 2.7 mM potassium chloride, 10 mM disodium phosphate, 1.8 mM monopotassium phosphate), prechilled at 4 °C before use.

2.2 Preparation of Protein Samples

2.2.1 Whole Cell Lysates

1. Trypsin-compatible buffer solution (40 mM HEPES/40 mM Tris/40 mM tetraethylammonium bromide (TEAB)/40 mM ammonium bicarbonate (ABC) [*see* **Note 4**]), pH 8.0. Prechilled at 4 °C before use and in the case of ammonium bicarbonate, prepare fresh.
2. Benzonase® nuclease (≥250 U/μL, ≥90%), store at 4 °C.
3. Protease inhibitor cocktail (e.g., cOmplete Mini, Roche Diagnostics).
4. Probe sonicator with micro-tip probe.

2.2.2 Membrane-Associated Proteins

1. Trypsin-compatible buffer solution (40 mM HEPES/40 mM Tris/40 mM TEAB/40 mM ABC [*see* **Note 4**]), pH 8.0. Prechilled at 4 °C before use and in the case of ammonium bicarbonate, prepare fresh.
2. Benzonase® nuclease (≥250 U/μL, ≥90%), store at 4 °C.
3. Sodium carbonate (0.1 M), prechilled at 4 °C before use.
4. Protease inhibitor cocktail (e.g., cOmplete Mini, Roche Diagnostics).
5. Probe sonicator with micro-tip probe.

2.3 Digestion of Samples

2.3.1 Trypsin Digestion of Complex Sample

1. Solubilization buffer: 6 M urea, 2 M thiourea, prepared fresh prior to use (*see* **Note 5**).
2. 100 mM ammonium bicarbonate (ABC) pH 8.0 in Milli-Q water. Prepared fresh, ensure pH ~8.0 prior to use.
3. 1 M dithiothreitol in 100 mM ABC pH 8.0, prepared fresh and stored at 4 °C.
4. 100 mM iodoacetamide in 100 mM ABC pH 8.0, prepared fresh and stored in the dark at 4 °C.
5. 0.1 μg/μL Lys-C in Milli-Q water. Store as single-use aliquots at −80 °C.
6. 1 μg/μL porcine sequencing grade trypsin in 100 mM ABC pH 8.0, store as single-use aliquots at −80 °C.

2.3.2 Digestion of Gel Electrophoresis Separates Proteins

1. SDS-PAGE gel with adequate polyacrylamide concentration to separate protein of interest.
2. SDS-PAGE fixative buffer: 10% methanol, 7% glacial acetic acid in Milli-Q water.
3. Coomassie blue stock: 0.1% (w/v) Coomassie Brilliant Blue G250, 10% ammonium sulfate, 5% phosphoric acid in 500 mL water.
4. 4X SDS-PAGE loading buffer: 280 mM Tris-HCl, pH 6.8, 4% lithium dodecyl sulfate (LDS) or sodium dodecyl sulfate (SDS), 40% (w/v) glycerol, 0.02% bromophenol blue.
5. 100% methanol.
6. 50 mM ammonium bicarbonate (ABC) pH 8.0.
7. 100% ethanol.
8. Reducing solution: DTT 10 mM in 50 mM ABC, prepared fresh.
9. Alkylation solution: iodoacetamide, 55 mM in 50 mM ABC pH 8.0, prepared fresh and stored in the dark at 4 °C.
10. Destaining solution: 50% ethanol, 50 mM ABC in Milli-Q water.
11. 20 ng/uL porcine sequencing grade trypsin: (prepared from trypsin stock) in 50 mM ABC pH 8.0.
12. Gel extraction buffer: 30% ethanol, 3% glacial acetic acid in Milli-Q water.

2.4 Desalting/Concentration of Peptides

1. For large amounts (>100 μg) of peptide material (from whole cell or membrane lysates), use C18 solid-phase extraction (SPE) cartridges.
2. For small amounts (<20 μg) of peptide material (gel digested samples), use STOP and GO tips (STAGE), using C18 Empore material, constructed according to the published method [58, 59].
3. ≥99.9% purity HPLC-grade acetonitrile (MeCN).

4. Buffer A*: 0.1 % trifluoroacetic acid (TFA, for HPLC ≥99.0 % purity,) in Milli-Q water (*see* **Note 6**).
5. Buffer B*: 0.1 % TFA in 80 % acetonitrile.
6. 10 % trifluoroacetic acid stock, aliquots and stored at −20 °C.
7. Vacuum centrifuge.

2.5 Construction of ZIC-HILIC STAGE Tips

1. 20 G, Kel-F Hub NDL blunt needle.
2. Plunger assembly N, RN, LT, LTN 1701/10 μL.
3. ZIC-HILIC material (ZIC-HILIC, 10 um, 200 Å).
4. C_8 material (Empore Octyl C_8 extraction disk, 3 M, available through Sigma-Aldrich; *see* **Note 7**).
5. P10 pipette tip.
6. 10 mL syringe.

2.6 ZIC-HILIC Enrichment of Glycopeptides

1. ZIC-HILIC buffer A: 0.1 % formic acid (FA, ~98 % purity) in Milli-Q water. Prepared fresh on day of the enrichment.
2. ZIC-HILIC buffer B: 5 % formic acid in 80 % acetonitrile. Prepared fresh on day of the enrichment (*see* **Note 8**).
3. ZIC-HILIC buffer C: 95 % acetonitrile. Prepared fresh on day of the enrichment.
4. Vacuum centrifuge.

2.7 LC-MS/MS

1. Mass spectrometer capable of resonance-based and beam-type collision-induced dissociation (referred to as CID and HCD fragmentation, respectively) e.g., LTQ Orbitrap or Fusion instruments (Thermo Scientific; *see* **Note 9**).
2. Nanoflow HPLC system (e.g., Agilent 1100/1200 series or Thermo Scientific EASY-nLC systems).
3. **Optional**: Pre/trapping column: 20 mm, 100 μm inner diameter (i.d.), 360 μm outer diameter (o.d.) fused silica (Polymicro Technologies™), packed with Aqua C_{18} reversed phase (RP) material (5 μm, Phenomenex) (*see* **Note 10**).
4. Analytical column with integrated emitter pulled in-house or purchased from New Objective: 150–500 mm, 75 μm i.d. 360 μm o.d fused silica packed with Reprosil-Pur C_{18}-AQ RP material (3 μm, Dr. Maisch).
5. Buffer A: 0.1 % formic acid in Milli-Q water.
6. Buffer A*: 0.1 % trifluoroacetic acid in Milli-Q water.
7. Buffer B: 0.1 % formic acid in 80 % MeCN.

2.8 Data Analysis

1. MASCOT (Matrix Science) search engine for peptide identification (proprietary software).
2. *C. jejuni* proteome database (such as NCTC11168 or 81–176 available from UNIPROT; *see* **Note 11**).

3. Xcalibur MS data viewer (available from Thermo Scientific).
4. MSconvert (part of the ProteoWizard software package) available from http://proteowizard.sourceforge.net/.
5. Protein Prospector MS-product website (http://prospector.ucsf.edu/ prospector/cgi-bin/msform.cgi?form=msproduct).

3 Methods

3.1 Bacterial Growth

1. Grow *C. jejuni* strains of interest to the required growth phase or density in the desired media, and harvest cells by centrifugation at 2000 × *g* for 10 min at 4 °C.
2. Wash cell pellets two times with ice-cold PBS, and collect cells by centrifugation at 2000 × *g* for 10 min at 4 °C to remove any media contamination.
3. For convenience cellular material can be freeze dried and stored at −80 °C until required.

3.2 Preparation of Complex Protein Samples

3.2.1 Whole Cell Lysates

Weigh out freeze-dried cells (10–20 mg of freeze-dried *C. jejuni* cells typically results in 4–8 mg of total protein).

1. Resuspend cells in 1 mL ice-cold trypsin-compatible buffering solution with protease inhibitor cocktail. Vortex to disperse cells in solution. Keep samples on ice.
2. Add 250 U of Benzonase® nuclease (*see* **Note 12**).
3. Lyse cells using a tip-probe sonicator with four rounds of 30 s with 2 min on ice between each round (*see* **Note 13**).
4. Remove cellular debris by centrifugation at 20,000 × *g* for 30 min at 4 °C.
5. Collect supernatant, quantify protein amount (*see* **Note 14**), and aliquot 1 mg protein lysate stocks.
6. Snap freeze and freeze dry aliquots; store at −80 °C for long-term storage.

3.2.2 Membrane-Associated Proteins

Weigh freeze-dried cells (30–40 mg of freeze-dried *C. jejuni* cells typically results in 3–4 mg of membrane protein).

1. Complete **steps 1–4** as in Subheading 3.2.1.
2. Resuspend cellular debris in 1 mL ice-cold trypsin-compatible buffering solution, and sonicate samples as in Subheading 3.2.1, **step 4**. Repeat five times (*see* **Note 15**).
3. Pool the supernatants, add six volumes of ice-cold 0.1 M sodium carbonate and stir mixture gently for 1 h at 4 °C.
4. Collect precipitated membranes by centrifugation at 35000 × *g* for 1 h at 4 °C.

5. Remove and discard the supernatant and wash the membrane pellet vigorously with 20 mL of ice-cold trypsin-compatible buffering solution.
6. Collect precipitated membranes by centrifugation at 35000 × *g* for 1 h at 4 °C.
7. Resuspend the membrane pellet in 10 mL ice-cold trypsin-compatible buffering solution by tip-probe sonication, quantify protein amount (*see* **Note 14**), and aliquot 1 mg stocks.
8. Snap freeze and freeze dry aliquots; store at −80 °C for long-term storage.

3.3 Digestion of Samples

3.3.1 Preparation and Trypsin Digestion of Low Complexity Protein Samples (See Note 16)

1. Prepare protein sample attempting to minimize potential protein modification (*see* **Note 17**).
2. Separate proteins using SDS-PAGE.
3. Fix gel using SDS-PAGE fixation buffer for 1 h.
4. Stain gel with 4:1 mixture of Coomassie blue stock/methanol overnight.
5. Destain gel in Milli-Q water for at least 4 h.
6. Excise protein of interest from the gel and cut into 1–2 mm^3 pieces to ensure optimal destaining/rehydration.
7. Destain gel pieces with ten volumes of destaining solution for 10 min with shaking; repeat till Coomassie is removed.
8. Dehydrate gel pieces with ten volumes of 100% ethanol for 10 min and dry in vacuum centrifuge.
9. Add ten volumes of reducing solution for 1 h at 56 °C with shaking.
10. Remove reducing solution and dehydrate gel pieces with ten volumes of ethanol for 10 min.
11. Remove excess solution and dry for 10 min in a vacuum centrifuge.
12. Add ten volumes of alkylation solution to dried gel plug and allow to rehydrate in the dark for 1 h at 37 °C.
13. Remove alkylation solution, wash gel slices with 50 mM ABC, and dehydrate gel pieces with ten volumes of ethanol for 10 min.
14. Remove excess solution; rehydrate gel slices with 20 volumes of 50 mM ABC for 10 min with shaking.
15. Remove excess solution; dehydrate gel pieces with ten volumes of ethanol for 10 min then dry in vacuum centrifuge till all liquid is gone.
16. Rehydrate gel slices with trypsin solution (20 ng/μL) for 1 h at 4 °C.

17. Remove excess trypsin solution, add ABC 50 mM till gel slices are covered, and digest overnight at 37 °C.
18. Collect supernatant form gel slices and incubate gel slices for 15 min in gel extraction buffer; collect supernatant and pool with previously collected supernatant.
19. Dehydrate gel slices with ethanol for 15 min then collect the supernatant and pool with previously collected supernatant.
20. Dry down collected supernatant in a vacuum centrifuge.

3.3.2 Trypsin Digestion of Complex Proteins Samples

1. Resuspend 1 mg dried protein lysate in 200 μL of 6 M urea, 2 M thiourea and vortex.
2. Add 2 μL of 1 M DTT (final concentration 10 mM); vortex and allow sample to incubate at room temp for 1 h in the dark.
3. Add 50 μL of 100 mM iodoacetamide (20 mM final concentration); mix and incubate in the dark for 1 h.
4. Quench excess iodoacetamide by addition of 2 μL 1 M DTT; incubate for 20 min in the dark.
5. Add 5 μg Lys-C (1/200 w/w) and incubate samples for 4 h at 25 °C.
6. Dilute sample 1:5 by addition of 1100 μL 100 mM ABC (final volume 1400 μL), and incubate for 24 h with 40 μg trypsin (1/25 w/w) at room temperature in the dark (*see* **Note 18**).

3.4 Desalting/Concentrating of Peptides

3.4.1 Desalting/Concentrating of Peptides from Low Complexity Samples (*See* Note 19)

1. Resuspend dried samples in Buffer A*.
2. Centrifuge at 10,000 × *g* for 10 min to remove insoluble material.
3. Prepare STAGE column from C18 Empore material according to protocol [58, 59], by washing with five bed volumes of 100% MeCN.
4. Wash column with five bed volumes of Buffer B* for STAGE tips.
5. Wash column with two rounds of five bed volumes Buffer A* for STAGE tips.
6. Load sample resuspended in Buffer A* on to prepared STAGE column.
7. Wash column with two rounds of five bed volumes Buffer A* for STAGE tips.
8. Elute peptides with five bed volumes Buffer B* for STAGE tips.
9. Dry sample by vacuum centrifugation.

3.4.2 Desalting/Concentrating of Peptides from Complex Samples (*See* Note 20)

1. Acidify digests to 0.1% TFA with 10% TFA stock; vortex to mix.
2. Centrifuge at 10,000 × *g* for 10 min to remove insoluble material.
3. Prepare SPE columns by washing with 1 mL 100% MeCN for C18 SPE cartridges.

4. Wash SPE column with two rounds of 1 mL Buffer B* for C18 SPE cartridges.
5. Wash SPE column with three rounds of 1 mL Buffer A* for C18 SPE cartridges.
6. Load acidified sample on to prepared SPE column.
7. Wash SPE column with three rounds of 1 mL Buffer A* for C18 SPE.
8. Elute peptides into an Eppendorf tube with 1 mL Buffer B* for C18 SPE cartridges.
9. Dry sample by vacuum centrifugation.

3.5 Construction of ZIC-HILIC STAGE Tips

1. Using a 20 G, Kel-F Hub NDL blunt needle, excise a C_8 disk.
2. Eject the C_8 disk into a P10/P200 tip using the Plunger assembly.
3. Resuspend the ZIC-HILIC resin in Milli-Q water and load resin on C_8 disk.
4. Pack ZIC-HILIC resin to a height of 0.5 cm by washing with Milli-Q water using pressure from a 10 mL syringe.

3.6 ZIC-HILIC Enrichment of Glycopeptides

1. Resuspend peptides in ZIC-HILIC buffer B, 200 μg of peptide to 100 μL of ZIC-HILIC buffer B and vortex. Store at 4 °C and use immediately after resuspension.
2. Prepare ZIC-HILIC column for glycopeptide binding, wash with four washes of 50 μL ZIC-HILIC buffer A, four washes of 50 μL ZIC-HILIC buffer C and four washes of 50 μL ZIC-HILIC buffer B (*see* **Note 21**).
3. Load peptides onto ZIC-HILIC STAGE tip column and wash ten times with 50 μL ZIC-HILIC buffer B (*see* **Note 22**).
4. Elute glycopeptides with 3×50 μL washes of ZIC-HILIC buffer A.
5. Dry eluted glycopeptides by vacuum centrifugation and store at −20 °C prior to analysis.

3.7 LC-MS/MS

1. Resuspend the dried samples from Subheading 3.6 in 12 μL Buffer A* and transfer to a 96-well autosampler plate.
2. Inject 5 μL of the glycopeptide sample onto the nanoflow LC system.
3. Separate glycopeptides using the analytical column by altering the gradient from 100% Buffer A to 40% Buffer B over 148 min, followed by 40% Buffer B to 80% Buffer B over 12 min. Wash the column for 10 min with 100% Buffer B to remove additional material retained on the column and to condition it for the next injection by washing for a further 10 min with 100% Buffer A. Use a constant flow of 250 nL/min for

all steps, and infuse the elution into the mass spectrometer with a coated fused-silica emitter.

4. Operate the mass spectrometer (e.g., LTQ Orbitrap Velos/Elite or Fusion) in data-dependent acquisition mode, automatically switching between MS and MS/MS selecting the 5–10 most intense ions. Acquire the full-scan MS scan in the Orbital trap at a resolution of R = 60 k/70 k (at 400 *m/z*) between 350 and 1800 *m/z* (*see* **Note 23**). Set the automatic gain control (AGC) target to 1,000,000 ions and a maximum injection time of 50 ms. Isolate the precursor ions of interest with an ion selection width of 2.0 *m/z*, and perform CID fragmentation followed by HCD fragmentation on the same ion (*see* **Note 24**). Dynamic exclusion should be enabled with the exclusion window reflecting the width and time of observed chromatography peaks.

3.8 Data Analysis

3.8.1 Manual Data Analysis Based on Diagnostic Ions

1. Using instrumentation with beam-type fragmentation ions corresponding to glycopeptides can be assessed by performing extracted ion chromatograms of a diagnostic carbohydrate ion, such as the oxonium 204.086 *m/z* ion of *N*-acetylhexosamines (Fig. 2a; *see* **Note 25**).
2. By manually inspecting scan events corresponding to these ions, glycopeptide can be identified by observing the following features: (1) the presence of intense diagnostic carbohydrate ions, (2) the presence of deglycosylated peptide ions, and (3) the presence of ions differing in spacing corresponding to individual carbohydrates (Fig. 2b).
3. Ions below the mass of the deglycosylated peptide ions may correspond to peptide fragmentation and can be used to assign the identity of the glycopeptide. All ions should be assigned in confident glycopeptide spectra (Fig. 2b). To ensure the correct assignment of ions, it is advised to use peptide fragmentation prediction tools, such as MS-product. To further confirm the assignment of glycopeptides, it is advised that a database search engine be used, such as MASCOT.

3.8.2 Data Analysis of ETD Fragmentation Data Using Database Searching (*See* Note 26)

1. Generate MASCOT generic files (.mgf) for each .RAW file using the MSconvert module of ProteoWizard.
2. Search the .mgf files against the appropriate database and the decoy database using MASCOT. Set fixed modifications to carbamidomethylation and variable modifications to methionine oxidation, deamidation of asparagine/glutamine, and the potential glycan modification expected (*see* **Note 24** for information on glycan masses). Allow two missed cleavages, a precursor mass tolerance of 10 ppm, and an MS/MS fragment mass tolerance of 20 ppm for fragmentation analyzed within

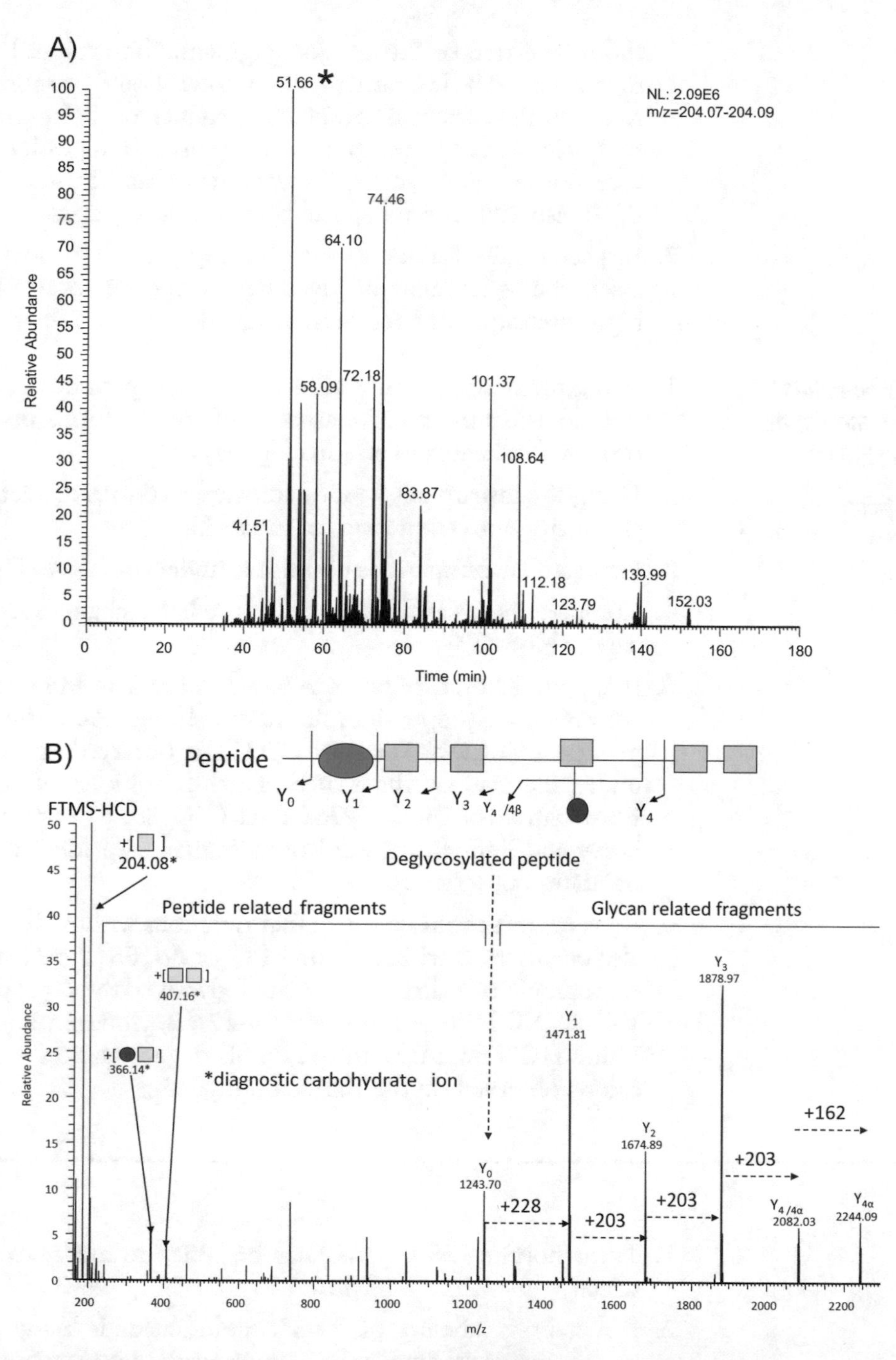

Fig. 2 Assessing beam-type fragmentation of glycopeptides. (**a**) Using extracted ion chromatograms of the diagnostic 204.086 *m/z* carbohydrate ion enables the identification of potential glycopeptides scan from HILIC-enriched samples. (**b**) Assessing the scan containing the most intense diagnostic ion, denoted with *asterisk* in Panel (**a**), the characteristic properties of a glycopeptide can be observed

the orbital trap or 0.6 Da for fragmentation analyzed within the ion trap. The instrumentation setting should be set to ETD to enable the correct assignment of product ions. The enzymes' specificity should be set to the enzyme used for digestion; however, nonspecific or semi-specific searches can be used to identify nonspecific or semi-specific peptide cleavage products.

3. Inspect results for identified glycopeptides. It is advised to assess the assignment of glycopeptide spectra to confirm all high-intensity ions have been assigned.

3.9 Glycopeptide/ Peptide Quantitation (*See* Note 27)

3.9.1 Manual Quantitation of Glycopeptides (*See* Note 28)

1. Using the observed m/z of identified glycopeptides, it is possible to assess the relative amount of specific forms using the observed total area of an ion (Fig. 3a, b).
2. Using the instrument vendor software to create extracted ion chromatograms of the ions of interest.
3. Generate the estimations of the area under the curve (Fig. 3b).
4. Compare the areas of ions to assess relative changes in abundance (**Note 27**).
5. It is good laboratory practice to assess the data MS and MS/MS data for glycopeptides of interest. In Fig. 3a, b the glycopeptide VVLKDNGTER of Cj0371 is observed to decrease under the one of the experimental conditions of interest. Examination of the MS/MS data (Fig. 3d, e) confirms the heavy and light forms are isotopologues with identical fragmentation observed.
6. Researchers should be aware that variations within glycans can also be observed within strains of *C. jejuni* [55] (*see* **Note 29**). An example of a further acetylated glycan of the glycopeptide VVLKDNGTER within strain 81–176 is shown in Fig. 3c, f. Using HCD fragmentation, Fig. 3f, the acetylation of glycan can be observed on the third carbohydrate.

4 Notes

1. Trimethoprim (25 μg/mL) can be added to media to ensure selective growth of *C. jejuni*.
2. If metabolic labeling of specific amino acids is being undertaken, ensure defined media is utilized such as Dulbecco's Modified Eagle Medium (Life Technologies) and the required amino acid, such as arginine or lysine isotopologues, is added accordingly.
3. Vancomycin, trimethoprim, polymyxin B, and defibrinated horse blood is added to Skirrow media after autoclaving to preserve the activity of the selection agents and prevent lysis of the blood. Add reagents when media is cool enough to handle using aseptic conditions.

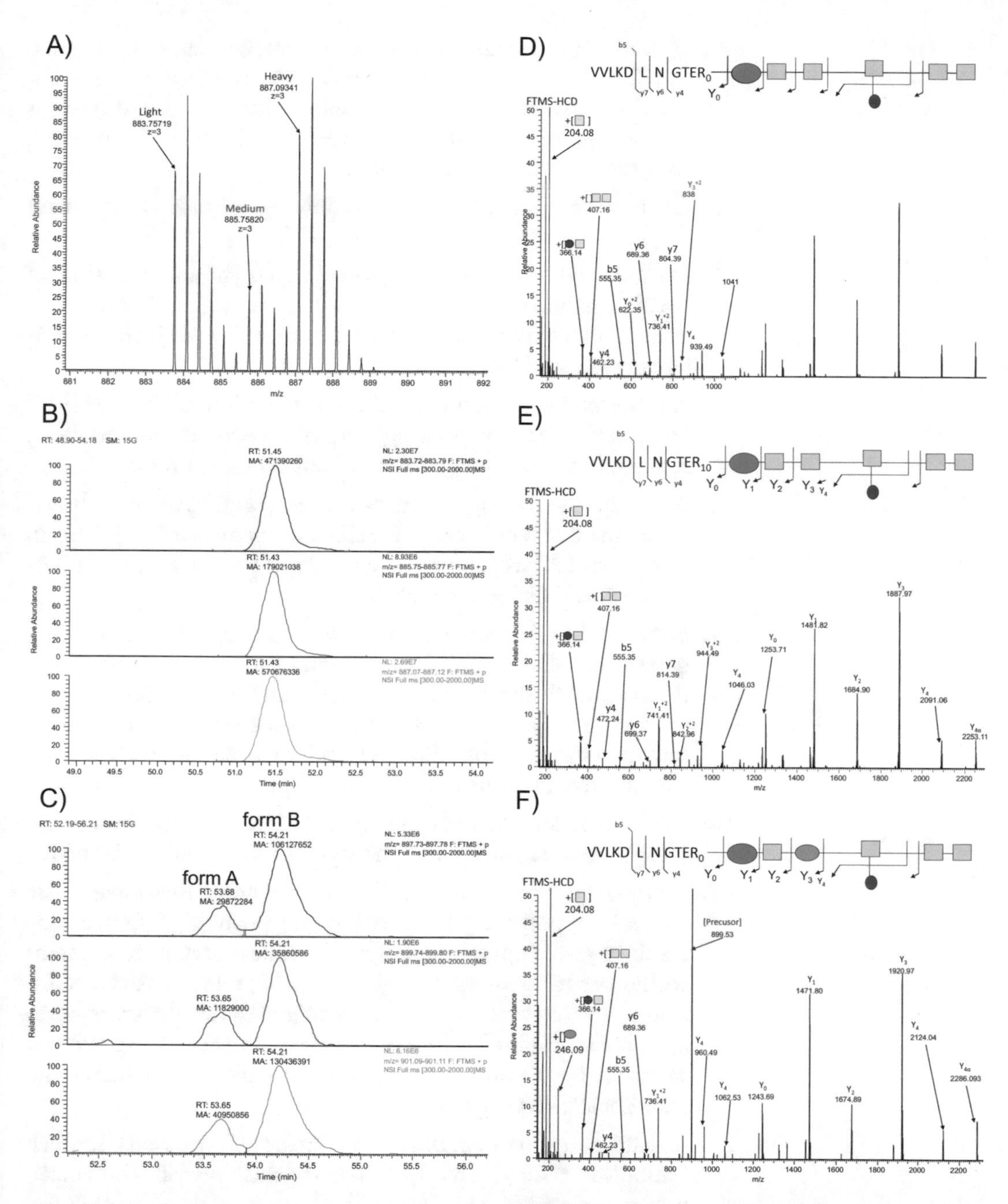

Fig. 3 Quantitative analysis of the VVLKDNGTER glycopeptide within *C. jejuni* 81–176. (**a**) MS spectra of SILAC forms of the glycopeptide VVLKDNGTER. Using SILAC labeling three conditions can be monitored within the same experiment enabling the comparison relative glycopeptide abundance. (**b**) Extracted ion chromatograms enable the determination of the area under the curve for quantitation of SILAC isotopologues. A single species of each glycopeptide is observed which co-elute. (**c**) Extracted ion chromatograms of isobaric forms of the acetylated N-linked glycan, two forms are observed, (**a**) and (**b**) that differ subtlety in elution time. (**d**) and (**e**) MS/MS for the light and heavy isotopologues of the VVLKDNGTER glycopeptide, identical fragmentation is observed with the exception of the mass of the heavy isotopologue being 10 Da heavier due to the presence of heavy arginine. (**f**) MS/MS for the light isotopologue of the acetylated glycan, form A. Using HCD fragmentation, the identity of the peptide was confirmed and modification in the glycan localized to the third carbohydrate

4. If non-metabolic labeling via amine labeling approaches such as iTRAQ/TMT [60] or dimethylation [61] are to be used, avoid the use of reagents with amine such as Tris and ABC as these reagents can reduce the labeling efficiency if not effectively removed prior to labeling.
5. Avoid heating urea/thiourea-containing solutions to decrease the risk of protein carbamylation [62].
6. Decant TFA or use glass pipettes; avoid the introduction of plastic tips into the stock solution as strong acids can leach contaminates from plastics that can interfere with the downstream MS analysis.
7. An alternative method for the construction of ZIC-HILIC columns instead of packing columns onto of C8 has been reported [63] and can be used based on users preferences.
8. Alterative ion pairing agents can be used and have been shown to augment the efficiency of HILIC enrichment [64]. Within our hands FA has shown to be the most robust for the enrichment of bacterial glycopeptides.
9. For complete/accurate assignment of glycopeptides, both the generation of high mass accuracy data and multiple fragmentation approach should be considered. Instrumentation such as the Orbitrap family of MS instruments offers multiple types of fragmentation options and the ability to analyze both precursor and product ions at high resolution.
10. For information on setting up a UPLC trapped based LC setup, users are referred to the work of Cristobal et al. [65].
11. If non-sequenced strains of *C. jejuni* are being investigated, be aware that polymorphism within the protein sequence may result in glycopeptide being unable to be matched to proteins within the database of a single *C. jejuni* strain. A partial solution to overcome this problem is to generate a database composed of multiple *C. jejuni* proteomes. However, the preferred situation is to only use a single proteome that matches the strain being investigated.
12. An alternative to Benzonase® nuclease is Cryonase (Clontech Laboratories Inc, Takara Bio Group), which provides enhanced nuclease digestion at 4 °C.
13. Ensure the sample remains chilled during sonication and is at 4 °C before beginning another round of sonication.
14. Protein/peptide quantification can be accomplished using multiple techniques; however, we recommend using tryptophan fluorescence as described by Wiśniewski and Gaugaz [66], as in our hands this provides an easy and accurate way to monitor protein and peptide amounts.

15. With each round of sonication, the amount of extracted protein should decrease. For *C. jejuni* strains, we have found four rounds of sonication typically result in effective solubilization of membrane material.
16. When preparing protein samples, avoid introducing reagents and processes that may lead to chemical modifications [62, 67], as these modifications will prevent optimal peptide identification.
17. For purified protein or low complexity samples, gel separation and digestion are a robust approach for the preparation of peptides for MS analysis. In addition to requiring reagents utilized within molecular biology laboratories, this approach is extremely effective at removing and reducing detergent and polymer contamination that can interfere with MS analysis.
18. Alternative enzymes may also be useful for generation of glycopeptides of suitable length conducive to their analysis (e.g., GluC, AspN). Importantly, due to the steric hindrance provided by larger glycans, nonspecific proteinases such as thermolysin, pepsin, or even proteinase-K can also be used to generate peptides of adequate length to enable glycopeptide identification [68]. If these enzymes are used, the corresponding digestion buffer may also require alteration to ensure maximum efficiency.
19. For effective enrichment of peptides, a 1:20 ratio of peptide to C18 resin should be used.
20. The presence of salts, urea, and thiourea severely compromises the enrichment of glycopeptides using ZIC-HILIC. For optimal results, ensure the effective cleanup of peptide samples.
21. Do not allow the ZIC-HILIC resin to become dry during the glycopeptide enrichment process, as this appears to seriously compromise the final glycopeptide yield.
22. For complex peptide lysates, ten 50 μL (total voulme 500 μl) washes with ZIC-HILIC buffer B are effective for the removal of the majority of non-glycosylated peptides. For low complexity samples such as single proteins, three 50 μL (total voulme 150 μl) washes are typically sufficient. The number of wash steps should be varied to assess the effect on the enrichment of glycopeptides for any new sample type. Washes can be collected and analyzed to assess the effectiveness of the washing procedure.
23. The addition of a glycan increases the mass of a peptide but does not increase the observed charge state. To increase the chances of selecting a glycopeptide, the lower threshold of the MS1 mass window can be increased to 500 m/z, thus 500–1,800 m/z.
24. Alternative combinations of fragmentation, such as ETD, may be used to gain information about the site of glycosylation. In addition to these fragmentation approaches, dynamic data-dependent methods, such as using the presence of diagnostic ions within HCD scans to trigger ETD, can also be used [69, 70].

25. As the *C. jejuni N*-linked glycan [15] contains five Gal*N*Ac, the oxonium 204.086 *m/z* ion is a reliable diagnostic ion for N-linked glycopeptides. For the identification of O-linked glycan, a range of oxonium ions can be investigated and readers should refer to the list of potential O-linked glycan mass in the work of Ulasi et al. 2015 [53].
26. ETD fragmentation is adversely effected by the presence of negatively charged glycans [71], such as the O-linked glycans observed on *C. jejuni*, glycopeptides, and displays poor efficiency on low charge density ions [72], such as typical N-linked *C. jejuni* glycopeptides. In cases where ETD is required, readers are suggested to augment the charge state of the observed ion using peptide labeling [73].
27. In cases where >10 ions are to be quantified, it is advised that users use tools such as the Skyline quantitative MS software package (https://skyline.gs.washington.edu/) to preform quantitation.
28. Only ions of similar or identical composition (such as isotopologues) can accurately be compared via MS-based analysis. This is due to the differences in ionization potential of different ions, which effects which ions will preferentially be observed [74]. In cases where chemically different species are compared, such as different glycoforms, extreme care should be taken to assess if such analysis is truly informative.
29. Extreme care must be taken to ensure the correct assignment of variation, as multiple artifacts are known to contribute to incorrect glycopeptide glycan assignment [75–77].

Acknowledgments

I would like to thank Beverley and Meowcroft Phillips for their tireless support and proofreading of this manuscript. N.E.S. is supported by a National Health and Medical Research Council (NHMRC) of Australia Overseas (Biomedical) Early Career Fellowship (APP1037373) and a Michael Smith Foundation for Health Research Trainee Postdoctoral Fellowship (award # 5363).

References

1. Spiro RG (2002) Protein glycosylation: nature, distribution, enzymatic formation, and disease implications of glycopeptide bonds. Glycobiology 12(4):43R–56R
2. Eichler J, Adams MW (2005) Posttranslational protein modification in Archaea. Microbiol Mol Biol Rev 69(3):393–425
3. Abu-Qarn M, Eichler J, Sharon N (2008) Not just for Eukarya anymore: protein glycosylation in bacteria and Archaea. Curr Opin Struct Biol 18(5):544–550. doi:10.1016/j.sbi.2008.06.010, S0959-440X(08)00098-5 [pii]
4. Szymanski CM, Wren BW (2005) Protein glycosylation in bacterial mucosal pathogens. Nat Rev Microbiol 3(3):225–237
5. Nothaft H, Szymanski CM (2010) Protein glycosylation in bacteria: sweeter than ever. Nat Rev Microbiol 8(11):765–778. doi:10.1038/nrmicro2383

6. Nothaft H, Szymanski CM (2013) Bacterial protein N-glycosylation: new perspectives and applications. J Biol Chem 288(10):6912–6920. doi:10.1074/jbc.R112.417857
7. Iwashkiw JA, Vozza NF, Kinsella RL et al (2013) Pour some sugar on it: the expanding world of bacterial protein O-linked glycosylation. Mol Microbiol 89(1):14–28. doi:10.1111/mmi.12265
8. Iwashkiw JA, Seper A, Weber BS et al (2012) Identification of a general O-linked protein glycosylation system in *Acinetobacter baumannii* and its role in virulence and biofilm formation. PLoS Pathog 8(6):e1002758. doi:10.1371/journal.ppat.1002758
9. Lithgow KV, Scott NE, Iwashkiw JA et al (2014) A general protein O-glycosylation system within the *Burkholderia cepacia* complex is involved in motility and virulence. Mol Microbiol 92(1):116–37
10. Szymanski CM, Burr DH, Guerry P (2002) Campylobacter protein glycosylation affects host cell interactions. Infect Immun 70(4):2242–2244
11. Howard SL, Jagannathan A, Soo EC et al (2009) *Campylobacter jejuni* glycosylation island important in cell charge, legionaminic acid biosynthesis, and colonization of chickens. Infect Immun 77(6):2544–2556
12. Pearson JS, Giogha C, Ong SY et al (2013) A type III effector antagonizes death receptor signalling during bacterial gut infection. Nature 501(7466):247–251. doi:10.1038/nature12524
13. Breitling J, Aebi M (2013) N-linked protein glycosylation in the endoplasmic reticulum. Cold Spring Harb Perspect Biol 5(8):a013359. doi:10.1101/cshperspect.a013359
14. Jensen PH, Kolarich D, Packer NH (2010) Mucin-type O-glycosylation--putting the pieces together. FEBS J 277(1):81–94. doi:10.1111/j.1742-4658.2009.07429.x
15. Young NM, Brisson JR, Kelly J et al (2002) Structure of the N-linked glycan present on multiple glycoproteins in the Gram-negative bacterium, *Campylobacter jejuni*. J Biol Chem 277(45):42530–42539
16. Morrison MJ, Imperiali B (2014) The renaissance of bacillosamine and its derivatives: pathway characterization and implications in pathogenicity. Biochemistry 53(4):624–638. doi:10.1021/bi401546r
17. Scott NE, Kinsella RL, Edwards AV et al (2014) Diversity within the O-linked protein glycosylation systems of Acinetobacter species. Mol Cell Proteomics. doi:10.1074/mcp.M114.038315
18. Nothaft H, Scott NE, Vinogradov E et al (2012) Diversity in the protein N-glycosylation pathways within the Campylobacter genus. Mol Cell Proteomics 11(11):1203–1219. doi:10.1074/mcp.M112.021519
19. Deeb SJ, Cox J, Schmidt-Supprian M et al (2014) N-linked glycosylation enrichment for in-depth cell surface proteomics of diffuse large B-cell lymphoma subtypes. Mol Cell Proteomics 13(1):240–251. doi:10.1074/mcp.M113.033977
20. Anugraham M, Jacob F, Nixdorf S et al (2014) Specific glycosylation of membrane proteins in epithelial ovarian cancer cell lines: glycan structures reflect gene expression and DNA methylation status. Mol Cell Proteomics 13(9):2213–2232. doi:10.1074/mcp.M113.037085
21. Yang Z, Halim A, Narimatsu Y et al (2014) The GalNAc-type O-Glycoproteome of CHO cells characterized by the SimpleCell strategy. Mol Cell Proteomics 13(12):3224–3235. doi:10.1074/mcp.M114.041541
22. Vester-Christensen MB, Halim A, Joshi HJ et al (2013) Mining the O-mannose glycoproteome reveals cadherins as major O-mannosylated glycoproteins. Proc Natl Acad Sci U S A 110(52):21018–21023. doi:10.1073/pnas.1313446110
23. Parker BL, Thaysen-Andersen M, Solis N et al (2013) Site-specific glycan-peptide analysis for determination of N-glycoproteome heterogeneity. J Proteome Res 12(12):5791–5800. doi:10.1021/pr400783j
24. Schirm M, Schoenhofen IC, Logan SM et al (2005) Identification of unusual bacterial glycosylation by tandem mass spectrometry analyses of intact proteins. Anal Chem 77(23):7774–7782
25. Thibault P, Logan SM, Kelly JF et al (2001) Identification of the carbohydrate moieties and glycosylation motifs in *Campylobacter jejuni* flagellin. J Biol Chem 276(37):34862–34870
26. McNally DJ, Aubrey AJ, Hui JP et al (2007) Targeted metabolomics analysis of Campylobacter coli VC167 reveals legionaminic acid derivatives as novel flagellar glycans. J Biol Chem 282(19):14463–75
27. Logan SM, Kelly JF, Thibault P et al (2002) Structural heterogeneity of carbohydrate modifications affects serospecificity of Campylobacter flagellins. Mol Microbiol 46(2):587–597
28. Goon S, Kelly JF, Logan SM et al (2003) Pseudaminic acid, the major modification on Campylobacter flagellin, is synthesized via the Cj1293 gene. Mol Microbiol 50(2):659–671

29. Ewing CP, Andreishcheva E, Guerry P (2009) Functional characterization of flagellin glycosylation in *Campylobacter jejuni* 81-176. J Bacteriol 191(22):7086–7093
30. Guerry P, Ewing CP, Schirm M et al (2006) Changes in flagellin glycosylation affect Campylobacter autoagglutination and virulence. Mol Microbiol 60(2):299–311
31. Karlyshev AV, Everest P, Linton D et al (2004) The *Campylobacter jejuni* general glycosylation system is important for attachment to human epithelial cells and in the colonization of chicks. Microbiol 150(Pt 6):1957–1964
32. Logan SM, Trust TJ, Guerry P (1989) Evidence for posttranslational modification and gene duplication of Campylobacter flagellin. J Bacteriol 171(6):3031–3038
33. Doig P, Kinsella N, Guerry P et al (1996) Characterization of a post-translational modification of Campylobacter flagellin: identification of a sero-specific glycosyl moiety. Mol Microbiol 19(2):379–387
34. Champion OL, Gaunt MW, Gundogdu O et al (2005) Comparative phylogenomics of the food-borne pathogen *Campylobacter jejuni* reveals genetic markers predictive of infection source. Proc Natl Acad Sci U S A 102(44): 16043–16048. doi:10.1073/pnas.0503252102
35. Parkhill J, Wren BW, Mungall K et al (2000) The genome sequence of the food-borne pathogen *Campylobacter jejuni* reveals hypervariable sequences. Nature 403(6770):665–668. doi:10.1038/35001088
36. Schoenhofen IC, McNally DJ, Brisson JR et al (2006) Elucidation of the CMP-pseudaminic acid pathway in Helicobacter pylori: synthesis from UDP-N-acetylglucosamine by a single enzymatic reaction. Glycobiology 16(9):8C–14C. doi:10.1093/glycob/cwl010
37. Chou WK, Dick S, Wakarchuk WW et al (2005) Identification and characterization of NeuB3 from *Campylobacter jejuni* as a pseudaminic acid synthase. J Biol Chem 280(43):35922–35928
38. Schoenhofen IC, McNally DJ, Vinogradov E et al (2006) Functional characterization of dehydratase/aminotransferase pairs from Helicobacter and Campylobacter: enzymes distinguishing the pseudaminic acid and bacillosamine biosynthetic pathways. J Biol Chem 281(2):723–732. doi:10.1074/jbc.M511021200
39. Schoenhofen IC, Vinogradov E, Whitfield DM et al (2009) The CMP-legionaminic acid pathway in Campylobacter: biosynthesis involving novel GDP-linked precursors. Glycobiology 19(7):715–725. doi:10.1093/glycob/cwp039
40. Zampronio CG, Blackwell G, Penn CW et al (2011) Novel glycosylation sites localized in *Campylobacter jejuni* flagellin FlaA by liquid chromatography electron capture dissociation tandem mass spectrometry. J Proteome Res. doi:10.1021/pr101021c
41. Scott NE, Nothaft H, Edwards AV et al (2012) Modification of the *Campylobacter jejuni* N-linked glycan by EptC protein-mediated addition of phosphoethanolamine. J Biol Chem 287(35):29384–29396. doi:10.1074/jbc.M112.380212
42. Scott NE, Parker BL, Connolly AM et al (2011) Simultaneous glycan-peptide characterization using hydrophilic interaction chromatography and parallel fragmentation by CID, higher energy collisional dissociation, and electron transfer dissociation MS applied to the N-linked glycoproteome of *Campylobacter jejuni*. Mol Cell Proteomics 10(2):M000031–MCP000201. doi:10.1074/mcp.M000031-MCP201
43. Scott NE, Marzook NB, Cain JA et al (2014) Comparative proteomics and glycoproteomics reveal increased N-linked glycosylation and relaxed sequon specificity in *Campylobacter jejuni* NCTC11168 O. J Proteome Res 13(11):5136–5150. doi:10.1021/pr5005554
44. Szymanski CM, Yao R, Ewing CP et al (1999) Evidence for a system of general protein glycosylation in *Campylobacter jejuni*. Mol Microbiol 32(5):1022–1030
45. Szymanski CM, Michael FS, Jarrell HC et al (2003) Detection of conserved N-linked glycans and phase-variable lipooligosaccharides and capsules from campylobacter cells by mass spectrometry and high resolution magic angle spinning NMR spectroscopy. J Biol Chem 278(27):24509–24520
46. Wacker M, Linton D, Hitchen PG et al (2002) N-linked glycosylation in *Campylobacter jejuni* and its functional transfer into *E. coli*. Science 298(5599):1790–1793
47. Nita-Lazar M, Wacker M, Schegg B et al (2005) The N-X-S/T consensus sequence is required but not sufficient for bacterial N-linked protein glycosylation. Glycobiology 15(4):361–367
48. Kowarik M, Young NM, Numao S et al (2006) Definition of the bacterial N-glycosylation site consensus sequence. EMBO J 25(9)
49. Schwarz F, Lizak C, Fan YY et al (2011) Relaxed acceptor site specificity of bacterial oligosaccharyltransferase *in vivo*. Glycobiology 21(1):45–54. doi:10.1093/glycob/cwq130
50. Ielmini MV, Feldman MF (2011) Desulfovibrio desulfuricans PglB homolog possesses oligosaccharyltransferase activity with relaxed glycan

specificity and distinct protein acceptor sequence requirements. Glycobiology 6: 734–742

51. Lizak C, Gerber S, Michaud G et al (2013) Unexpected reactivity and mechanism of carboxamide activation in bacterial N-linked protein glycosylation. Nat Commun 4:2627. doi:10.1038/ncomms3627
52. Gerber S, Lizak C, Michaud G et al (2013) Mechanism of bacterial oligosaccharyltransferase: *in vitro* quantification of sequon binding and catalysis. J Biol Chem 288(13):8849–8861. doi:10.1074/jbc.M112.445940
53. Ulasi GN, Creese AJ, Hui SX et al (2015) Comprehensive mapping of O-glycosylation in flagellin from *Campylobacter jejuni* 11168: a multi-enzyme differential ion mobility mass spectrometry approach. Proteomics. doi:10.1002/pmic.201400533
54. Whitworth GE, Imperiali B (2015) Selective biochemical labeling of *Campylobacter jejuni* cell-surface glycoconjugates. Glycobiology. doi:10.1093/glycob/cwv016
55. Ding W, Nothaft H, Szymanski CM et al (2009) Identification and quantification of glycoproteins using ion-pairing normal-phase liquid chromatography and mass spectrometry. Mol Cell Proteomics 8(9):2170–2185. doi:10.1074/mcp.M900088-MCP200, M900088-MCP200 [pii]
56. Linton D, Allan E, Karlyshev AV et al (2002) Identification of N-acetylgalactosamine-containing glycoproteins PEB3 and CgpA in *Campylobacter jejuni*. Mol Microbiol 43(2): 497–508
57. Scott NE, Bogema DR, Connolly AM et al (2009) Mass spectrometric characterization of the surface-associated 42 kDa lipoprotein JlpA as a glycosylated antigen in strains of *Campylobacter jejuni*. J Proteome Res. doi:10.1021/pr900544x
58. Rappsilber J, Ishihama Y, Mann M (2003) Stop and go extraction tips for matrix-assisted laser desorption/ionization, nanoelectrospray, and LC/MS sample pretreatment in proteomics. Anal Chem 75(3):663–670
59. Rappsilber J, Mann M, Ishihama Y (2007) Protocol for micro-purification, enrichment, pre-fractionation and storage of peptides for proteomics using StageTips. Nat Protoc 2(8):1896–1906. doi:10.1038/nprot.2007.261
60. Thompson A, Schafer J, Kuhn K et al (2003) Tandem mass tags: a novel quantification strategy for comparative analysis of complex protein mixtures by MS/MS. Anal Chem 75(8):1895–1904
61. Boersema PJ, Raijmakers R, Lemeer S et al (2009) Multiplex peptide stable isotope dimethyl labeling for quantitative proteomics. Nat Protoc 4(4):484–494. doi:10.1038/nprot.2009.21, nprot.2009.21 [pii]
62. Kollipara L, Zahedi RP (2013) Protein carbamylation: *in vivo* modification or *in vitro* artefact? Proteomics 13(6):941–944. doi:10.1002/pmic.201200452
63. Dedvisitsakul P, Jacobsen S, Svensson B et al (2014) Glycopeptide enrichment using a combination of ZIC-HILIC and cotton wool for exploring the glycoproteome of wheat flour albumins. J Proteome Res. doi:10.1021/pr401282r
64. Mysling S, Palmisano G, Hojrup P et al (2010) Utilizing ion-pairing hydrophilic interaction chromatography solid phase extraction for efficient glycopeptide enrichment in glycoproteomics. Anal Chem 82(13):5598–5609. doi:10.1021/ac100530w
65. Cristobal A, Hennrich ML, Giansanti P et al (2012) In-house construction of a UHPLC system enabling the identification of over 4000 protein groups in a single analysis. Analyst 137(15):3541–3548. doi:10.1039/c2an35445d
66. Wisniewski JR, Gaugaz FZ (2015) Fast and sensitive total protein and peptide assays for proteomic analysis. Anal Chem 87(8):4110–4116. doi:10.1021/ac504689z
67. Means GE, Feeney RE (1990) Chemical modifications of proteins: history and applications. Bioconjug Chem 1(1):2–12
68. Larsen MR, Hojrup P, Roepstorff P (2005) Characterization of gel-separated glycoproteins using two-step proteolytic digestion combined with sequential microcolumns and mass spectrometry. Mol Cell Proteomics 4(2):107–119
69. Saba J, Dutta S, Hemenway E et al (2012) Increasing the productivity of glycopeptides analysis by using higher-energy collision dissociation-accurate mass-product-dependent electron transfer dissociation. Int Jurnal of proteomics 2012:560391. doi:10.1155/2012/560391
70. Wu SW, Pu TH, Viner R et al (2014) Novel LC-MS(2) product dependent parallel data acquisition function and data analysis workflow for sequencing and identification of intact glycopeptides. Anal Chem 86(11):5478–5486. doi:10.1021/ac500945m
71. Thaysen-Andersen M, Wilkinson BL, Payne RJ et al (2011) Site-specific characterisation of densely O-glycosylated mucin-type peptides using electron transfer dissociation ESI-MS/MS. Electrophoresis 32(24):3536–3545. doi:10.1002/elps.201100294
72. Good DM, Wirtala M, McAlister GC et al (2007) Performance characteristics of electron transfer dissociation mass spectrometry. Mol Cell Proteomics 6(11):1942–1951

73. Thingholm TE, Palmisano G, Kjeldsen F et al (2010) Undesirable charge-enhancement of isobaric tagged phosphopeptides leads to reduced identification efficiency. J Proteome Res 9(8):4045–4052. doi:10.1021/pr100230q
74. Schmidt A, Karas M, Dulcks T (2003) Effect of different solution flow rates on analyte ion signals in nano-ESI MS, or: when does ESI turn into nano-ESI? J Am Soc Mass Spectrom 14(5): 492–500, doi:S1044030503001284 [pii]
75. Darula Z, Medzihradszky KF (2015) Carbamidomethylation side-reactions may lead to glycan misassignments in glycopeptide analysis. Anal Chem. doi:10.1021/acs.analchem.5b01121
76. Medzihradszky KF (2014) Noncovalent dimer formation in liquid chromatography-mass spectrometry analysis. Anal Chem 86(18):8906–8909. doi:10.1021/ac502790j
77. Darula Z, Medzihradszky KF (2014) Glycan side reaction may compromise ETD-based glycopeptide identification. J Am Soc Mass Spectrom 25(6):977–987. doi:10.1007/s13361-014-0852-9

Chapter 19

Methods for Proteome Analysis of *Campylobacter jejuni* Using 2-D Electrophoresis

Ramila C. Rodrigues, Nabila Haddad, and Odile Tresse

Abstract

This chapter describes protocols used for two-dimensional electrophoretic analysis of the proteome or subproteome of *Campylobacter jejuni*, a major human food-borne pathogen. The following protocols, adapted to *Campylobacter* strains, include all the steps from cultivation to gel-support protein separation.

Key words 2-D electrophoresis, *Campylobacter*, Protein extraction, Protein purification, Isoelectric point, SDS-PAGE

1 Introduction

Organisms must constantly respond to external and internal cues and adapt their metabolism to survive. These metabolic alterations occur through various regulations induced at different cellular levels including DNA, RNA, protein, and enzymatic activities. The cellular processes involved in response to cell internal and external cues could be analyzed using targeted or holistic approaches. Proteomic analysis refers to the identification and quantification of the complete set of proteins (the proteome) of a biological system (e.g., cell, tissue, organ, biological fluid, or organism) at a specific point in time. Proteome analyses require several steps from protein extraction to protein identification [1]. The fundamentals of two-dimensional (2-D) electrophoresis were first reported by O'Farrell and Klose [2, 3]. Optimization of these steps depends on the organism and the addressed cellular compartment. For instance, extraction procedures will be different from Gram-positive and Gram-negative bacteria and protein purification will differ from cytosoluble and membrane proteins. Several technologies have been developed to analyze bacterial proteome or for subproteome fingerprintings. The following protocol focuses on gel-support techniques to separate proteins from *Campylobacter* cultured in

James Butcher and Alain Stintzi (eds.), *Campylobacter jejuni: Methods and Protocols*, Methods in Molecular Biology, vol. 1512, DOI 10.1007/978-1-4939-6536-6_19, © Springer Science+Business Media New York 2017

suspension as reported in several published manuscripts [4–7]. This method of protein separation is based on the protein charge and protein size. In 2-D electrophoresis, the first dimension separates proteins according to the protein charge under a horizontal electric field using strips with linear or nonlinear pH gradients. During this step, proteins migrate until they reach their uncharged isoelectric point (pI). The amphoteric feature of proteins depends on the primary structure (the amino acid chain) and on the pH. The second dimension is performed perpendicular from the first one using sodium dodecyl sulfate (SDS) polyacrylamide gel electrophoresis (SDS-PAGE). This second step is conducted in denaturing conditions and separates the proteins according to their molecular size under a vertical electric field. The compounds used to achieve gels for the second dimension are the two monomers acrylamide and bisacrylamide in presence of SDS. SDS is an anionic surfactant which, in an aqueous medium, forms micelles composed of 70–80 molecules. In the presence of SDS, proteins form complexes resulting in an overall negative charge around proteins to allow them to migrate under the electric field according to their size. Proteins with a lower molecular weight pass through the gel more rapidly than proteins with higher molecular mass. Thus, the 2-D electrophoresis results from two sequential chromatographic techniques.

Currently, the implementation of 2-D electrophoresis enables a high resolution of the various species proteins present in a biological sample in a reproducible manner. Depending on the size of the gel and the pH gradient used, 2-D electrophoresis can resolve over 5000 proteins simultaneously and may detect smaller quantities as less than 1 ng protein per spot [1].

This chapter provides an overview of the current status of 2-D electrophoretic technology for proteome analysis of *Campylobacter*, with special emphasis on protein extraction, fractionation, and purification. The main steps of the procedure are summarized in Fig. 1.

2 Materials

2.1 Solutions

Prepare all solutions using ultrapure water (prepared by purifying deionized water to reach an electrical resistivity of 18 MΩ cm at 25 °C) and analytical grade reagents. Prepare and store all reagents at 4 °C (unless otherwise indicated). Carefully follow all waste disposal regulations when disposing waste materials.

1. Karmali agar (composition: Columbia Agar Base 39 g/L, activated charcoal 4 g/L, hemin 0.032 g/L) supplemented with *Campylobacter* Selective Supplement (composition: sodium pyruvate 100 mg/L, cefoperazone 32 mg/L, vancomycin 20 mg/L, cycloheximide 100 mg/L).
2. Brain-heart infusion (BHI) broth: Available from various suppliers (e.g., Oxoid). Prepare according to manufacturer's instructions.

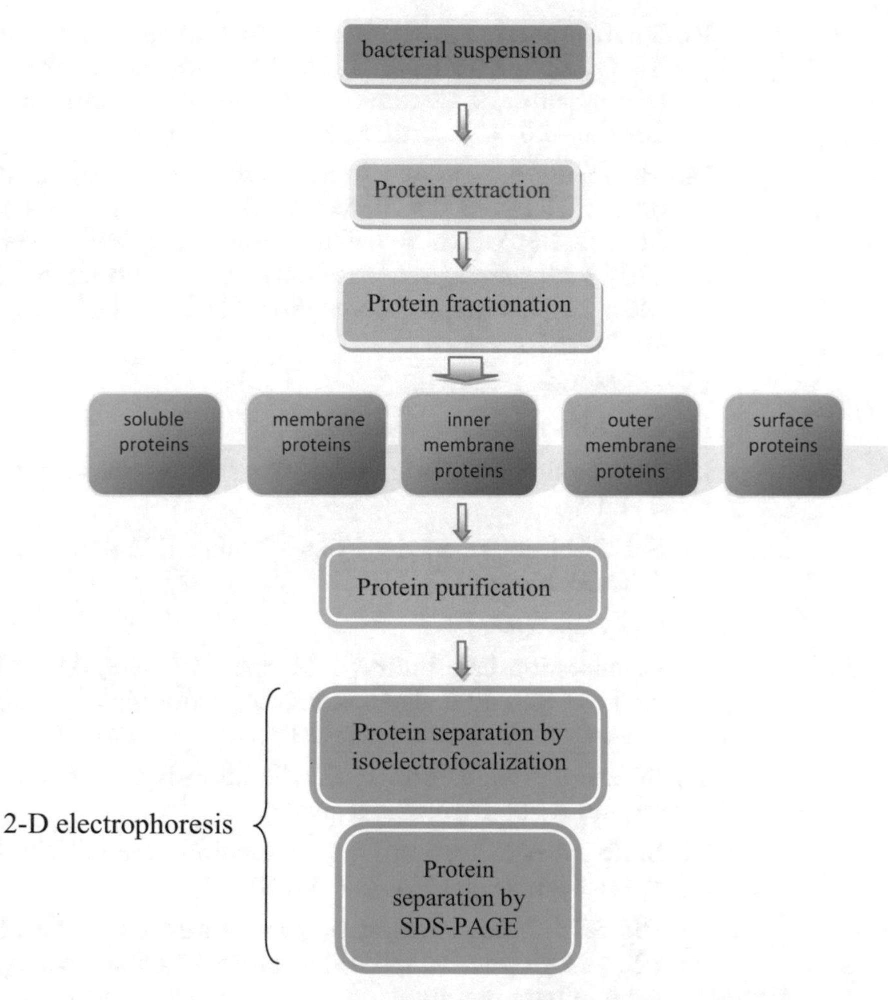

Fig. 1 Main steps of extraction and separation of *C. jejuni* proteins for 2-D electrophoresis analysis

3. Protease inhibitor tablet (e.g., complete protease inhibitor tablet, Roche Diagnostics. 1/2 tablet can be added to each sample).
4. 200 mM glycine pH 5.8 (*see* **Note 1**). Store at 4 °C.
5. 100 mM Tris–HCl pH 7.0 (*see* **Note 2**). Store at 4 °C.
6. 10 mM Tris–HCl pH 7.0 (*see* **Note 2**). Store at 4 °C.
7. 1 mM ETDA.
8. 0.5 % sodium lauroyl-sarcosinate in ultrapure water.
9. Nuclease buffer: 1.5 M Tris–HCl pH 7.0, 1.5 M Tris base, 1 M $MgCl_2$ in ultrapure water.
10. DNase I 1 U/μL.
11. 3 mg/mL RNase A in nuclease buffer.
12. Protein assay kit (e.g., Micro BCA Protein Assay Reagent Kit).

13. Soluble protein rehydration buffer: 6 M urea, 2 M thiourea, 4% CHAPS, 2% Biolyte 3/10 (Bio-Rad), bromophenol blue (some grains), 0.4% dithiothreitol (DTT), in ultrapure water. Store at –20 °C in 1 mL aliquots (*see* **Note 3**).
14. Membrane protein rehydration buffer: 7 M urea, 2 M thiourea, 4% CHAPS, 2% Biolyte 3/10 (Bio-Rad), 1% Coomassie Blue (R-250), 2 mM tributylphosphine (TBP) (ReadyPrep TBP Reducing agent, Bio-Rad), 1% amidosulfobetaine-14 (ASB-14), in ultrapure water. Store at –20 °C in 1 mL aliquots (*see* **Note 3**).
15. 1.5 M Tris–HCl pH 8.8.
16. 25% SDS in ultrapure water.
17. Ammonium persulfate: 10% w/v solution in water (*see* **Notes 5** and **6**).
18. N,N,N,N′-tetramethyl-ethylenediamine (TEMED). Store at 4 °C (*see* **Note 6**).
19. 95% isopropanol.
20. Equilibration base buffer: 6 M urea, 2% SDS, 0.05 M Tris–HCl pH 8.8, 30% glycerol, bromophenol blue (some grains) in ultrapure water. Store at –20 °C in 1 mL aliquots.
21. Buffer of equilibration I: Equilibration base buffer with 2% dithiothreitol (*see* **Notes 3** and **8**).
22. Buffer of equilibration II: Equilibration base buffer with 4% iodoacetamide (*see* **Notes** 7 and **8**).
23. 10× SDS-PAGE running buffer: 25 mM Tris–HCl pH 8.3, 192 mM glycine, 0.1% SDS (*see* **Note 9**). Dilute to 1× prior to use in ultrapure water.
24. Low melting point agarose.
25. Fixation solution: 50% ethanol (95%), 12% glacial acetic acid, 100 μL formaldehyde (35%), ultrapure water (final volume of 200 mL).
26. 50% ethanol and 70% ethanol.
27. Pretreatment solution: 0.02 g of sodium thiosulfate ($Na_2S_2O_3$), ultrapure water (final volume of 100 mL).
28. Impregnation solution: 0.1 g of silver nitrate ($AgNO_3$), 100 μL formaldehyde 35%, ultrapure water (final volume of 100 mL).
29. Developing solution: 6 g sodium carbonate (Na_2CO_3), 0.4 mg sodium thiosulfate ($Na_2S_2O_3$), 50 μL formaldehyde (35%), ultrapure water (final volume of 100 mL).

2.2 Equipment

1. Chemical fume hood.
2. Laminar flow hood.
3. Ultracentrifuge.

4. Centrifuge.
5. Rocker.
6. Sonicator.
7. Protean IEF Cell device for isofocalization with focalization supports (Bio-Rad) or alternative IEF focusing apparatus.
8. Protean II xi cell device for SDS-PAGE (Bio-Rad) or alternative SDS-PAGE system.
9. Power supply.
10. Glass container for gel staining.
11. Vortex.
12. Water bath at 60 °C.
13. SpeedVac Concentrator.
14. Micropipettes.
15. 12 mL Ultra-Clear Tubes (or equivalent for ultracentrifugation).
16. Sterile plastic tubes of 15 and 50 mL.
17. Sterile tips of 100 and 1000 μL.
18. Sterile pipettes of 5, 10, and 25 mL.
19. 12 kDa cutoff dialysis tubing cellulose membrane.
20. IEF strips of 17 cm.
21. Mineral oil.
22. Wicks papers.
23. Vacuum flask.
24. Plastic rack.
25. Spatula and tweezers (clean with alcohol prior to use).
26. Ice.

3 Timeline for Experiments

- Extraction of proteins: 1 day.
- Dialysis of proteins: 2 days.
- Protein concentration, rehydration, and isoelectrofocalization: 1 day.
- SDS-PAGE electrophoresis: 0.5 day.
- Protein staining with silver nitrate: 0.5 day.

4 Method

Carry out all procedures at room temperature unless otherwise specified.

4.1 Extraction and Fractionation of Proteins

1. Inoculate *C. jejuni* strain into Karmali agar and incubate for 48 h in microaerobic conditions (5% O_2, 10% CO_2 and 85% N_2) at 42 °C.
2. Inoculate one colony of *C. jejuni* in liquid BHI medium and incubate for 24 h in microaerobic conditions at 42 °C.
3. After the subculture incubation, inoculate 4 mL of cell suspension in 200 mL of BHI. Incubate in microaerobic conditions at 42 °C for overnight culture (~16 h).
4. After cultivation, harvest the cells by centrifuging at 6000 × *g* for 20 min at 4 °C.
5. Discard the supernatants and resuspend the pellets of bacteria in 50 mL (final volume) of 200 mM glycine solution at 4 °C. Incubate 20 min on ice on the rocker.
6. Centrifuge the suspension at 6000 × *g* for 20 min at 4 °C. Rinse the pellets with 100 mM Tris–HCl pH 7.0 and centrifuge with the same parameters. Resuspend the bacteria in 5.5 mL of 10 mM Tris–HCl pH 7.0, add ½ tablet of protease inhibitor, and transfer the suspension in a 15 mL sterile tube.
7. Sonicate bacteria on ice for 6 min (alternate 30 s sonication at 50 kHz and 30 s of rest). Let stand on ice for at least 5 min. Repeat three times.
8. Centrifuge the suspension at 6000 × *g* for 20 min at 4 °C to pellet the non-lysed cells. Collect the supernatant and store on ice.
9. Resuspend the pelleted bacteria in 5.5 mL of 10 mM Tris–HCl pH 7.2 and repeat **step 7**. Centrifuge the sonicated cells at 6000 × *g* for 20 min at 4 °C.
10. Pool the two supernatants obtained (**steps 8** and **9**) and centrifuge at 10000 × *g* for 20 min at 4 °C to ensure complete removal of the cell debris.
11. Transfer the supernatant into a 12 mL Ultra-Clear Tube.
12. Centrifuge at 188,000 × *g* for 1 h at 4 °C. Collect the supernatant containing the cytosoluble proteins and store at −80 °C in 2 mL aliquots.
13. Store the membrane proteins by resuspending the pellet in 1 mL of 1 mM EDTA at 4 °C, and store at −80 °C or proceed to **step 14** to separate the inner/outer membrane protein fractions.
14. To separate the inner membrane fraction from the outer membrane fraction, resuspend the pellet in 10 mL of 0.5% sodium lauroyl-sarcosinate at 4 °C; vortex and centrifuge at 188,000 × *g* for 1 h at 4 °C. Collect the supernatant containing proteins

from the inner membrane and store at –80 °C in 2 mL aliquots. Resuspend the pellet containing the outer membrane proteins in 1 mL of 1 mM EDTA and store at –80 °C.

4.2 Protein Dialyzation

1. Cut out more cellulose membrane dialysis tubing than needed for each sample (to accommodate expansion of the tubing). Wash the dialysis tubing cellulose membrane with hot water (60 °C) for 2 min.
2. Pipette 3 mL of protein sample into the tubing, 5 μL of DNase I (1.5 μL/mL of proteins), and 18 μL of RNase A (6 μL/mL of proteins), and close off the other end of the tubing with another closure. Try to avoid including too many air bubbles.
3. Insert the dialysis tubing containing the protein sample into ultrapure water in a large beaker. The volume of the buffer should be at least 100 times the original volume of the protein sample.
4. Stir the dialysis bag in ultrapure water slowly with a stir bar and magnetic stir plate overnight.
5. Discard the ultrapure water twice during the following day, and replace it with the same volume of fresh ultrapure water and dialyze for another night.
6. Remove the tubing from the beaker containing ultrapure water and carefully open one end of the tubing. Pipette the dialyzed protein solution into tubes. Store the proteins at –80 °C.
7. Quantify the concentration of proteins extracted (e.g., Micro BCA Protein Assay Reagent Kit) and store at –80 °C.

4.3 Protein Concentration and Rehydration

1. Place the equivalent of 100 μg of protein in an Eppendorf tube.
2. Evaporate the liquid with a SpeedVac to obtain no more than 15 μL ± 5 μL.
3. Add 275 μL of rehydration buffer (either for soluble or membrane proteins depending on application) to the concentrated protein sample. Vortex for 1 min to resuspend the proteins. Then centrifuge few seconds to pool the sample at the bottom of the tube.
4. Pipette and distribute gently and uniformly the sample in the focalization support, taking care not to make bubbles (*see* **Note 10**).
5. A strip of 17 cm (Bio-Rad) is used for each sample. Remove the protective tape using tweezers (cleaned with alcohol). Gently place the strip into the support rails, ensuring that the sample is completely under the strip (if necessary slightly turn the focalization support at the time of depositing the strip and reinject the sample on the strip using a pipette).

6. Uniformly coat the strip with 1 mL mineral oil and close the focalization support.
7. Install the support in the focusing machine and begin rehydration step using the active rehydration program default to 50 V.
8. Set the rehydration step for at least 12 h.

4.4 Isoelectro focalization

The electrofocalization parameters are given for a 17-cm IEF strip. For other lengths, refer to the manufacturer's procedure and alter accordingly.

1. Humidify two Wicks papers using ultrapure water, and, after removing the excess water, put the paper in the focusing support rail, between the electrodes and the strip (*see* **Note 11**).
2. Isoelectrofocalization is performed at 20 °C with the following parameters: **step 1**, voltage ramping from 0 V to 250 V for 3 h; **step 2**, voltage ramping from 250 to 6000 V for 3 h; **step 3**, constant voltage at 6000 until 54000 V h; and **step 4**, constant voltage at 500 V for a maximum of 20 h.
3. Following focalization, transfer the strips (gel side up) in a plastic rack and store them at –80 °C until the second dimension.

4.5 Preparation of the Two-Dimensional Gels for SDS-PAGE

The proportions are given for a 18.5 × 20 cm gel and IEF strips of 17 cm. For other sizes refer to the manufacturer's procedure and alter accordingly.

1. Clean the glass plates with water and then with 70% ethanol. It is critical to remove all dust and small particles, especially any bits of leftover polyacrylamide. Rinse with ultrapure water and assemble the plates.
2. For the preparation of a gel of 1.5 mm thick and 12% of acrylamide/bisacrylamide, mix 19.8 mL of acrylamide/bisacrylamide (37:1), 16.5 mL of 1.5 M Tris–HCl pH 8.8, and 29.08 mL ultrapure water in a 50 mL vacuum flask. Cap the flask and apply a vacuum for 10 min while swirling the flask.
3. Add 317 μL of 25% SDS, 396 μL of 10% ammonium persulfate, and 10 μL of TEMED. Gently swirl the flask to mix, being careful not to generate bubbles.
4. Use a pipette to deliver the solution into one corner of the plate, taking care not to introduce any air pockets. Allow space for the gel strip.
5. Using a pipettor or syringe, add 95% isopropanol (*see* **Note 12**). Let it polymerize for about 1 h.
6. After polymerization rinse the top of gel three times carefully with 1× SDS-PAGE running buffer. Discard the running buffer and remove the excess of running buffer solution using an absorbing paper.

4.6 Equilibration of Strips for the Second Dimension

1. Following focalization, rinse the IEF strip with ultrapure water and remove excess water with an absorbing paper.
2. Transfer the strips (gel side up) in a plastic rack and add 1.5 mL of equilibration I buffer.
3. Place it on a rocker and equilibrate for 20 min.
4. Rinse the strip gel with ultrapure water and remove excess water with an absorbing paper.
5. Transfer the strips (gel side up) in a plastic rack and add 1.5 mL of equilibration II buffer.
6. Place the plastic rack on a rocker and equilibrate for 20 min.
7. Rinse the strip gel with the running buffer solution.
8. Place the strip with the acidic end (+) to the left and the plastic side backing against one of the glass plates. With a spatula, gently push the IPG strip down so that the entire lower edge of the IPG strip is in contact with the top surface of the slab gel. Ensure that no air bubbles are trapped between the IPG strip and the slab gel surfaces or in between the gel back and the glass plate.
9. Seal the strip with 1% agarose. Pipette slowly to avoid introducing bubbles. Wait until the agarose solidifies before proceeding.

4.7 SDS-PAGE

1. Start the migration step (*see* **Note 13**). The electrophoresis is firstly performed at 20 mA/gel and until the blue line comes completely out of the strip to stack protein sample and then continue at 40 mA/gel. Stop migration when the dye front line is approximately 1 mm from the bottom of the gel.
2. Following electrophoresis, pry apart the gel plates using a spatula. The gel usually remains stuck on one of the glass plates. Carefully transfer the gel to a glass container with 125 mL of fixation solution.
3. Leave the gel in fixation solution overnight.
4. Stain the gel with silver nitrate or Coomassie Blue (*see* **Note 14**).

4.8 Protein Staining with Silver Nitrate

1. Fix the proteins using 100 mL of the fixation solution. Place the gel on a rocker for 24 h (*see* **Note 15**).
2. Wash the gel in 100 mL of 50% ethanol. Place the gel on a rocker for 20 min. Repeat this step three times.
3. Pretreat the gel in 100 mL of pretreatment solution for 1 min.
4. Wash the gel in 100 mL of ultrapure water for 20 s. Repeat this step three times.
5. Impregnate the gel with the impregnation solution. Place the gel on a rocker for 20 min (*see* **Note 16**).

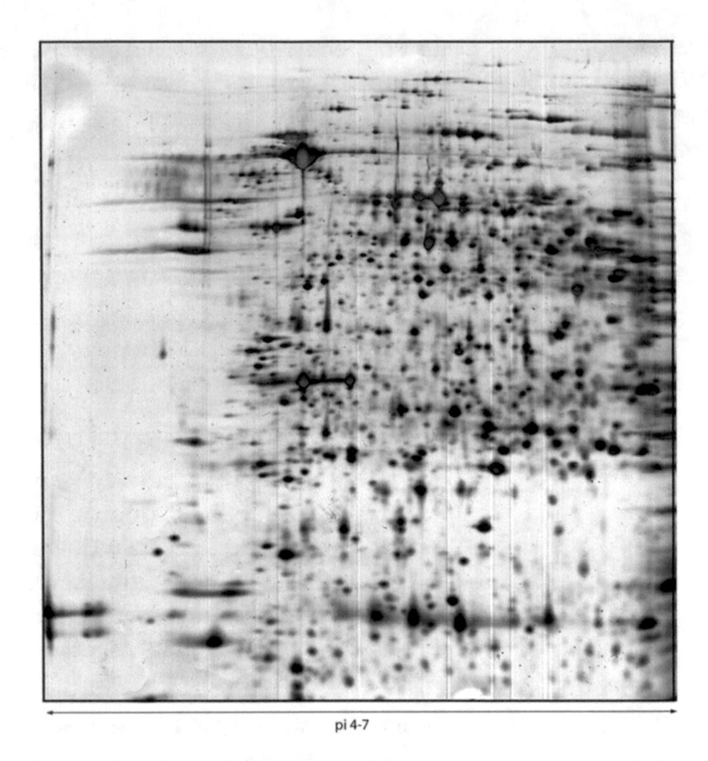

Fig. 2 Electrophoretic profile of cytosoluble proteins separated by 2-D electrophoresis using 17 cm, pH 4–7 linear IEF strips and 12 % acrylamide/bisacrylamide gels for SDS-PAGE. The silver stained gel was scanned using a GS 700 densitometer (Bio-Rad)

6. Wash the gel in 100 mL of ultrapure water for 20 s. Repeat this step two times.
7. Reveal the gel using a developing solution to obtain the desired staining.
8. Wash the gel quickly in 100 mL of ultrapure water and stop the stain development using 100 mL of fixation solution for at least 10 min (Fig. 2).
9. Store the gel at 4 °C in ultrapure water until image acquisition.

5 Notes

1. Concentrated HCl (10 N) can be used at first to narrow the gap from the starting pH to the required pH. It is advisable to use series of HCl concentrations (e.g., 6 and 1 N) with lower ionic strengths to avoid a sudden drop in pH below the required pH.

2. The volume of Tris–HCl must be prepared according to the number of protein samples to be extracted. For bacterial pellet washing step, the volume used is approximately 50 mL of 100 mM Tris–HCl pH 7 for each sample. For bacterial pellet suspension step, the volume used is approximately 11 mL of 10 mM Tris–HCl pH 7.2 for each sample.
3. The dithiothreitol should be kept at 4 °C and added only at the time of use.
4. Acrylamide should be stored at 4 °C and filtered using Whatman filter paper to discard crystals.
5. Ammonium persulfate should be prepared at the time of use.
6. Ammonium persulfate and TEMED allow the polymerization of the gel. Once these two reagents are added, the gel preparation should be rapidly poured into the casting apparatus.
7. Iodoacetamide should be kept at 4 °C and added only at the time of use.
8. Buffers I and II should be prepared at the time of use.
9. Simple method of preparing running buffer: Prepare 10× native buffer (25 mM Tris–HCl pH 8.3, 192 mM glycine, SDS 0.1%). Weigh 3.03 g of Tris and 14.41 g of glycine, then mix them in 1 L ultrapure water. Dilute 100 mL of 10× native buffer to 990 mL with water and add 10 mL of 10% SDS. Care should be taken to add SDS solution since it makes bubbles.
10. The focalization support rails should always be occupied with numerical order, without leaving empty rails in between samples.
11. The Wicks papers are changed to remove salt residues.
12. The overlay of isopropanol prevents contact with atmospheric oxygen (which inhibits acrylamide polymerization) in addition to helping to level the resolving gel solution.
13. The migration step must occur at low temperature (14 °C for two gels).
14. The methods most commonly used for the staining used silver nitrate and organic dyes such as Coomassie Blue G-250 and R-250. In general, the Coomassie Blue staining is sufficiently sensitive to detect 38 ng of a generic protein in a spot on the gel. When using Bio-Safe Coomassie Blue (Bio-Rad) is not necessary to perform the gel fixation step. The staining using Coomassie Blue is done in about 1 h [8].
15. All the staining steps should be carried out using gloves.
16. To avoid degradation of impregnating solution due to the exposure to light, it is necessary to cover the container with aluminum foil.

References

1. Gorg A, Weiss W, Dunn MJ (2004) Current two-dimensional electrophoresis technology for proteomics. Proteomics 4(12):3665–3685. doi:10.1002/pmic.200401031
2. Klose J (1975) Protein mapping by combined isoelectric focusing and electrophoresis of mouse tissues. A novel approach to testing for induced point mutations in mammals. Humangenetik 26(3):231–243
3. O'Farrell PH (1975) High resolution two-dimensional electrophoresis of proteins. J Biol Chem 250(10):4007–4021
4. Bieche C, de Lamballerie M, Chevret D et al (2012) Dynamic proteome changes in *Campylobacter jejuni* 81-176 after high pressure shock and subsequent recovery. J Proteome 75(4):1144–1156. doi:10.1016/j.jprot.2011.10.028
5. Bieche C, de Lamballerie M, Federighi M et al (2010) Proteins involved in *Campylobacter jejuni* 81-176 recovery after high-pressure treatment. Ann N Y Acad Sci 1189:133–138. doi:10.1111/j.1749-6632.2009.05180.x
6. Haddad N, Tresse O, Rivoal K et al (2012) Polynucleotide phosphorylase has an impact on cell biology of *Campylobacter jejuni*. Front Cell Infect Microbiol 2:30. doi:10.3389/fcimb.2012.00030
7. Sulaeman S, Hernould M, Schaumann A et al (2012) Enhanced adhesion of *Campylobacter jejuni* to abiotic surfaces is mediated by membrane proteins in oxygen-enriched conditions. PLoS One 7(9), e46402. doi:10.1371/journal.pone.0046402
8. Bradford MM (1976) A rapid and sensitive method for the quantitation of microgram quantities of protein utilizing the principle of protein-dye binding. Anal Biochem 72:248–254

Chapter 20

Analyzing Prokaryotic RNA-Seq Data: A Case Study Identifying Holo-Fur Regulated Genes in *Campylobacter jejuni*

Sophie Bérubé, James Butcher, and Alain Stintzi

Abstract

In recent years, RNA-seq has become an important method in the process of measuring gene expression in various cells and organisms. This chapter will detail all the bioinformatic steps that should be undertaken to determine differentially expressed genes from a typical RNA-seq experiment. Each step will be clearly explained in "non-bioinformatic" terminology so that readers embarking on RNA-seq analysis will be able to understand the rationale and reasoning behind each step. Moreover, the exact command lines used to process the data will be presented along with a description of the various flags and commands.

Key words RNA-seq, Bioinformatics, *Campylobacter jejuni*, Differential expression

1 Introduction

1.1 General Introduction

The ability to rapidly and accurately profile an organism's complete transcriptional response to an environmental change (the transcriptome) has proven to be a powerful tool for identifying the function of regulatory proteins and generating novel regulatory hypotheses. The classical method for doing these transcriptome analyses has been the use of microarray technologies, where each transcript that is expected to be transcribed is printed on the microarray slide. Equal amounts of cDNA from RNA extractions under the conditions studied are then labeled and cross hybridized onto the microarray slide for comparison. While microarray studies are instructive, they are being increasingly surpassed by RNA sequencing using next-generation sequencing technologies (RNA-seq). Several studies have suggested RNA-seq can determine gene expression level more accurately than microarrays. In addition, as RNA-seq is not limited by a priori knowledge of the transcriptional content to be studied, they also offer unbiased insight into the transcripts that are present. The increasing affordability of

James Butcher and Alain Stintzi (eds.), *Campylobacter jejuni: Methods and Protocols*, Methods in Molecular Biology, vol. 1512, DOI 10.1007/978-1-4939-6536-6_20,

performing RNA-seq has allowed more researchers to use this tool. However, it can be initially daunting for researchers to analyze the data generated from these experiments. RNA-seq data analysis typically requires the use of various bioinformatic programs for different steps, including quality control, genomic alignment, and differential gene identification. While these programs often have their own stand-alone tutorials or instructions, it would be useful for researchers to have a resource that details the exact bioinformatic steps used to complete an analysis with a reference set (preferably published and publically available) for comparison.

The following is a step-by-step analysis of RNA sequencing data from a study with *Campylobacter jejuni* [1]. The wild-type strain NCTC11168 and an isogenic deletion mutant of the primary regulator for iron homeostasis, Δ*fur*, were grown in MEMα + 40 μM $FeSO_4$, and total RNA was extracted at mid-log phase. The RNA was purified, depleted of rRNA, and subjected to paired-end high-throughput sequencing on an Illumina HiSeq 2500 [1]. The raw sequencing results obtained were processed and analyzed according to the following steps, and ultimately, a list of differentially expressed genes in the Δ*fur* as compared to the wild-type strain was assembled. The raw data and all the results for this experiment have been published [1] and are publically available for download from the NCBI SRA archive (http://www.ncbi.nlm.nih.gov/bioproject/PRJNA255798). The following sections in the introduction list the various programs used to complete the analysis along with a description their function.

1.2 SRA Toolkit

SRA toolkit is a package from NCBI for downloading data that has been deposited in the SRA archive. This can be used to download either raw or aligned data from various bioinformatic studies. For more information on SRA toolkit, access the following website: http://trace.ncbi.nlm.nih.gov/Traces/sra/sra.cgi?view=software.

1.3 FastQC

FastQC [2] processes the reads from a sequencing run to generate an overall quality report in an HTML format. In particular, FastQC reports the quality of each base in the sequence, the average quality score per read, %GC content of reads, average length of the reads, and overrepresented sequences. The program automatically flags elements in the report that appear abnormal and thus serves as an indicator of the overall quality of the next-generation sequencing (NGS) run. In some cases, flagged categories such as base pair content of the report can be rationalized based on previously known characteristics of a particular organism's genome, such as a high overall AT content or the nature of the sample. For example, highly repetitive sequences are expected in Amplicon libraries and RNA-seq applications. However, certain flagged elements can indicate a problem with the sequencing run and explain incongruous data. For more information on FastQC, access: http://www.bioinformatics.babraham.ac.uk/projects/fastqc/.

1.4 Bowtie2

Bowtie2 [3] is a popular program for aligning NGS reads to a reference genome. The program uses a Burrows–Wheeler transformation to store compressed version of the *C. jejuni* genome making the alignment of individual sequences to the genome faster. Bowtie2 can output the information about an alignment in the SAM format (a defined format for storing sequence alignment information). For more information about Bowtie2, access the following website: http://bowtie-bio.sourceforge.net/bowtie2/index.shtml.

1.5 SAMtools

SAMtools [4] contains a number of useful tools for working with alignments in SAM format. In particular, SAMtools can convert SAM files (e.g., those created by Bowtie2) into BAM files which are compressed and indexed for random access. This allows quick access to random regions in an aligned genome without needing to read the entire file. Note that for this manuscript, SAMtools 0.1.19 was used. For more information about SAMtools, access the following website: http://samtools.sourceforge.net.

1.6 Bedtools

Bedtools [5] is a suite of different scripts for working with genomic alignments and information stored in the BED format. Bedtools scripts can be used for extracting the reads aligning to a genomic interval, comparing two sets of sequences and several other functions. In this case, we will use Bedtools to count the number of reads aligning to each gene in the *C. jejuni* NCTC genome. For more information about Bedtools, access the following website: http://bedtools.readthedocs.org/en/latest/.

1.7 EdgeR

EdgeR [6] is an R package for performing statistical analysis on RNA-seq data to determine differentially abundant transcripts. In this case, the number of reads aligning to each gene in the wild-type strain is compared to the number of reads in the Δ*fur* strain to yielding a list of genes that are differentially abundant due to Fur deletion. EdgeR also normalizes the different sequencing runs to account for a differing number of reads between samples. Various thresholds for determining differentially expressed genes can be used, for this manuscript fold changes ≥ 1.5 or ≤ -1.5 with a false discovery rate (Benjamini–Hochberg FDR) corrected $p \leq 0.05$ were considered statistically significant in this set of data.

1.8 Command Line Conventions

The methods section details the commands used to perform the analyses. With the exception of the commands used in R for the edgeR analysis, all commands are performed in a Linux/Unix-like command line environment. Several of these programs can also be natively installed on a Mac. Working on the command line can appear quite daunting at first, so this manuscript attempts to clearly illustrate how the commands are executed and what the user should expect as output. The command prompt in your "shell" or terminal is represented by '~$' (this will normally be your username and directory location) with the R command prompt represented as '>'. For those unfamiliar with command line actions, it is important to note that capitalization and spacing are important. Files such as "myfile.txt" or "myFile.txt" are distinct and must be referenced precisely. Avoid using spaces in file names at all times. All commands should be written as a single line. Where the length of the command precludes it being present as a single line in this document, we have used a '\' to indicate where the command was broken, with the rest of the command continued on a second line with indenting (do not include the "\" in the command and do not insert a space before the text on the second line). For example, the following commands are equivalent:

```
~$ fastqc --no-group filename.fastq.gz
~$ fastqc --no-group \
     filename.fastq.gz
```

2 Materials

2.1 Equipment

1. Computer with ideally multiple CPU cores, ≥2 GB RAM, and ≥40 GB hard disk space.
2. Linux/Unix (or Mac) operating system (OS) with the required software packages installed. BioLinux [7] is an excellent Ubuntu-based OS for bioinformatics analysis as it comes with many useful applications preinstalled and configured.

2.2 Software Packages (Versions Refer to the Ones Used During This Specific Analysis)

1. SRA toolkit (2.5.5).
2. FastQC (v0.10.1).
3. Bowtie2 (v2.1.0).
4. SAMtools (v0.1.19).
5. Bedtools (v2.17.0).
6. R (v3.1.2) with edgeR package (v3.2.4) installed.
7. RStudio (optional).

3 Methods

3.1 Acquire Raw Data from NCBI SRA Archive

1. Download the appropriate version of SRA toolkit for the operating system that you are using from http://trace.ncbi.nlm.nih.gov/Traces/sra/sra.cgi?view=software.
2. Uncompress the SRA toolkit file, and using the command line, navigate to the bin subdirectory where the SRA toolkit scripts are located (the name of the main folder will also depend on your operating system).
```
~$ cd uncompressed_sra_toolkit_folder/bin
```

3. As the data was deposited as unaligned paired-end fastq files, the "fastq-dump" script will be used to download the files from NCBI using the accession numbers for the wild-type (SRR1525673, SRR1525674, SRR1525675) and Δ*fur* (SRR1525676, SRR152677, SRR1525678) iron-replete experiments. The accession numbers for each experiment can be accessed through the "SRA experiments" link on the main project page (http://www.ncbi.nlm.nih.gov/bioproject/PRJNA255798). The following command will download the paired-end fastq data for the experiment and save the two paired-end fastq files (denoted as _1 and _2, respectively). The files will be downloaded as compressed files to minimize the amount of storage space required (*see* **Note 1**).

```
~$ fastq-dump --split-files --gzip SRR1525673
```

4. Repeat this command for each SRR entry to obtain all the raw sequencing data. The files can be moved to another folder if desired. There is no need to decompress the fastq.gz files.

3.2 Assess the Quality and Overall Number of Reads in a Raw Sequencing File of the Reads

1. Navigate to the correct folder (where the fastq.gz files are stored) by typing in the command.

```
~$ cd foldername
```

2. The following command will run FastQC on a single fastq.gz file (*see* **Note 2**).

```
~$ fastqc --no-group filename.fastq.gz
```

3. When the command is finished running, uncompress the zipped output file and open the corresponding index.html file to view the result (*see* **Note 3**).

3.3 Align the Reads to the C. jejuni Genome and Count How Many Reads Align to Each Gene

1. Download the *C. jejuni* NCTC11168 genome in FASTA format from NCBI (http://www.ncbi.nlm.nih.gov/nuccore/NC_002163), and place in the index folder.
2. Create an index for the *C. jejuni* genome by running the following command (this assumes that the genome is named nctc11168.fa; change as appropriate) (*see* **Note 4**).

```
~$ bowtie2-build nctc11168.fa nameOfIndex
```

3. Align the reads to the *C. jejuni* genome using the following Bowtie2 command (*see* **Note 5**, **6**).

```
~$ bowtie2 --no-head -p 2 name of index\
-1 filenameR1.fastq.gz -2 filenameR2.fastq.gz
-S filename.sam
```

4. Convert the text-delimited SAM file into a BAM file using the following command (*see* **Note 7**).

```
~$ samtools view -b -t nctc11168.fa.fai\
-S filename.sam > filename.bam
```

5. Create a BAM file that is sorted by coordinate location on the genome (*see* **Note 8**), and index the sorted BAM file for random access.

```
~$ samtools sort filename.bam filename.sorted
~$ samtools index filename.sorted.bam
```

6. Repeat Subheading 3.3, **steps 3–5** for each sequencing run to be analyzed.
7. Count the number of reads aligning to each gene using Bedtools (*see* **Note 9**). The number of reads aligning to each gene will be in the seventh column of filename.bed.

```
~$ bedtools multicov D bed gene_names.bed
-bams filename.sorted.bam \
> filename.bed
```

3.4 Determine Differentially Expressed Genes

1. Format the output of Bedtools so that it can be imported into R for edgeR analysis (*see* **Note 10**)
   (a) Construct a table named "counts.txt" whose leftmost column contains the names of all the genes in the *C. jejuni* genome, the adjacent column containing the gene length and every subsequent column the data from each RNA-seq experiment.
   (b) The first few rows of the table should look something like the following:

```
geneslen wt_1 wt_2 wt_3 fur_1 fur_2 fur_3
dnaA 1325 223 270 341 188 169 242
dnaN 1070 1038 1071 1121 872 561 949
```

2. In order to define the comparison between the wild-type strain and the Δ*fur* mutant, another table describing the contents of the merged Bedtools results needs to be created. In this particular case, the wild-type samples are the control and the Δ*fur* samples are the test. Name this file "experiment.txt." The table for the published results looks like the sample below (*see* **Note 11**). The columns are separated by tabs not spaces and there is an empty tab before "lane" in the first row.

```
lane treatment label
wt_1 1 control wt1
wt_2 2 control wt2
wt_3 3 control wt3
fur_1 4 test fur1
fur_2 5 test fur1
fur_3 6 test fur1
```

3. Launch R (all following commands are run in the R terminal) (*see* **Note 12**).

```
~$ R
```

4. Load the edgeR library:
```
>library(edgeR)
```
5. Use the following command to read the targets file associating treatments with samples (*see* **Note 13**):
```
> targets <- readTargets("experiment.txt")
```
6. Use the following command to read the file of counts (*see* **Note 14**):
```
> count_data<-read.delim("counts.txt", row.names=1,
stringsAsFactors = FALSE)
```
7. Use this command to put the counts and other information into a DGEList Object named "gene_counts" (*see* **Note 15**):
```
> gene_counts<-DGEList(counts=count_
data[,3:8] , group = targets$treatment,
genes= data.frame(Length= count_data [,2]))
> colnames(gene_counts) <- targets$label
```
8. Use this command to normalize the data:
```
> gene_counts <- calcNormFactors(gene_counts)
```
9. Use this command to verify the labeling of the samples has been completed appropriately. The output is the first few lines of a table showing the counts associated with each gene for the targeted group of samples.
```
> head(gene_counts$counts)
```
10. Use this command to create a simple plot of the samples using multidimensional scaling (MDS) (*see* **Note 16**):
```
> plotMDS(gene_counts)
```
11. Use this command to estimate the dispersion of the data (*see* **Note 17**):
```
> gene_counts <- estimateCommonDisp(gene_
counts, verbose=TRUE)
```
12. Use the following command to estimate gene-specific dispersions within samples.
```
> gene_counts <- estimateTagwiseDisp(gene_
counts)
```
13. Use the following command to compute differential expression:
```
> et <- exactTest(gene_counts)
```
14. This command performs and stores the differential results in the R object named "top". By default, the topTags function will only output the most differentially expressed genes. This modified command will output the results for every gene tested. Note that fold changes output by edgeR are in a Log2 scale, and the *p* values have been FDR corrected using the Benjamini–Hochberg method by default.

```
> top <- topTags(et, n=nrow(et$table))$table
```

15. We can also get a summary of the results using the summary() command. This will list the number of significantly differentially expressed genes (Log2 fold change >1 or<−1 with an FDR-corrected $p \leq 0.05$).

```
> summary(decideTestsDGE(et))
```

16. Use the following command to export the list of differentially expressed genes.

```
> write.table(top,file="name_of_your_file.
txt",sep="\t")
```

17. The resulting text file can be opened in any spreadsheet program for plotting and/or further analysis. Note that the column header will be offset by one column (move the headers to the right by one to get the correct headings).

4 Notes

1. Raw data from next-generation sequencing projects can be quite large. In this case, downloading all six raw fastq files (even though they are compressed) requires ~12 GB of hard drive storage space. The downstream analyses will also require ~20 GB of storage space (assuming that every intermediary file is kept). If storage space is an issue, the following command will only download the first 1 million sequences for each run. Running the analysis only using this subsetted data will obviously give slightly different results; however, the results are highly comparable for this particular dataset.

```
$ fastq-dump --split-files --gzip -X 1000000
SRR1525673
```

2. The --no-group flag indicates that each report generated by the program shows quality data for each base in the read. Without the --no-group flag, FastQC presents averaged results for bases beyond 10 nt. In addition, the following command below will run FastQC on every fastq.gz file present in the directory. The -t flag and the number following it indicates the number of cores on the computer that will be used to process the data (i.e., the number of runs analyzed simultaneously, four in this example), while the *.fastq.gz wildcard allows for every file to be processed without further user input.

```
~$ fastqc -t 4 --no-group *.fastq.gz
```

3. The FastQC report generated for this particular RNA-seq data shows acceptable results in most categories. The per base sequence content category, however, shows an uneven distribution of bases at various locations. In particular the sequences, on average, have a high AT content, which is a known

characteristic of the *C. jejuni* genome. The per sequence GC content does not show a normal distribution among all reads but instead shows two major peaks, suggesting two groupings of sequences. For *C. jejuni*, which frequently incorporates sequences from other organisms into its genome, the grouping observed in per sequence GC content plot is not unexpected. One peak likely corresponds to native *C. jejuni* sequences, while another peak likely corresponds to acquired sequences. The high sequence duplication levels can be explained by certain overrepresented sequences in the samples. Since the sequenced data comes from RNA extractions, overrepresented regions of the genome correspond to highly transcribed regions of the genome, for instance, the 16S rRNA, which is expressed at high levels in bacterial cells. While the total RNA extracted was depleted of rRNA sequences, this process was not 100% successful. Since the 16S gene is expressed at such high levels in *C. jejuni*, even after depletion, the sample still contains disproportionate amounts of the ribosomal region. The previously noted overrepresented regions of the genome also explain large peaks in the Kmer content plot.

4. Building the Bowtie2 index via bowtie2-build requires a full sequence of the *C. jejuni* genome, starting at a preestablished base in the sequence. The program uses a Burrows–Wheeler transformation to store compressed version of the *C. jejuni* genome making the alignment of individual sequences to the genome faster.

5. There are two major types of alignment: paired and unpaired. Unpaired reads are generated when the sequencing reaction is only performed for one end of the DNA fragment, whereas paired-end reads are generated from sequencing performed on both ends. Paired-end reads are particularly useful when the species being sequenced is unknown because the sequencing libraries can be constructed to generate overlapping paired-end reads. Using the overlapping reads, the genome of an organism can more readily be assembled from scratch. However, paired-end reads raise the cost of sequencing.

6. The --no-head flag in the command indicates the SAM file does not contain a header making it easier for some of the other programs later in the pipeline to process the reads. The -p flag and the number following it indicate the number of cores used to align the reads to the genome. The -1 and -2 flags number the reads that, when combined, form a paired-end read. In order to perform an unpaired alignment, use the following command:

```
~$ bowtie2 --no-head -p 4 nameOfIndex \
-U filename.fastq.gz -S filename.sam
```

7. SAM files are tab-delimited text formats, while BAM files display the alignment information for a particular read in a binary format. The binary format allows the computer to sort and index the reads rapidly. The -b flag indicates that the output of the view command is in a BAM format, while the –t flag is for adding the correct genome annotation. The '*.fa.fai' file can be generated by running '~$ samtools faidx filename.fa' on the reference genome. This is useful when different downstream programs use different genome naming conventions and you want to be able to convert between them. The –S flag indicates that the input for the view command is a SAM file.
8. The aligned reads were sorted in order of the leftmost base pair of the read. Indexing the BAM files allows certain overlapping alignments to be retrieved quickly without needing to go through the entire genome to retrieve the aligned reads. This is done through a process called binning. SAMtools puts each alignment into a "bin" so that when other programs search through the aligned reads, they do not need to search through the entire genome but can instead search through bins sorted in the order that genes and sequences appear in the genome.
9. Bedtools uses a previously constructed list of genes in the *C. jejuni* genome and uses the indexed alignment of each read to pair each aligned read to a gene. Bedtools then counts the number of reads associated with each gene and outputs these counts in a tab-delimited text format, which can be viewed in spreadsheet applications (e.g., Microsoft Excel). The -D flag indicates that the duplicate reads are counted. This is particularly important for RNA-seq data where counting each read is crucial for determining the differential expression of a particular gene. The -bed flag indicates that the output is in a BED format.
10. EdgeR is a program that performs statistical analysis on the data generated by Bedtools. In this case, the number of reads aligned to each gene in the wild-type strain is compared to the number of reads in the Δ*fur* strain, thus yielding a list of differentially expressed genes. However, before a comparison is made, the data must be normalized to account for a differing number of reads between samples. False discovery rates and log fold changes are calculated, and edgeR can also apply certain filters to the data, to varying degrees of stringency.
11. It is very important that the names/locations of the columns detailed in the target.txt file be correct. Mistakes in the names (names are case sensitive) will lead to errors when importing into R.
12. RStudio is a companion program for working in R. It provides a GUI that allows one to monitor the creation/type of R objects.

13. This function allows R to associate samples or groups of reads with a label and classifies them as control samples or treatment samples. The "experiment.txt" is the tab-delimited text file that labels samples as treatment or control.
14. It is possible to filter out genes that are very lowly expressed in all three biological replicates, using various levels of stringency. Filtering was not used for this analysis as almost no genes (<5) were expressed at such low levels.
    (a) Use the following command to filter data at varying levels of stringency. The following keeps all genes that are expressed at greater than one count per million in at least three of the samples:
    ```
    > keep <- rowSums(cpm(gene_counts)>1) >= 3
    > gene_counts <- gene_counts[keep,]
    ```
    (b) Use this command to recompute the library sizes prior to the normalization:
    ```
    > gene_counts$samples$lib.size <- colSums(gene_
    counts$counts)
    ```
15. The R command creates a DGEList R object from our gene count information (count_data) and experimental information (targets). We are telling R that our count information is in the third to eighth columns in the count_data object (counts= count_data [,3:8]), to take our groups from the "treatment" column in the targets object (group= targets$treatment), and that the gene lengths are in the second column of the count_data object (genes=data.frame(Length=count_data[,2])). Finally, the second command adds in the names of each run to the count data so that we can differentiate each experiment (e.g., the MDS plot).
16. This step shows the dispersion of the groups (the three biological replicates of wild type and Δ*fur*) of samples on a plot. This shows whether or not the biological replicates of a sample are clustered together and can highlight any outlying biological replicates.
17. In order to calculate the dispersion of the data, a biological coefficient of variation is measured between the samples, using the common dispersion of the data averaged over all of the genes in the *C. jejuni* genome. This command is particularly important for estimating the variation between biological replicates and identifying any outlying biological or technical replicates.

Acknowledgments

This work was supported by CIHR (MOP#84224).

References

1. Butcher J, Handley RA, van Vliet AH et al (2015) Refined analysis of the *Campylobacter jejuni* iron-dependent/independent Fur- and PerR-transcriptomes. BMC Genomics 16:498. doi:10.1186/s12864-015-1661-7
2. Andrews S (2010) FastQC: a quality control tool for high throughput sequence data. Available online at http://www.bioinformatics.babraham.ac.uk/projects/fastqc
3. Langmead B, Trapnell C, Pop M et al (2009) Ultrafast and memory-efficient alignment of short DNA sequences to the human genome. Genome Biol 10(3):R25. doi:10.1186/gb-2009-10-3-r25
4. Li H, Handsaker B, Wysoker A et al (2009) The Sequence Alignment/Map format and SAMtools. Bioinformatics 25(16):2078–2079. doi:10.1093/bioinformatics/btp352
5. Quinlan AR, Hall IM (2010) BEDTools: a flexible suite of utilities for comparing genomic features. Bioinformatics 26(6):841–842. doi:10.1093/bioinformatics/btq033
6. Robinson MD, McCarthy DJ, Smyth GK (2010) edgeR: a Bioconductor package for differential expression analysis of digital gene expression data. Bioinformatics 26(1):139–140. doi:10.1093/bioinformatics/btp616
7. Field D, Tiwari B, Booth T et al (2006) Open software for biologists: from famine to feast. Nat Biotechnol 24(7):801–803. doi:10.1038/nbt0706-801

Chapter 21

Generation and Screening of an Insertion Sequencing-Compatible Mutant Library of *Campylobacter jejuni*

Jeremiah G. Johnson and Victor J. DiRita

Abstract

The advent of next-generation sequencing technology has enabled experimental approaches to characterize large, complex populations of DNA molecules with high resolution. Included among these are methods to assess populations of transposon insertion libraries for the fitness cost of any particular mutant allele after applying selection to a population. These approaches have proven invaluable for identifying genetic factors that influence survival of bacterial pathogens within different environments, including animal hosts. One such method, termed insertion-site sequencing (INSeq), was designed to generate a 16 bp fragment of transposon-flanking genomic DNA captured during the protocol, which then serves as the substrate for massively parallel sequencing. Here we describe the generation of a transposon mutant library of *Campylobacter jejuni* amenable to INSeq and its use in identifying colonization determinants in a day-of-hatch chicken colonization model.

Key words Transcription factor, Fur, Strep-tag affinity chromatography, Apo-metalloregulator, Bacterial metalloregulator

1 Introduction

A long-standing interest among microbial pathogenesis investigators is the identification of factors essential for microbes to colonize and cause disease within a host. Several technologies have emerged for this purpose including, most notably, in vivo expression technology (IVET), transcriptional microarrays, signature-tagged mutagenesis (STM), and transposon in situ hybridization (TraSH) [1–3]. While these techniques offer genome-level interrogation of the pathogenicity traits, each has its limitations, often around the scale with which the analysis can be carried out or due to inefficient hybridization methods. With the development of rapid and cheap next-generation sequencing methods, it is now possible to generate exceptionally high-throughput coverage of complex DNA mixtures, circumventing many limitations of other approaches.

James Butcher and Alain Stintzi (eds.), *Campylobacter jejuni: Methods and Protocols*, Methods in Molecular Biology, vol. 1512, DOI 10.1007/978-1-4939-6536-6_21, © Springer Science+Business Media New York 2017

One protocol for simultaneous discovery of virulence determinants from transposon insertion libraries—for which STM was originally developed—is insertion sequencing (INSeq) or transposon sequencing (TnSeq) [4, 5]. These methods were developed to apply massively parallel sequencing approaches to identify and enumerate transposon mutants within a population of insertions. The INSeq protocol relies on engineering a MmeI restriction site within the Himar1 insertion sequence at the ends of the *mariner* transposon [6]. MmeI digests DNA approximately 20 bp away from its recognition site, so digestion of genomic DNA harboring the engineered transposon results in a short fragment that includes chromosomal DNA within 16 bases from the site of insertion. After adding barcoded adapters, these fragments can then be sequenced, mapped, and quantified to determine the prevalence of any insertion site in multiple environments, including animal hosts.

Campylobacter jejuni is a leading cause of foodborne infection due to its ability to asymptomatically colonize the gastrointestinal tracts of birds, most notably chickens, from which it can be disseminated to meat products during processing [7]. Humans become infected following consumption of undercooked poultry or other food that has contacted a contaminated surface [8]. As such, it is of interest to better understand bacterial and host factors that influence colonization of the chicken gastrointestinal tract in the hope of identifying potential points of intervention that can reduce bacterial loads and improve the safety of meat products.

To that end we developed an INSeq-compatible transposon for use in *C. jejuni* and constructed a transposon library of approximately 8500 *C. jejuni* mutants. After determining the heterogeneity of the collection, we subsequently introduced the library into day-of-hatch white leghorn chicks and determined the populations present within cecal contents at seven days post-inoculation. Comparison of the known population in the inoculum to that of the cecal contents made it possible to identify determinants that influence *C. jejuni*'s ability to efficiently colonize and reside within the chicken gastrointestinal tract [9].

2 Materials

2.1 Amplification and Cloning of INSeq-Compatible Transposon

1. Plasmid pRY109 [10].
2. Commercial miniprep kit (e.g., QIAprep Spin Miniprep).
3. PvuII.
4. Agarose.
5. Tris–acetate–EDTA buffer (TAE; 10×): 242 g Tris, 57.1 mL acetic acid, 100 mL 0.5 M EDTA, add ddH_2O to 1 L.
6. Commercial gel extraction kit (e.g., QIAquick Gel Extraction Kit).

7. Taq DNA polymerase.
8. Deoxynucleoside triphosphates (dNTPs).
9. pGEM-T Easy Vector Systems (Cat. No. A1360) (Promega).
10. JM109 cells (Cat. No. P9751) (Promega).
11. NotI.
12. 5′cat_INSeq: 5′-ACGCGTCCTAACAGGTTGGATGATAAG TCCCCGGTCTTCGTATGCCGTCTTCTGCTTGGCGCG CCCTCGAGCAATTGTGCTCGGCGGTGTTCCTTTC CAA-3′.
13. 3′cat_INSeq: 5′-ACGCGTCCTAACAGGTTGGATGATAAG TCCCCGGTCTTCGTATGCCGTCTTCTGCTTGGCGC GCCCTGCAGTCTAGTGCGCCCTTTAGTTCCTAAAG GGT-3′.
14. LB + chloramphenicol (25 μg/mL) plates: autoclave 7.5 g agar, 5 g tryptone, 2.5 g yeast extract, and 2.5 g NaCl in 500 mL ddH_2O. When cool to the touch, add 500 μL 25 mg/mL chloramphenicol in 70% ethanol, and pour approximately 20 mL into standard petri dishes.
15. T7 oligonucleotide: 5′-TAATACGACTCACTATAGGG-3′.
16. SP6 oligonucleotide: 5′-ATTTAGGTGACACTATAG-3′.

2.2 Construction of INSeq-Compatible C. jejuni Mutant Library

1. Genomic DNA from wild-type *C. jejuni* DRH212 [11].
2. 5× salt buffer: 50% glycerol, 125 mM HEPES (pH 7.9), 500 mM NaCl, 25 mM $MgCl_2$.
3. Bovine serum albumin (BSA) (10 mg/mL).
4. MarC9 transposase (purified in-house) [12].
5. 100 mM dithiothreitol.
6. Genomic lysis buffer: 50 mM Tris, 50 mM EDTA, 1% SDS, 10 mM NaCl, pH 7.5.
7. MPC Protein Precipitation Reagent (Cat. No. MMP03750) (Epicentre).
8. 2-propanol (isopropanol).
9. SpeedVac.
10. TE buffer, pH 7.0.
11. NEBuffer 2 (Cat. No. B7002S) (NEB).
12. Deoxynucleoside triphosphates (dNTPs).
13. T4 DNA polymerase.
14. T4 DNA ligase.
15. 0.025 μm nitrocellulose filter.
16. Mueller-Hinton (MH) agar with 10% sheep's blood and trimethoprim: 19 g of Mueller-Hinton agar (Difco) was added to 500 mL ddH_2O and autoclaved. Media was allowed to cool,

and 50 mL of pre-warmed sheep's blood (Hemostat Labs) and 500 μL of trimethoprim (10 mg/mL in DMSO) were added to media. Approximately 20 mL was added individually to standard petri dishes.

17. Wild-type *C. jejuni* DRH212 [11].
18. Disposable inoculating loops.
19. Tissue culture incubator (at 5 % CO_2) for inducing competence in *Campylobacter* (Thermo).
20. Tri-gas water-jacketed incubator (at 5 % O_2, 10 % CO_2, 85 % N_2) for growing *Campylobacter*.
21. Mueller-Hinton broth: 10.5 g Mueller-Hinton broth (Difco) in 500 mL ddH_2O and autoclave.
22. Non-tissue culture-treated 96-well plates (Falcon).
23. Sterile deep-well 96-well plates (1.0 mL).
24. 96-pin replicator (Boekel).
25. Mueller-Hinton broth with 40% glycerol: 10.5 g Mueller-Hinton broth and add 300 mL ddH_2O and 200 mL glycerol (Sigma). Autoclave.

2.3 Characterization of *C. jejuni* Transposon Mutant Populations

Many of the reagents listed below, including the various oligonucleotides, were ordered exactly to the specifications of Goodman et al. [4, 13].

1. Sterile cotton swabs with wooden applicator (Fisherbrand).
2. Mueller-Hinton broth with 20% glycerol: 10.5 g Mueller-Hinton broth and add 400 mL ddH_2O and 100 mL glycerol (Sigma). Autoclave before use.
3. Cryogenic tubes (Corning).
4. RNase A (100 mg/mL) (Qiagen).
5. Buffer EB (Qiagen).
6. 3 M sodium acetate, pH 5.5.
7. Kit for measuring DNA concentrations (i.e., Qubit HS DNA assay kit).
8. Platinum Pfx polymerase (Cat. No. 11708-021) (Invitrogen).
9. BioSamA primer: 5′-Bio-TEG-CAAGCAGAAGACGGCATACGAAGACC-3′.
10. Commercial PCR purification kit (e.g., QIAquick PCR Purification).
11. Streptavidin-coated M-280 Dynabeads (Cat. No. 112-05D) (Invitrogen).
12. Magnetic particle concentrator (MPC) (magnetic rack) for 1.5 mL tubes and for PCR tubes.

13. 2× binding and wash buffer (B&W): 2 M NaCl, 10 mM Tris, 1 mM EDTA, pH 7.5.
14. 1× B&W buffer: dilute 2× buffer 1:1 with ddH_2O.
15. Low-TE buffer: 3 mM Tris and 0.2 mM EDTA, pH 7.5.
16. 2.5 mg/mL random hexamers (N_6).
17. Klenow fragment.
18. M12_top primer: 5′-CTGTCCGTTCCGACTACCCTCCC GAC-3′.
19. M12_bot primer: 5′-GTCGGGAGGGTAGTCGGAACG GACAG-3′.
20. NEBuffer 4 (Cat. No. B7004S) (NEB).
21. S-adenosylmethionine.
22. MmeI (Cat. No. R0637S) (NEB).
23. Barcoded adapters (Table 1).
24. LIB-PCR5 primer: 5′-CAAGCAGAAGACGGCATACGAA GACCGGGGACTTATCATCCAACCTGT-3′.
25. LIB-PCR3 primer: 5′-AATGATACGGCGACCACCGAAC ACTCTTTCCCTACACGACGCTCTTCCGATCT-3′.
26. Xylene cyanol (Sigma).
27. GelRed (Cat. No. 41002) (Biotium).
28. Handheld UV source.

2.4 Identification of Colonization Factors Using the Chicken Colonization Model

1. Sterile PBS.
2. Phenol (Sigma).
3. White leghorn chicken eggs (Charles River).
4. *Campylobacter* selective media: 19 g of Mueller-Hinton agar into 500 mL of ddH_2O and autoclave. Allow to cool, and add 50 mL of pre-warmed sheep's blood and 500 μL each of 40 mg/mL cefoperazone, 100 mg/mL cycloheximide, 10 mg/mL trimethoprim, and 100 mg/mL vancomycin.

3 Methods

3.1 Amplification and Cloning of INSeq-Compatible Transposon

1. Isolate plasmid pRY109 using a commercial miniprep kit.
2. Digest the purified pRY109 with PvuII for 2 h at 37 °C, and run products at 130 V on a 1% agarose/1× TAE gel.
3. Use a commercial gel extraction kit to purify the approximately 1 kb *cat* gene-containing fragment. Elute in 32 μL of supplied buffer.
4. Adenylate the ends of the PvuII cut *cat* gene fragment to allow for cloning into pGEM-T Easy Vector: Mix the following components for the adenylation reaction:

Table 1

Barcoded adapter sequences [4, 13]

Adapter oligonucleotide	Sequence (5′–3′) (barcode in bold)
LIB_AdaptT_A	TTCCCTACACGACGCTCTTCCGATCT**TTTT**NN
LIB_AdaptT_B	TTCCCTACACGACGCTCTTCCGATCT**ACCA**NN
LIB_AdaptT_C	TTCCCTACACGACGCTCTTCCGATCT**ACGT**NN
LIB_AdaptT_D	TTCCCTACACGACGCTCTTCCGATCT**AGGA**NN
LIB_AdaptT_E	TTCCCTACACGACGCTCTTCCGATCT**AGTC**NN
LIB_AdaptT_F	TTCCCTACACGACGCTCTTCCGATCT**ATCG**NN
LIB_AdaptT_H	TTCCCTACACGACGCTCTTCCGATCT**CATG**NN
LIB_AdaptT_I	TTCCCTACACGACGCTCTTCCGATCT**CCAA**NN
LIB_AdaptT_J	TTCCCTACACGACGCTCTTCCGATCT**CCCC**NN
LIB_AdaptT_K	TTCCCTACACGACGCTCTTCCGATCT**CGAT**NN
LIB_AdaptT_O	TTCCCTACACGACGCTCTTCCGATCT**GCTA**NN
LIB_AdaptT_P	TTCCCTACACGACGCTCTTCCGATCT**GGAA**NN
LIB_AdaptT_Q	TTCCCTACACGACGCTCTTCCGATCT**GGGG**NN
LIB_AdaptT_R	TTCCCTACACGACGCTCTTCCGATCT**GTAC**NN
LIB_AdaptT_S	TTCCCTACACGACGCTCTTCCGATCT**GTCA**NN
LIB_AdaptT_U	TTCCCTACACGACGCTCTTCCGATCT**TATA**NN
LIB_AdaptT_V	TTCCCTACACGACGCTCTTCCGATCT**TCAG**NN
LIB_AdaptT_W	TTCCCTACACGACGCTCTTCCGATCT**TCGA**NN
LIB_AdaptT_Y	TTCCCTACACGACGCTCTTCCGATCT**TTAA**NN
LIB_AdaptT_Z	TTCCCTACACGACGCTCTTCCGATCT**AAAA**NN
LIB_AdaptT_AA	TTCCCTACACGACGCTCTTCCGATCT**GAAG**NN
LIB_AdaptT_CC	TTCCCTACACGACGCTCTTCCGATCT**CCTT**NN
LIB_AdaptT_DD	TTCCCTACACGACGCTCTTCCGATCT**AACC**NN
LIB_AdaptT_EE	TTCCCTACACGACGCTCTTCCGATCT**TTGG**NN
LIB_AdaptB_A	**AAAA**AGATCGGAAGAGCGTCGTGTAGGGAA
LIB_AdaptB_B	**TGGT**AGATCGGAAGAGCGTCGTGTAGGGAA
LIB_AdaptB_C	**ACGT**AGATCGGAAGAGCGTCGTGTAGGGAA
LIB_AdaptB_D	**TCCT**AGATCGGAAGAGCGTCGTGTAGGGAA
LIB_AdaptB_E	**GACT**AGATCGGAAGAGCGTCGTGTAGGGAA
LIB_AdaptB_F	**CGAT**AGATCGGAAGAGCGTCGTGTAGGGAA
LIB_AdaptB_H	**CATG**AGATCGGAAGAGCGTCGTGTAGGGAA
LIB_AdaptB_I	**TTGG**AGATCGGAAGAGCGTCGTGTAGGGAA
LIB_AdaptB_J	**GGGG**AGATCGGAAGAGCGTCGTGTAGGGAA
LIB_AdaptB_K	**ATCG**AGATCGGAAGAGCGTCGTGTAGGGAA
LIB_AdaptB_O	**TAGC**AGATCGGAAGAGCGTCGTGTAGGGAA
LIB_AdaptB_P	**TTCC**AGATCGGAAGAGCGTCGTGTAGGGAA
LIB_AdaptB_Q	**CCCC**AGATCGGAAGAGCGTCGTGTAGGGAA
LIB_AdaptB_R	**GTAC**AGATCGGAAGAGCGTCGTGTAGGGAA
LIB_AdaptB_S	**TGAC**AGATCGGAAGAGCGTCGTGTAGGGAA
LIB_AdaptB_U	**TATA**AGATCGGAAGAGCGTCGTGTAGGGAA
LIB_AdaptB_V	**CTGA**AGATCGGAAGAGCGTCGTGTAGGGAA
LIB_AdaptB_W	**TCGA**AGATCGGAAGAGCGTCGTGTAGGGAA
LIB_AdaptB_Y	**TTAA**AGATCGGAAGAGCGTCGTGTAGGGAA
LIB_AdaptB_Z	**TTTT**AGATCGGAAGAGCGTCGTGTAGGGAA
LIB_AdaptB_AA	**CTTC**AGATCGGAAGAGCGTCGTGTAGGGAA
LIB_AdaptB_CC	**AAGG**AGATCGGAAGAGCGTCGTGTAGGGAA
LIB_AdaptB_DD	**GGTT**AGATCGGAAGAGCGTCGTGTAGGGAA
LIB_AdaptB_EE	**CCAA**AGATCGGAAGAGCGTCGTGTAGGGAA

(a) 0.5 μL 10 mM dNTPs.
(b) 1 μL 50 mM $MgCl_2$.
(c) 2.5 μL Taq buffer.
(d) 0.5 μL Taq.
(e) 21.5 μL purified *cat* gene from gel extraction.

5. Incubate for 10 min at 72 °C.
6. Ligate fragment into pGEM-T Easy Vector. Mix the following reagents for the ligation reaction:

(a) 3 μL adenylated *cat* gene fragment.
(b) 5 μL ligase buffer.
(c) 1 μL ligase.
(d) 1 μL pGEM-T Easy Vector.

7. Incubate for 1 h at room temperature.
8. Transform the ligation product into *E. coli* JM109 cells according to the manufacturer's instructions, and select overnight on LB + chloramphenicol (25 μg/mL) plates.
9. Isolate ligated plasmids, and confirm the insertion of the *cat* gene using T7 and SP6 primers and Sanger sequencing.
10. Store the pGemTINSeq-*cat* plasmid DNA until further use.

3.2 Construction of INSeq-Compatible *C. jejuni* Mutant Library

1. Perform the transposition reaction by mixing the following components together [11]:
   (a) 16 μL 5× salt buffer.
   (b) 2.0 μL 10 mg/ml BSA.
   (c) 1.6 μL 100 mM DTT.
   (d) 2.0 μg DRH212 genomic DNA.
   (e) 0.6 μg pGemTINSeq-*cat* DNA (*see* **Note 1**).
   (f) 500 ng MarC9 transposase.
   (g) Use ddH_20 to bring volume to 80 μL.
2. Incubate transposition reactions at 30 °C for 4 h.
3. Purify the resulting mutagenized genomes (*see* **Note 2**):
   (a) Add 300 μL Genomic Lysis Buffer and 150 μL Epicentre MPC Protein Precipitation Reagent to the transposition reactions.
   (b) Vortex for 20 s.
   (c) Place on ice for 10 min.
   (d) Spin down proteins at 16,800 × *g* for 15 min at room temp.
   (e) Remove supernatant and transfer into a clean 1.5 ml tube.

(f) Add 750 μL of isopropanol and invert 35–40×.

(g) Spin down DNA at 16,800 × *g* for 5 min at 4 °C.

(h) Wash pellet 2× in 70% ethanol.

(i) Decant ethanol and dry for 10 min using a SpeedVac.

(j) Resuspend DNA pellet in 40 μL TE buffer.

4. Repair transposed DNA—step 1 (filling in gaps due to transposition):

   (a) 40 μL precipitated DNA from transposition reaction.

   (b) 6 μL NEB buffer 2.

   (c) 4.8 μL 1.25 mM dNTPs.

   (d) 7.7 μL ddH_2O.

   (e) 1.5 μL diluted T4 DNA polymerase (1 U/μL concentration).

5. Incubate at 11 °C for 20 min.

6. Heat inactivate the polymerase by incubating at 75 °C for 15 min.

7. Repair transposed DNA—step 2 (ligating sections back together):

   (a) 60 μL from **step 6** of Subheading 3.2.

   (b) 12 μL NEB T4 ligase buffer.

   (c) 1.5 μL NEB T4 ligase.

   (d) 46.5 μL ddH_2O.

8. Incubate the DNA at 16 °C overnight.

9. Dialyze the DNA for 20 min on a nitrocellulose membrane floating on 20 mL ddH_2O.

10. Grow wild-type *C. jejuni* DRH212 at 37 °C for 48 h under microaerobic conditions (85% N_2, 10% CO_2, 5% O_2) on Mueller-Hinton (MH) agar with 10% sheep's blood plus trimethoprim (10 μg/ml).

11. Harvest cells using an inoculating loop, streak onto another Mueller-Hinton (MH) agar with 10% sheep's blood plus trimethoprim (10 μg/ml), and grow under the microaerobic conditions overnight.

12. Prepare a solution of bacteria at an $OD_{600} = 0.05$ in MH broth from the plates.

13. Prepare biphasic tubes by adding 1 mL MH agar to a culture tube and allow to solidify.

14. Once the biphasics have solidified, add 0.5 mL of the 0.5 OD_{600} bacterial suspension prepared in **step 12** of Subheading 3.2.

15. Incubate the biphasic tubes at 37 °C for 3 h under tissue culture conditions (5% CO_2).

16. Add the dialyzed DNA (~120 μL) from **step 19** of Subheading 3.2 to the liquid phase of a biphasic tube and incubate at 37 °C for an additional 4 h under tissue culture conditions.
17. Transfer 0.5 mL liquid phase to a microcentrifuge tube and pellet for 1 min at 16,800 × *g*. Resuspend pellet in 110 μL MH broth, and spread 100 μL and 10 μL onto separate MH blood plus chloramphenicol (30 μg/ml) agar plates.
18. Grow plates at 37 °C for 2–3 days under microaerobic conditions (*see* **Note 3**).
19. Pick chloramphenicol-resistant colonies using sterile toothpicks, and inoculate into 100 μL of MH broth in 96-well plates (*see* **Note 4**).
20. Grow bacteria in 96-well plates at 37 °C for 48 h under microaerobic conditions.
21. Replicate 96-well plates into deep-well plates containing 500 μL MH broth using a 96-pin replicator. Add an equal volume of MH broth + 40% glycerol (100 μL) to each well of the 96-well plates, and store at −80 °C.
22. Grow bacteria in deep-well plates at 37 °C for 48 h under microaerobic conditions.
23. After 48 h, add an equal volume of MH broth + 40% glycerol to each well of the deep-well plates (500 μL), and store at −80 °C.

3.3 Characterization of *C. jejuni* Transposon Mutant Populations

1. Stamp all wells from the prepared transposon library onto MH blood + chloramphenicol plates using a 96-pin replicator, and grow at 37 °C for 48 h under microaerobic conditions.
2. Harvest the replicated colonies using a sterile cotton swab, and place into 30 mL of MH broth + 20% glycerol. Store 2 mL aliquots of this suspension in cryogenic tubes at −80 °C.
3. Thaw three aliquots of pooled transposon libraries, and pipette 100 μL of each onto individual MH blood + chloramphenicol plates. Grow these cultures for approximately 4 h at 37 °C under microaerobic conditions.
4. Scape cells from each plate, and dilute into 4× 1 mL suspensions of 10^9, 10^8, 10^7, 10^6, 10^5, and 10^4 cfu/mL.
5. Extract genomic DNA as in **step 3** of Subheading 3.2 except resuspend the DNA in 200 μL of TE buffer.
6. Add 0.5 μL RNase A (100 mg/mL) to 100 μL of purified genomic DNA, and incubate for 2 min at room temperature.
7. Genomic DNA was ethanol precipitated from these reactions as follows:
   (a) Add 300 μL 100% ethanol and 10 μL 3 M sodium acetate to each 100 μL RNase reaction. Incubate for 2 h at −80 °C.

(b) Centrifuge at 16,800 × *g* for 15 min at 4 °C.
(c) Wash DNA pellets with 500 μL of 70% ethanol.
(d) Decant 70% ethanol and dry DNA pellets for approximately 10 min using a SpeedVac.
(e) Resuspend DNA pellets in 52 μl of buffer EB.
(f) Store samples at –20 °C.

8. Measure the DNA concentration of each sample (e.g., using a Qubit 2.0 and High Sensitivity dsDNA reagents) (*see* **Note 5**).

3.4 Generate Fragments Flanking the Transposon Insertion Site

1. The protocol presented here to generate the fragments flanking the transposons is based on that of Goodman et al. [4, 13]:
2. Assemble the linear PCR reactions on ice as follows:
   (a) 10 μL Pfx buffer.
   (b) 2 μL 10 mM dNTPs.
   (c) 2 μL 50 mM $MgCl_2$.
   (d) 5 μL of 1 pmol/μL BioSamA primer.
   (e) 0.5–2 μg genomic DNA.
   (f) 1 μL Pfx polymerase.
   (g) ddH_2O to 100 μL.
3. Split into two 50 μL aliquots and place into separate PCR tubes. These were run in a thermocycler using the following program: 94 °C for 2 min followed by 50 cycles of 94 °C for 15 s and 68 °C for 1 min.
4. Pool tubes, and clean using a commercial PCR purification kit (e.g., QIAquick PCR cleanup) according to the manufacturer's instructions. Elute samples in 50 μL of provided elution buffer.
5. Transfer streptavidin-coated beads (32 μL per sample) to a microcentrifuge tube, and concentrate in a magnetic particle concentrator (MPC). Wash the beads 3× by resuspending (by inversion) in 1 mL of 1× B&W buffer and concentrating in the MPC (*see* **Note 6**).
6. After the final wash is removed, resuspend the streptavidin beads in 2× B&W buffer (52 μL per sample), and aliquot into PCR tubes.
7. Add the cleaned PCR products (~50 μl) from **step 4** of Subheading 3.4 to a streptavidin bead aliquot and incubate at room temperature on a rolling drum.
8. Concentrate the beads using the MPC and wash 3× with 100 μL of 1× B&W buffer.
9. After the final wash, resuspend the beads in 100 μL of low-TE buffer and store overnight at 4 °C (*see* **Note 7**).

10. Denature the samples in a thermocycler for 2 min at 95 °C and snap chill to 4 °C.
11. Prepare the following second-strand synthesis master mix for each sample:
    (a) 2 μL 2.5 mg/mL random hexamers.
    (b) 1 μL 10 mM dNTPs.
    (c) 1 μL Klenow (exo$^-$).
    (d) 16 μL ddH_2O.
12. Place the denatured sample tubes from **step 10** of Subheading 3.4 on the MPC, and discard excess low-TE buffer.
13. Add 20 μL of the second-strand synthesis master mix to each sample and incubate on a thermocycler at 37 °C for 30 min. Mix the samples every 15 min by flicking the tubes.
14. Using the MPC, wash the beads 3× with 100 μL low-TE buffer.
15. Prepare 50 μM of double-stranded M12 oligonucleotide by mixing 15 μL of 100 μM M12_top, 15 μL 100 μM M12_bot, and 1 μL 1 M NaCl and annealing the mixture in a thermocycler by heating at 95 °C for 5 min followed by cooling to 4 °C at a rate of 0.1 °C s^{-1}:
16. Prepare the MmeI restriction mixture on ice by mixing the following reagents for each sample:
    (a) 2 μL 10× NEB buffer #4.
    (b) 0.08 μL 32 mM S-adenosylmethionine.
    (c) 0.2 μL 50 μM M12 double-stranded oligonucleotide.
    (d) 16.8 ddH_2O.
17. Resuspend the beads in 19 μL of MmeI restriction mixture and add 1 μL MmeI. Incubate reactions in a thermocycler for 1 h at 37 °C, with inversion mixing every 15 min.
18. Using the MPC, wash the streptavidin beads 3× with 100 μL low-TE buffer.
19. Thaw barcoded adapters (50 μM) on ice and dilute to 5 μM by adding nine volumes of cold 1× T4 DNA ligase buffer.
20. Prepare ligation master mixtures for each sample by mixing 16.4 μL ddH_2O and 2 μL of 10× T4 DNA ligase buffer. After adding the ligation master mix to each sample, add a unique barcoded adapter (0.6 μL) and 1 μL of T4 DNA ligase. Incubate reactions for 1 h at 16 °C.
21. Wash beads 2× with 100 μL low-TE buffer and resuspend in 100 μL of low-TE buffer. Store samples overnight at 4 °C (*see* **Note 8**).
22. Prepare the PCR master mixture on ice:
    (a) 10 μL 10× Pfx buffer.
    (b) 2 μL 10 mM dNTPs.

(c) 2 μL 50 mM MgCl2.

(d) 2 μL each of 5 μM LIB-PCR5 and 5 μM LIB-PCR3.

(e) 0.5 μL Pfx polymerase.

(f) 31.5 μL ddH_2O.

23. Concentrate the beads from **step 21** of Subheading 3.4 using the MPC and resuspend in 50 μl of the PCR mixture. Run the following PCR program: 2 min at 94 °C, followed by 18 cycles of 94 °C for 15 s, 60 °C for 1 min, and 68 °C for 2 min with a final extension at 68 °C for 4 min.

24. Collect beads using the MPC and transfer the supernatant to a clean tube. **Save the supernatant. This is the part to be sequenced**.

25. Add xylene cyanol to the supernatants to a 1× concentration, and load the samples onto a 2% agarose gel pre-stained with GelRed (add 10 μL of 10,000× GelRed to 100 mL of molten agarose and allow to cool). Run the samples for approximately 1 h at 70 V.

26. Visualize gels using a handheld long-wave UV light (*see* **Note 9**). Excise bands of approximately 125 bp from the gel and purify using a commercial gel purification kit (e.g., QIAquick gel purification kit), according to the manufacturer's directions. Elute fragments in 32 μL of supplied elution buffer.

27. Quantify DNA present in samples (e.g., Qubit). Typically, we observe yields of between 0.8 and 2.4 ng/μL.

28. Normalize each barcoded sample to 10 nM and pool samples together. Sequence using a single Illumina HiSeq 2000 lane formulated for single-end, 50 bp reads.

29. Trim resulting reads, and map the reads to the *Campylobacter jejuni* 81-176 genome using the Burrows–Wheeler Aligner (BWA) [14]. Eliminate reads that (a) are not within 4 nt of a corresponding read on the opposite strand and that (b) exhibit a disparity in sequencing the opposite strand by at least one order of magnitude—legitimate insertion sequences should result in reads from both strands within a few nucleotides that are relatively equal in abundance. Additionally, we eliminated mutants from our analysis that had insertions occur in the last 10% of a gene, since these mutations may not interrupt gene function.

3.5 Identification of Colonization Factors Using Chicken Colonization Model

1. Thaw a pooled transposon mutant library stored at −80 °C on ice, and pipette 100 μL of cell suspension onto a MH blood + chloramphenicol plate. Incubate the plate for 4 h at 37 °C under microaerobic conditions.

2. Harvest cells into sterile PBS and measure the OD_{600} by spectrophotometry.

3. Collect three samples of approximately 10^9 cfu from this suspension, and extract genomic DNA as described in **step 5** of Subheading 3.3. This represents the "input" for chicken colonization studies (*see* **Note 9**).
4. Generate a cell suspension of approximately 10^7 cfu/mL in sterile PBS (where an OD_{600} of 1.0 equals 10^9 cells) using the OD_{600} reading in **step 2** of Subheading 3.5.
5. Inoculate day-of-hatch white leghorn chickens with 100 μL (10^6 cfu) of this suspension by oral gavage with a ball-tipped syringe. Additionally, inoculate positive control birds with either wild-type *C. jejuni* DRH212 and/or negative control birds with sterile PBS.
6. House the birds for 7 days before euthanizing and collecting the cecal contents.
7. Resuspend the cecal contents in sterile PBS by vigorous vortexing at a 10^{-1} dilution (e.g., 100 mg cecal content in 1 mL of sterile PBS).
8. Serially dilute a 100 μL aliquot of the resuspended cecal contents, and plate on selective media to enumerate total *C. jejuni* present.
9. Dilute the remaining cecal suspension 1:1 in a sterile PBS—1 % phenol solution and vortex briefly.
10. Centrifuge the mixture at 800 × *g* for 3 min at 4 °C and transfer supernatant to a clean tube.
11. Centrifuge the supernatant at 9260 × *g* for 5 min at 4 °C and decant the supernatant.
12. Extract genomic DNA as described in **step 5** of Subheading 3.3 from the *C. jejuni* pellet (*see* **Note 10**).
13. Using the number of bacteria present within the ceca (from **step 8** of Subheading 3.5), calculate the amount of DNA from each sample required to give an equivalent of 10^9 cfu *C. jejuni*. In many cases, this may require pooling bacterial DNA from between 2 and 5 birds and represents the "output" for chicken colonization studies (*see* **Note 11**).
14. The "input" and "output" extracted genomic DNA is then subjected to the same protocol and analysis described in Subheading 3.4.
15. To determine an approximate "fitness ratio," take the pooled number of raw reads from each gene represented in the output and divide by the number of reads from the same gene in the input. If a gene is required for colonization, it would be expected to have a lower abundance in the output pools, yielding a fitness ratio <1. Conversely, if a factor is not required for colonization, it will yield a fitness ratio ≥ 1.

4 Notes

1. The amount of transposon DNA will need to be varied based on the size of the delivery vector since this will affect the molar ratio of recipient:transposon DNA—the molar ratio used in our lab is approximately 1:120, respectively. Overabundance of transposon DNA leads to strains that contain multiple insertions, which would ruin the library. This can be tested for by restriction digestion of mutant chromosomes followed by Southern blot analysis or by preparing a number of mutants using the INSeq protocol above and identifying multiple loci.
2. The original protocol called for a phenol–chloroform-based purification of the mutagenized DNA, but this was found to result in low yields. Instead, we opted for the protocol that we routinely use for preparing genomic DNA from cells. In our hands, this led to significantly better yields.
3. We generally obtain 200–500 colonies per in vitro transposition and transformation.
4. To maintain heterogeneity of the library, we often only pick 100–200 colonies from each transformation. At approximately 8500 mutants, this means our library is the aggregate of at least 42 separate in vitro transposition reactions and transformations.
5. Quantification of DNA following purification indicated that samples containing less than 10^8 cfu of *C. jejuni* did not yield appreciable amount of genomic DNA. While this is a limitation with our genomic DNA isolation protocol, other investigators may have more efficient means and should verify their protocol's efficiency prior to using the entire protocol.
6. The original protocol by Goodman et al. stated that the magnetic beads were to be washed by pipetting. We noticed that beads were retained by the pipette tips, and following the entire protocol, very few remained. As such, we eliminated the pipetting steps and substituted inversion and "flicking" the tubes at all steps.
7. This is a "pause point" in the protocol.
8. This is a "pause point" in the protocol.
9. We were only able to generate detectable bands on genomic samples that represented at least 10^9 cfu *C. jejuni*. Repeated attempts to use less bacteria never yielded bands that could be visualized—again, this may be due to our inability to get appreciable amounts of genomic DNA from low numbers of *C. jejuni*.
10. The pelleted material at this stage has previously been shown to consist primarily of *C. jejuni* cells [15].

11. Due to the inability to generate the PCR fragments that contain the transposon-flanking sequence on samples that contain less than 10^9 cfu of *C. jejuni*, we were often required to pool birds in order to cross this threshold. Day-of-hatch chicks often maintain colonization loads of 10^8–10^9 cfu/gram of cecal contents at 1 week post-inoculation, but due to their small size, will often only have 100–500 mg of cecal contents. This represents 10^7–5×10^8 total cfu of *C. jejuni* making it an unfortunate limitation that only birds with higher colonization loads could be used.

Acknowledgments

This work was supported by grants from the US Department of Agriculture National Institute for Food and Agriculture (awards 2010-65201-20594 and 2013-67012-21136), the NIH National Institute of Allergy and Infectious Diseases (R01 AI069383), and the Molecular Mechanisms of Microbial Pathogenesis Training Program (T32 AI007528).

References

1. Darwin AJ (2005) Genome-wide screens to identify genes of human pathogenic Yersinia species that are expressed during host infection. Curr Issues Mol Biol 7(2):135–149
2. Dudley EG (2008) *In vivo* expression technology and signature-tagged mutagenesis screens for identifying mechanisms of survival of zoonotic foodborne pathogens. Foodborne Pathog Dis 5(4):473–485. doi:10.1089/fpd.2008.0104
3. Le Breton Y, Mistry P, Valdes KM et al (2013) Genome-wide identification of genes required for fitness of group A *Streptococcus* in human blood. Infect Immun 81(3):862–875. doi:10.1128/IAI.00837-12
4. Goodman AL, McNulty NP, Zhao Y et al (2009) Identifying genetic determinants needed to establish a human gut symbiont in its habitat. Cell Host Microbe 6(3):279–289. doi:10.1016/j.chom.2009.08.003
5. Fu Y, Waldor MK, Mekalanos JJ (2013) Tn-Seq analysis of *Vibrio cholerae* intestinal colonization reveals a role for T6SS-mediated antibacterial activity in the host. Cell Host Microbe 14(6):652–663. doi:10.1016/j.chom.2013.11.001
6. Lampe DJ, Churchill ME, Robertson HM (1996) A purified mariner transposase is sufficient to mediate transposition *in vitro*. EMBO J 15(19):5470–5479
7. Young KT, Davis LM, Dirita VJ (2007) *Campylobacter jejuni*: molecular biology and pathogenesis. Nat Rev Microbiol 5(9):665–679. doi:10.1038/nrmicro1718
8. Silva J, Leite D, Fernandes M et al (2011) Campylobacter spp. as a foodborne pathogen: a review. Front Microbiol 2:200. doi:10.3389/fmicb.2011.00200
9. Johnson JG, Livny J, DiRita VJ (2014) High-throughput sequencing of *Campylobacter jejuni* insertion mutant libraries reveals *mapA* as a fitness factor for chicken colonization., J Bacteriol
10. Yao R, Alm RA, Trust TJ et al (1993) Construction of new *Campylobacter* cloning vectors and a new mutational *cat* cassette. Gene 130(1):127–130
11. Hendrixson DR, Akerley BJ, DiRita VJ (2001) Transposon mutagenesis of *Campylobacter jejuni* identifies a bipartite energy taxis system required for motility. Mol Microbiol 40(1):214–224
12. Lampe DJ, Akerley BJ, Rubin EJ et al (1999) Hyperactive transposase mutants of the Himar1 mariner transposon. Proc Natl Acad Sci U S A 96(20):11428–11433

13. Goodman AL, Wu M, Gordon JI (2011) Identifying microbial fitness determinants by insertion sequencing using genome-wide transposon mutant libraries. Nat Protoc 6(12):1969–1980. doi:10.1038/nprot.2011.417
14. Li H, Durbin R (2009) Fast and accurate short read alignment with Burrows-Wheeler transform. Bioinformatics 25(14):1754–1760. doi:10.1093/bioinformatics/btp324, btp324 [pii]
15. Jerome JP, Bell JA, Plovanich-Jones AE et al (2011) Standing genetic variation in contingency loci drives the rapid adaptation of *Campylobacter jejuni* to a novel host. PLoS One 6(1):e16399. doi:10.1371/journal.pone.0016399

ERRATUM TO

Chapter 12
Assays to Study the Interaction of *Campylobacter jejuni* with the Mucosal Surface

Marguerite Clyne, Gina Duggan, Ciara Dunne, Brendan Dolan, Luis Alvarez, and Billy Bourke

James Butcher and Alain Stintzi (eds.), *Campylobacter jejuni: Methods and Protocols*, Methods in Molecular Biology, vol. 1512, DOI 10.1007/978-1-4939-6536-6_12, © Springer Science+Business Media New York 2017

DOI 10.1007/978-1-4939-6536-6_22

The original version of this chapter was inadvertently published with incorrect author name. The correct name should be Luis Alvarez.

The updated original online version for this chapter can be found at
DOI 10.1007/978-1-4939-6536-6_12

James Butcher and Alain Stintzi (eds.), *Campylobacter jejuni: Methods and Protocols*, Methods in Molecular Biology, vol. 1512, DOI 10.1007/978-1-4939-6536-6_22, © Springer Science+Business Media New York 2017

Index

James Butcher and Alain Stintzi (eds.), *Campylobacter jejuni: Methods and Protocols*, Methods in Molecular Biology, vol. 1512, DOI 10.1007/978-1-4939-6536-6, © Springer Science+Business Media New York 2017

M

N

P

R

S

T

V

W

Printed in the United States
By Bookmasters